D1053401

PATERNITY

PATERNITY

THE ELUSIVE QUEST FOR THE FATHER

NARA B. MILANICH

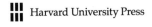 Harvard University Press

CAMBRIDGE, MASSACHUSETTS

LONDON, ENGLAND | 2019

First printing

Library of Congress Cataloging-in-Publication Data
Names: Milanich, Nara B., 1972– author.
Title: Paternity : the elusive quest for the father / Nara B. Milanich.
Description: Cambridge, Massachusetts : Harvard University Press, 2019. |
    Includes bibliographical references and index.
Identifiers: LCCN 2018052382 | ISBN 9780674980686 (alk. paper)
Subjects: LCSH: Paternity testing. | Paternity. | Fatherhood—
    Social aspects.
Classification: LCC RA1138 .M55 2019 | DDC 362.82—dc23
LC record available at https://lccn.loc.gov/2018052382

To Nino, with love and gratitude

CONTENTS

WHO'S YOUR DADDY?

> It is hard to grasp, given the immense and visible parcel in which man-made history has packaged the idea of paternity, that paternity is in fact an abstract idea.
>
> —Mary O'Brien, *The Politics of Reproduction*, 1981, p. 29

THE PROTAGONISTS OF THE courtroom drama consisted of a young mother, a putative father, and an adorable red-haired baby. In the early 1940s, as war raged abroad, the inquest unfolded in a packed Los Angeles courtroom. This was not just any paternity suit. The mother was Joan Berry, a twenty-three-year-old aspiring actress; the baby, her daughter Carol Ann; and the accused father Charlie Chaplin, Hollywood celebrity.

Berry was Chaplin's onetime protégé, and in happier times, the two had read Shakespeare together and practiced drama. Now the fifty-four-year-old actor, whose penchant for much younger women was well known, stood accused of fathering Berry's baby. He admitted the romance but vehemently denied the paternity charge. A week after the case broke, the actor wed his fourth wife, the eighteen-year-old daughter of playwright Eugene O'Neill. Thanks to his British citizenship and leftist political leanings, Chaplin's ideological proclivities were for some sectors of the American public as questionable as his romantic ones. For her part, Berry was portrayed as a hapless ingénue, "aflame with the glamour of Hollywood," possibly mentally unstable, "pretty to the eye," but, by her own lawyer's account, "of limited intelligence."[1]

But it was the baby who was the star of the show. Not yet born when the suit over her paternity was filed, over the course of the saga Carol Ann blossomed into an amiable toddler. She was a courtroom fixture, sitting on the wooden table in front of her mother's lawyer. The press gleefully reported on her colorful pinafores and penchant for patty-cake. Still, the legal proceedings were serious business. At stake was the identity of a child. Would she face a life of penury or comfort? Would she have a name, a patrimony, a father? The paternity suit, her mother's lawyer declared and the newspapers would repeat, was the baby's "day in court."[2]

The drama featured other players, too: witnesses like Chaplin's handyman and butler, who testified about the couple's trysts; and the members of the jury, ordinary women and men—housewives, an interior decorator, a retired property manager—who arrived in court carefully coiffed in anticipation of the cameras. There was Berry's attorney, himself something of a courtroom thespian, who in an especially memorable three-hour closing statement denounced the actor as a "cheap cockney cad" and a "lecherous hound."[3] (Chaplin's own lawyer responded by comparing his client to Christ crucified on the cross.) Finally, there were the newspapermen, and a few newspaperwomen, who breathlessly broadcast the story to the public. Their daily reports from the courtroom included descriptions of the protagonists' outfits (Joan's chartreuse coat) and moods (Charlie's grimaces). The heady spectacle of sex, celebrity, and scandal not only reached American readers but, thanks to the global wire services, was beamed out to a world at war.

The two-year inquest had numerous twists and turns. In a related criminal charge, Chaplin was tried and acquitted of trafficking Berry across state lines for immoral purposes. He briefly faced deportation charges as an alien. As for the proceedings concerning Carol Ann's paternity, the first suit ended in a mistrial when the jury deadlocked, and a new trial was called. The Chaplin–Berry saga had begun as President Roosevelt ordered striking coal miners back to wartime production and Allied troops amassed in the Mediterranean readying to invade Italy. By the time it ended, Roosevelt was dead and the Allied victory in Europe was weeks away. Yet for all its protracted drama, "the case is a simple one," the judge reminded the courtroom as the proceedings drew to a close. It revolved around one question: "Is the defendant the father?"[4]

His question was less simple than it appeared. Paternity is a question of long-standing cultural, legal, political, and scientific interest and, according to a long Western tradition, an intractable one. Whereas a mother's identity can be known by the fact of birth, the father has always been maddeningly uncertain. The quest to identify him animated medical experts at least since Hippocrates and preoccupied jurists of Roman, Islamic, and Jewish law. Literary fathers have brooded over their paternity in the works of Homer and Shakespeare, Hardy and Machado de Assis. Theorists from Friedrich Engels to Sigmund Freud posited paternal uncertainty as a primordial foundation of human society and the human psyche. For a generation of early-twentieth-century anthropologists, cross-cultural beliefs about paternity were "the most exciting and controversial issue in the comparative science of man."[5]

But paternity is not just a subject of intellectual rumination. As the Chaplin–Berry inquest suggests, it matters to men and women, to children and families, for reasons that are patrimonial, practical, and existential. Questions of paternity have historically arisen in the context of disputes over child support and inheritance. The orphaned and the adopted have asked this question in relation to lost identities. More recently, assisted reproductive technologies—gamete donation, surrogacy—have raised old issues in new ways.

Paternity's stakes are public as well as private: they matter to states and societies and not just individuals. This is why the dispute over Carol Ann's father took place in a courtroom and was governed by rules set by law. For while kinship is often seen as a "pre-modern" or "non-Western" form of association, it is at the heart of modern social and economic citizenship and a key symbol demarcating public and private spheres. Family ties are important to states because they confer access to war pensions and social security, to nationality and the right of noncitizens to settle in a country. Historically, children bereft of kin ties have become public charges. The question of the father raises questions about the balance of rights and responsibilities between individuals and societies.

Of course, it wasn't Carol Ann's parentage generally but her paternity specifically that was in dispute. Significantly, the question "who is the father?" has no parallel query concerning the mother. Paternity has been understood as naturally uncertain, whereas maternity is obvious and unproblematic. Paternal identity is posed as a question, in short, because the answer is considered potentially unknown. Moreover, in patriarchal societies, the most important resources it has traditionally conferred—economic support, patrimony, nationality, a patronym, an "identity"—are not transmitted by maternity. When Joan Berry's lawyer exhorted the jury to pronounce Chaplin to be Carol Ann's father in order to "give this baby a name," he assumed that only a father, not a mother, had the power to do so.[6] The question of paternal identity reflects paternity's distinctive economic, political, and cultural stakes.

If the quest for the father has a long history, the Chaplin case reflected several modern twists on the story. The idea that a fatherless baby like Carol Ann was a citizen entitled to her day in court contrasted with an earlier era when children were objects of charity, not subjects of rights, and gave greater urgency to the dispute over her parentage. The role of the press was also new. As novelists and playwrights had long known, mysteries of identity were the stuff of melodrama. In the twentieth century, the mass media began to tell those stories to a fascinated public. The Chaplin affair had instant star power, but such stories did not need a Hollywood celebrity to rivet audiences the world over.

Above all, the Chaplin–Berry drama featured a new protagonist in the perennial quest for the father: the scientist. In fact, it featured three of them. Accompanied by his attorney, in February 1944 Chaplin visited a local laboratory, where a thimbleful of his blood was drawn. An hour later, Berry and her baby arrived for the procedure. Three medical specialists tested the samples and later presented their findings in court, assisted by what one observer described as "a maze of alphabetical designations, long words and big charts."[7] The test they had conducted was an analysis of hereditary blood groups, and the three experts agreed unanimously about what it revealed: Joan Berry had blood type A and baby Carol Ann had type B, which by the laws governing blood group heredity meant that her paternal progenitor must have type B or AB blood. Chaplin, however, had type O. The actor might be a notorious cad who had admitted to a romance with Berry. But he could not be Carol Ann's biological father.

Hereditary blood grouping was just one of many scientific methods that, beginning in the 1920s, promised a potentially revolutionary solution to the timeless quest for the father. "Medical experts hope that in the blood that is transferred from parent to child, down through the ages, there exists some as yet unknown but vital element that links them inevitably."[8] They sought that vital element in blood types but also in other, long forgotten methods involving the electronic vibrations of blood, its crystallization patterns, and its chromatic characteristics. They also looked beyond the veins, in the inheritance of nose shape, to similarities in the conformation of the teeth, and at the bumps and ridges of the palate. Anthropometric analyses of the body and especially the face attempted to make objective the unmistakable yet ambiguous phenomenon of family resemblance. Perhaps the secret of paternity was tucked into the intricate folds of the human ear, the delicate whorls and loops of the fingerprints, or the patterning of eye, hair, and skin color.

The scientific methods were myriad, but the central assumption of all of them was that the truth of kinship was located somewhere on the physical bodies of father and child. Such an approach implied not just a new method for revealing paternity but a broader set of claims: that paternity was a knowable quality, that it was in the public interest that it should be known, and that the scientific expert could discover it. Most fundamentally, it implied a belief about what paternity was in the first place: a physical relationship rather than a social one.

Such an understanding of paternity is familiar in the era of DNA. Today we routinely dispatch finger pricks and cheek swabs to faraway labs to expose the recondite mysteries of our identity. We understand kinship as a physical fact, the body as a source of truth, and science as the means to reveal it. But such ideas are relatively recent. In an older tradition, biological paternity was considered an ineffable enigma of nature, not just unknown but indeed unknowable. Paternity was less physical than metaphysical, a relationship deduced from behaviors and social conventions. In many legal traditions, it was marriage that made paternity: the father was the husband of the mother. As for a child born outside of marriage like Carol Ann Berry, the father was revealed in other ways: he was the man who cohabitated with the mother or kissed the baby in public, the man whom the neighbor had seen paying the wet nurse. Paternity was not primarily a natural fact deriving from the act of procreation; it was a social fact brought into being through a man's words and deeds and the observations of the community.

Following this social logic, in medieval tradition, if a widow remarried quickly and then gave birth, the child could choose its father, depending on whether it was more advantageous to be the youngest child of the first husband or the oldest child of the second one. Other legal traditions provided for partible paternity. Scandinavian law, for example, held that if two men had a relationship with the mother, support for her child might be split between them. It could also be partial: a man might be responsible for supporting a child economically but not give that child his name or inheritance. When paternity was disputed, those called to elucidate it were not scientists or doctors but friends, associates, the neighbors, the mother, or the man himself.

Some children simply had no father. In Anglo-American law, the illegitimate child was historically deemed a *filius nullius*, a child of nobody. If many circumstances demanded a father, in others "who's your daddy" was deliberately left unanswered. In slave societies, the father of the enslaved child might well be the mother's owner. And what about the debauched priest, or the case in which the husband was not the father of his wife's child? Colonizers and soldiers deployed in foreign lands have often been excused from responsibility for the children they engender there. Because paternity is embedded in social relations of power, it is also potentially disruptive. Politics, morality, and the public purse might require a father in some situations but demand something else—discretion, suppression, invention—in others.

If the understanding of paternity distilled in a blood test is deeply familiar to us, it involves a series of assumptions about what paternity is, the need to know it, and how it can be known, that are by no means universal and indeed are surprisingly recent. These ideas became increasingly powerful in the first decades of the twentieth century, not only in the United States but across the Americas, north and south, as well as in Europe. Tentatively at first, and then with increasing enthusiasm, these beliefs and techniques found practical application from Buenos Aires to Berlin to Los Angeles. As they did, they spawned boundless fascination among transatlantic publics and shaped the way states and societies thought about kinship, identity, and belonging.

But like any new technology, paternity science raised a host of practical and ethical questions. It raised questions about the circumstances in which tests of parentage should be performed, who should have access to the results, and whether revelation was always a good thing. If the community, the judge, the mother, and the man himself had traditionally defined who was a father, paternity science now vested this power in a new authority, the biomedical expert. What happened when the expert's assessment was at odds with older social and legal notions of paternity?

The dispute over Carol Ann Berry's paternity captures such tensions. By the 1940s, blood group heredity was well-established scientific doctrine, and scientists considered the results of a test that excluded an impossible father to be conclusive and incontrovertible. Because he was of an incompatible blood type, Chaplin could not have fathered Carol Ann. "The law of heredity," his lawyer reminded the jury, "is as certain as nature itself. If a child does not have the blood of a certain man in its veins, then that man cannot be the father."[9]

Yet while nature might have been certain, the law was more ambiguous. The judge admitted the blood test into evidence but explained to the court that in the state of California, it was not considered conclusive. A blood test was just one more piece of evidence to be weighed alongside others, like witness testimony or the word of the mother. For his part, Joan Berry's lawyer rejected the blood group analysis outright, calling it an "abomination" because it could only exclude an impossible father but never positively identify the real one. "In no way could Chaplin lose and in no way could the baby win," he thundered.[10] He urged the jury to pay heed to the true stakes of their decision. "Nobody can stop Chaplin and his lecherous conduct all through these years—nobody but you, ladies and gentlemen of the jury!"[11]

As for the eleven ladies and one gentleman of the jury in the second and final trial, they had their own thoughts about the quest for the father. After deliberating for three hours, they reached a startling conclusion: Charlie Chaplin was the father of Carol Ann. The courtroom erupted in applause and cheers, but many observers regarded the verdict with incredulity and outrage. "California has in effect decided that black is white, two and two are five and up is down," wrote one editorialist. In fact, such outcomes were not uncommon in U.S. courts. Critics attributed them to the ignorance of juries or the inherent conservatism of the law. A seasoned paternity lawyer summed up the Chaplin fiasco as "contrary to science, nature and truth."[12]

But which truth? While the defense attorney had urged the jury to remember that "by the cold, scientific proof of blood tests Chaplin could not be the father of this child," Berry's lawyer ascribed to the test a different meaning.[13] "To hold the blood test binding," he told the court, "would be to say, in effect, 'You little tramp, get out of here,' and let the rich father do as he pleases."[14] In his vision, the paternity inquest was less about biology than about morality and justice: the power of a rich, famous man to seduce hapless young women. Chaplin's paternity derived not from his biological link to Carol Ann Berry but from his relationship to her mother. That logic was hardly peculiar to the much-maligned Los Angeles jury. It is the same logic that in multiple legal traditions held the husband to be the father of his wife's children. It is the same logic that declared the father of an illegitimate child unknowable and perhaps nonexistent. It is, in short, the logic of the social rather than the biological.

The judge had suggested that the case was simple. The jury needed only decide whether the comedian was the baby's father, and the blood test was supposed to help them do that. But rather than revealing the answer, it had raised an even more basic question: what was a father in the first place? Instead of exposing the truth, the test laid bare the tensions between different possible truths. It reified rather than resolved the distinctions between the social and the biological, revelation and suppression, truth and morality.

A paternity suit like the Chaplin case was the most obvious place to debate such issues, but in the twentieth century the vexed question of the father arose in a surprising variety of contexts. To many observers, the modern world had actually intensified the millennial enigma of identity. Urbanization, immigration, demographic growth, changing sexual and family mores, and social heterogeneity had unmoored primordial and intimate ties of kith and kin. Luckily, however, modern science promised an antidote. In addition to dispensing with the sordid he said / she said of paternity disputes, it would resolve inheritance suits and ferret out offspring born of adulterous liaisons. It would reconstitute families fractured by the vicissitudes of modern life, whether in baby mix-ups in new maternity wards or due to world war. The new science of paternity was used to investigate not just identity but sex. It was deployed in investigations of rape and deflowering, where its subject was not the child or the father but the mother and her sexual behavior. Kinship science drew on universal laws of heredity in the service of incontrovertible truths, but its practical applications and social meanings were decidedly local.

If the new science of paternity sought to reknit ties, it could also rend them asunder. As in the Chaplin case, it was often better equipped to demonstrate that a given man could not be the father of a given child than at positively identifying whom the progenitor actually was. Blood testing did not find a father for baby Carol Ann; it threatened instead to deprive her of one. Proof of nonpaternity could also expose the abiding legal fiction that wives always gave birth to their husbands' children. Given its power to orphan children, imperil marriage, and upend public morality, this new science was potentially perverse, and this could serve as a powerful reason to restrict it.

In all these contexts, paternity science helped to define, defend, and sometimes destabilize kinship, sex, and marriage. But paternity belongs not only to the family; it has also been inextricably entwined with the history of race and nation. The science of the father germinated in the noxious soil of race science and eugenics, and its practical applications have often had racial objectives. Across radically different political contexts, nation-states have used paternity tests to set racial boundaries and defend the nation from racial outsiders. In the 1930s, Nazi authorities rewrote the law and science of paternity in an attempt to find Jews hidden in Aryan genealogies. Cold War–era U.S. immigration officials enlisted paternity science to bar Chinese immigrants by challenging their alleged kinship with Chinese American citizens. If paternity denoted a form of relatedness and belonging, its science was also deployed in the service of discrimination and exclusion.

Today the quest for the father has taken what appears to be a radical new turn, perhaps even reaching a definitive conclusion. The emergence of DNA fingerprinting in the 1980s made it possible for the first time in human history to know the father with 99.9 percent certainty. That promise has spawned a multibillion-dollar global industry, and today an infallible test of paternity—once the stuff of science fiction—has become ubiquitous and the promise of certainty banal.

Yet despite the unprecedented power of modern genetic science, paternity remains ensnarled in a thicket of unresolved social, economic, and political questions. Kinship technologies may have changed dramatically over the past hundred years, but the questions they raise have remained surprisingly constant. It remains unclear whether paternity science should be regulated and who should have access to its truths. Who, in other words, has the power to decide who is the father: individuals, communities, the state, or, most recently, profit-making companies? These techniques also force societies to consider whose interests they serve, whether those of men or women, children or adults, public or private good. They further raise the specter of racial exclusion, given that wealthy countries today routinely require DNA testing of nonwhite immigrants from the global south. Technology lays bare the existence of multiple possible paternities—social, affective, legal, biological—and asks which should prevail when they are in contradiction. Indeed, it begs the question of what, ultimately, paternity is.

The history of the quest for the father reveals precedents and patterns, lessons and cautions that can illuminate our relationship to genetic and reproductive technologies in the present. This history speaks not just to new technologies; it is germane to new social practices as well, from transnational adoption to same-sex unions. is germane to new social practices as well, from transnational adoption to same-sex unions. It helps us appreciate the extent to which paternity as well as maternity, family, and identity have always been malleable categories, made and remade over time and across place. Tracing these creative reformulations in the past helps us take stock of seemingly unprecedented practices and relationships in the present.

History also illustrates the ambiguous impact of new knowledge and new truths. If science has come to play a particularly powerful role in elucidating the vexed quest for the father, it has not resolved that quest but merely complicated it. Today, after a century of scientific advances, we are no closer to answering the millennial question, who is the father? Indeed, perhaps we are farther away than ever from doing so.

1

LOOKING FOR THE FATHER

Paternity is as mysterious as the source of the Nile.

—French legislator, 1883

OF THE MANY MIRACLES for which Saint Anthony is revered, one is a paternity test. While preaching in Ferrara, the Franciscan was approached by a weeping gentlewoman who pleaded for his help. Her jealous husband had become convinced that the child born to her a few months earlier was not his and had threatened to kill them both. Anthony comforted the distraught woman, assuring her that God never abandoned the innocent. A short time later, he happened upon the husband walking with some friends, followed by his scorned wife and the infant. Anthony stopped them, caressed the baby, and said, "tell me, child, who is your father?" The infant turned and fixed its eyes on the jealous husband. Then he called the man by name and announced, "that is my father." The stunned man burst into tears, embraced his child, and declared his love and esteem for his wife.[1] The miracle of the talking babe and his saintly paternity test was later commemorated in Renaissance frescoes, marble, and bronze reliefs.

The fascination with proving paternity is an ancient one. Hippocrates commented on the uses of physical resemblance to identify the father, and pre-Islamic Arab societies recognized a special physiognomic method for doing so. One ancient Jewish sage claimed that the blood of one individual

In a miraculous paternity test, depicted in this 1511 fresco by Tiziano Vecellio
(Titian), Saint Anthony caused a baby to speak and identify its doubting father.
Titian, *The Miracle of the Newborn Child,* 1511, Scuola del Santo, Padua, Wikimedia.

dropped on the bones of another would be absorbed if the two were kin. A
forensic text from thirteenth-century China described a similar test, in which
the blood of two people was mixed in a vessel of water. If the blood blended,
they were related; if it clumped, they were not. On the basis of this procedure, one Japanese scientist in the 1930s claimed an ancient Asian origin
for paternity testing based on ABO blood types.

But before the twentieth century, the question of the father was most
frequently and systematically treated as a problem not for miracles or medicine but rather for law. Paternity has preoccupied jurists both religious and
secular, ancient and modern. Perhaps the best known and most globally influential formulation is the one put forth in Roman law. *Pater semper
incertus est,* the father is always uncertain, because paternity cannot be
observed, merely intuited, or presumed, whereas *mater certissima est,* the
mother is very certain, thanks to the observable fact of birth. The Roman
law of parentage is typical of the strong asymmetry that has historically characterized maternity and paternity in Western thought.

Roman law also establishes a third principle: *pater est quem nuptiae
demonstrant,* the father is he whom marriage indicates. That is, a woman's
husband is always, according to the law, the father of her children. Paternity may be intrinsically uncertain, but marriage makes the father. The cor-
ollary, of course, is that the father necessarily remains uncertain in the case
of an unmarried mother. If paternity is made knowable by marriage, it is
unknowable outside of it. This basic constellation of principles informed

not only Roman law but also Islamic, Jewish, and canon law. These are hardly the archaic principles of a pre-genetic world: to a remarkable degree, they endure in modern legal regimes today, including Anglo-American common law and continental civil law.

A fixture of law, the uncertain father also appears in literature from Homer to Shakespeare. "'Tis a wise child that knows his own father," Telemachus tells Athena in *The Odyssey*, while in *The Merchant of Venice* Launcelot reverses the dictum, telling Gobbo, "it is a wise father that knows his own child." Physical paternity—powerful yet always presumptive— recurs in Renaissance poetry and drama.[2] Fathers in *King Henry IV, King John,* and *A Winter's Tale* all muse on the problem of paternity and the tricks of filial resemblance. Enlightenment thinkers likewise explored patterns of physical similarity as part of a broad fixation with questions of inheritance and generation, nature and nurture.[3]

IN THE NINETEENTH CENTURY, the unknown father became an object of even more elaborate fetishization. Anglo-American law traditionally sought to identify the father because the fatherless child would have to be supported by the parish, but the New Poor Law of 1834 challenged this imperative, ex- empting men from responsibility for their illegitimate progeny.[4] As one mag- istrate explained, previously "a woman of dissolute character" could "pitch upon any unfortunate young man . . . and swear that child to him."[5] The reform shifted parental responsibility squarely onto the mother in order to discourage women from indulging in immoral behavior. Within a decade, however, the reform was reversed under an avalanche of public protest. Erasing the father had not only contributed to a rise in bastardy and infanticide; it had forced the community to shoulder the cost of illegitimate children. The imperative to identify the father for economic reasons thus reasserted itself.

On the Continent, a similar moral and economic calculus governed illegitimate paternity. In eighteenth-century France, the unmarried mother had been pressured for a *déclaration de grossesse* in which she identified the author of her pregnancy. But nineteenth-century law elevated unknown paternity to an ontological truth—and unlike in English law, there would be no reversals. The 1804 Napoleonic Code prohibited paternity suits and in so doing erased the identity of the extramarital father. Unmarried women could no longer bring charges against the authors of their pregnancies; children born outside marriage had no right to economic support or the paternal name. Jurists charged that paternity suits caused scandal, allowed licentious women to profit from their own immorality, and encouraged imposters to persecute upstanding families with spurious paternity claims. Even more to the point, investigating paternity was nonsensical because it could never truly be known. Maternity was "a material fact, visible, subject to the domination of anyone's senses." But paternity was a "mystery of nature," "an act for which it is impossible to give clear proof of any kind." The most common metaphor invoked by nineteenth-century jurists was that Nature had concealed fatherhood by an "impenetrable veil."[6] No human could glimpse behind it.

In succeeding decades, the Napoleonic idea that paternity was unknowable and therefore should not be the subject of legal investigation spread across continental Europe, Latin America, and then parts of the colonial empires. It became more difficult, when not impossible, to legally establish an illegitimate child's father. Law defined paternity not as an empirical fact that derived from procreation but as an act of volition that came into being only when a man freely recognized it. In the absence of the man's will, there could be no paternity. This vision of paternity favored men at the expense of mothers and children, who had no right to paternal succor.

Significantly, the nineteenth-century law of paternity was at variance with contemporary scientific orthodoxies. In an era when science and society were fixated on biological heredity as a determinant of human life, jurists declared the inheritance of the father unknowable. The idea that maternity and paternity were ontologically different, which was at the heart of Napoleonic law, was also at odds with an emerging scientific consensus that male and female parents made equal contributions to generation.[7]

Paternity law, in other words, reflected not prevailing scientific doctrine but social and political ideas. In nations emerging from the upheaval of revolution, from France to the newly independent republics of Latin America, the architects of the social order understood the patriarchal family as key to their task. The insistence on paternal unknowability fortified the family by reinforcing the power of legitimate families over illegitimate interlopers and men's control over women, children, and family patrimony. Ideas of paternal uncertainty also bore the mark of economic and political liberalism. Leaving up to men the choice of whether to recognize their children reflected liberal ideas of individual freedom, privacy, and property rights.[8] The architects of new national communities sought to defend order, morality, patriarchy, and patrimony. Identifying all fathers undermined those objectives.

As for paternity within marriage, nineteenth-century civil law tended to reinforce the old Roman presumption of marital legitimacy, making it more difficult for a husband to challenge the paternity of his wife's child. Marriage and the legitimate family must be protected, even if doing so sometimes required quiet fictions about who had sired whom. Here too volitional paternity was at work, for in contracting marriage to a woman a man assented to be the father of her children. Even as jurists invoked nature to justify their erasure of illegitimate paternity, the distinct treatment of marital and extramarital paternity reveals that it was society, not nature, that determined the father. His identity, indeed whether or not he could be discovered in the first place, depended on the social circumstances surrounding procreation.

The conceit of paternal uncertainty also flourished in nineteenth-century literature. The husband tortured by doubts about his wife's children provided plots and subplots for authors from Balzac, Thomas Hardy, and August Strindberg to Guy de Maupassant and Machado de Assis.[9] In these literary reckonings, men's uncertainty about their paternity was a corrosive obsession: it destroyed men and their relationships to women and to each other. Unknown parentage also appeared in stories about orphans. Nineteenth-century literature abounds with plucky youngsters with absent or unknown parents.[10] For Oliver Twist, Tom Sawyer, Rémi (of the French novel *Sans Famille*), and Horatio Alger's characters, the very absence of progenitors catalyzes the novelistic journey: it is what forces the young heroes to make their way in the world, taking the reader with them.

In perhaps no story did paternal uncertainty play a more central role than in the one that nineteenth-century social theorists told about the origins of humanity. For Victorian thinkers, the unknowable father helped explain human social evolution, the rise of modern economic relations, changing gender roles, and the deepest recesses of the human psyche. Johann Bachofen, Friedrich Engels, Lewis Henry Morgan, and others told a story that went something like this: in primitive human societies, promiscuity and "group marriage" were the rule. As a result, paternity could never be known, and social organization was therefore matrilineal and matriarchal. The fact that in primitive society "the mothers [were] the only certain parents of their children," Engels famously argued, "secured for . . . their whole sex, a higher social status than women have ever enjoyed since." But this primitive matriarchy—or what the Swiss theorist Bachofen called "mother-right"—was fated to collapse. With the advent of monogamous marriage, men became certain of their offspring, making patrilineal descent and inheritance possible. The rise of private property and patriarchy thus flowed directly from paternal confidence. In social theory as in law, marriage made the father. Engels identified the Napoleonic Code as "the final result of three thousand years of monogamous marriage." He characterized this transformation from promiscuity to monogamy, uncertainty to certainty, and matriarchy to patriarchy as "one of the most decisive revolutions ever experienced by humanity." In the story that he and his contemporaries told, paternity helped determine the development of human civilization: knowledge of the father distinguished one evolutionary stage from the next.[11]

Paternity was also central to theories of human psychology. For Freud, a crucial turning point of psychic development occurs when the child learns the facts of procreation, realizes that the identity of its father is necessarily uncertain, and begins to imagine alternative progenitors as part of the "family romance" of wishful projection. The notion of paternity as an intellectual inference rather than an empirical fact—something presumed but never really known—recurs throughout his work.[12] Freud thus recast as a psychic process what his nineteenth-century predecessors had narrated as a historical one: the child passes through developmental stages that recapitulate those of human civilization.[13] In both accounts, the unsolvable problem of paternal uncertainty is the motor that propels development, whether of society or the individual.

THE PREOCCUPATION WITH PATERNAL uncertainty continued to evolve in the twentieth century. By the 1920s and 1930s, the theory of primitive matriarchy had largely fallen out of favor, but a new generation of theorists relocated the problem of paternal uncertainty from the distant human past to the sexual lives of contemporary "savages." In these years, the discipline of anthropology was roiled by a debate over whether certain "primitive" peoples understood physiological paternity. On one side of the debate were those like Bronislaw Malinowski, a founding father of modern anthropology, who asserted that they did not. Drawing on ethnographic work in the Trobriand Islands of the South Pacific, Malinowski claimed that the "natives" had no understanding of the male parent's biological role in procreation. His account challenged the Victorian theorists who had claimed that paternal uncertainty inevitably produced matriarchy and precluded the development of family and private property. Among the Trobrianders, according to Malinowski, ignorance of biological paternity coexisted with marriage, family, and a social idea of paternity (a man was a father not because of a biological relationship to a child but because of his social relationship to its mother). Of course, the enduring scholarly fixation with paternity tells us more about Euroamericans' ideas about kinship than about those of the people they studied.[14]

Malinowski's was an important revision of the nineteenth-century theories of group marriage and primitive matriarchy. Yet like earlier theorists, he placed paternity at the center of his account of Melanesian social life and indeed of anthropology generally. Examining how primitives understood paternity led "directly into the study of kinship and social organization, religious beliefs, systems of totemism, and magical ritual"—in short, into the very heart of anthropological inquiry.[15] For all of these thinkers, paternal knowledge helped to define human cultural development itself: savages past and present lived in wanton promiscuity, perhaps not even understanding how babies were made.

Such ideas seeped out of rarefied scholarly circles and into the popular imagination. Science fiction in particular took up the relationship of sex and society, asking: if procreation shapes human society, how might technology alter both? The most famous answer was Aldous Huxley's *Brave New World* (1932). Huxley imagined a future dystopia that provocatively reversed sexual mores, such that promiscuity was civilized and the nuclear family savage. One of the novel's characters contrasted the "appalling dangers of family life"—monogamy, chastity, fathers, mothers—with the more salubrious sex lives of primitives. In a reference to Malinowski, a character describes the bucolic order of the Trobriands, where "conception was the work of ancestral ghosts; nobody had ever heard of a father."[16] In the future that Huxley imagined, technology had obliterated paternity (as well as maternity) and monogamy, dissolving the biological, affective, and social bonds of filiation. In other words, science had delivered humankind back to Malinowski's South Pacific and Engels's primitive group marriage.

Twenty years later, Arthur C. Clarke imagined another scenario in which technology revolutionized sex and social order. In *Childhood's End* (1953), considered by many to be Clarke's master work, future human society has discovered an "infallible method—as certain as fingerprinting, and based on a very detailed analysis of the blood"—to identify the father.[17] Together with fail-safe contraception (Clarke was writing in the decade before the pill), this paternity test had "swept away the last remnants of the Puritan aberration," obliterating earlier sexual mores as well as marriage. If in Engels's imagined past, monogamy resolved the problem of uncertain paternity, in Clarke's fanciful future, technology did so. Both saw the uncertain father as determining the organization of marriage, family, and society. Making him certain was therefore socially transformative. Clark's vision also reflects how, as late as the 1950s, a conclusive test of biological paternity was still the stuff of science fiction.

Decades later, paternal uncertainty reemerged yet again, now in the theories of second-wave feminists. They echoed an earlier generation of feminists who had drawn on the idea of primitive matriarchy to explain and critique the rise of modern patriarchy. In a 1972 essay, Gloria Steinem invoked a time when "paternity had not yet been discovered and it was thought (as it still is in some tribal cultures) that women bore fruit like trees." Echoing the nineteenth-century thinkers, she compared the momentous discovery of paternity to the taming of fire and the invention of nuclear energy.[18] The most thoroughgoing feminist critique of paternity was Mary O'Brien's 1981 classic *The Politics of Reproduction,* which argued that patriarchy had its origins in men's relationship to physical reproduction.[19] Because procreation involved "the alienation of the male seed," men experienced it as estrangement, negation, and uncertainty. To resolve this alienation, they created not just marriage but the whole division between public and private spheres. "Whether mud hut or extended household," wrote O'Brien, "the private realm is a necessary condition of the affirmation of particular paternity" because men could be certain of their fatherhood only when it took place within a clearly defined domestic sphere. At the heart of the "huge and oppressive structure" of patriarchy "lies the intransigent reality of [paternal] estrangement and uncertainty." In other words, patriarchy exists so that men can know their children.[20]

O'Brien invoked a biological determinism that fell largely out of favor in later feminist thought. Yet theories that place biological paternity at the heart of patriarchy have found an incongruous echo among contemporary sociobiologists and primatologists. Whether one is a monkey or a human, "the trouble with being a male primate," primatologist Sarah Blaffer Hrdy has observed, is the inability to know for certain one's offspring. This basic biological fact presents a formidable evolutionary challenge and gives rise to a series of morphological and behavioral adaptations among male primates. These include testicle size (large; to ensure adequate sperm delivery and increase the chances of impregnation), lousy fathering (common; because why invest in offspring if you can't know they're yours?), and newborns' propensity to physically resemble their fathers (so that the man finds affirmation of his paternity in the baby's face, though this much-studied phenomenon is disputed by recent studies).[21]

RUNNING THROUGHOUT THIS LONG history of paternal uncertainty—across law, literature, social theory, ethnography, science fiction, feminism, and sociobiology—is the persistent assumption that biological facts and social facts are distinct, and the biological determines the social. The natural fact of unknowable paternity profoundly shapes our most primal social and cognitive structures. Through mechanisms by turns historical, psychic, or evolutionary, uncertainty molds our minds, bodies, behaviors, and social structures.

The sheer persistence of this idea is powerful inducement to accept its main premise: that the uncertain father is a timeless and immutable fact of nature. Yet he is less a biological truth than a historical idea. The trope of uncertainty obscures the extent to which paternity has always been more "certain" and "knowable" in social practice. Societies have historically had clear, authoritative, and enduring social and legal rules for identifying the father. The presumption of marital legitimacy is of course one such rule (for "fictitious" or not, it definitively and unambiguously gives a father to every child born to a married mother). Even in the absence of marriage, in diverse societies from medieval England to pre-revolutionary France to early nineteenth-century Chile, paternity was routinely established through the sorts of empirical evidence that established all disputed facts. When a man paid a midwife's bill, contracted a wet nurse, or showed affection for the baby, communities and courts read these as social acts of paternal recognition.

Intimacy or cohabitation with the mother could likewise serve as evidence of paternity. So too could witness testimony by the local priest, the servant, a neighbor, or a mother's own declaration. The trope of the essential and inevitable uncertainty of paternal identity obscures the extent to which, according to rules that varied cross-culturally and over time, social facts defined parentage unambiguously. It is not just that courts, families, and communities relied on social methods to discover a natural truth but that paternity was itself a social truth.

What is more, the trope of paternal uncertainty obscures the fact that some fathers have always been more uncertain than others. This is because the inability to know the father is not always a "problem": in fact, it may be strategically advantageous to the social order. In many colonial and postcolonial societies, marriage across color, caste, and class lines was stigmatized or forbidden. Sex across these divisions was often quite another matter, not just tolerated but woven into the very fabric of domination. But what about the children born of such relationships? Legal and social norms that erased the identity of their fathers helped protect patriarchal prerogatives, patrimonial interests, and the socioracial order.

Consider the emblematic case of Atlantic slave societies, where the principle of *partus sequitur ventrem* held that enslaved mothers passed on their status as chattel to their children. Sally Hemings's children would be slaves like their mother, while the identity of their father Thomas Jefferson was considered formally unknown and, in any event, legally irrelevant. This strategic public fiction endured for almost 200 years despite the Hemings descendants' private knowledge to the contrary (not to mention numerous studies over the years by mostly African American journalists and historians). The law of slavery erased the father to the benefit of masters, their white kin, and the system of bondage generally. Jefferson's relatives, allies, and biographers abetted this erasure long after slavery ended. Their persistent denials that Thomas Jefferson could possibly have fathered Hemings's children, as well as others' equally dogged attempts to expose his paternity, reflect the immense social and political stakes of the quest for the father.[22]

The logic of strategic erasure was especially powerful in slave societies, but it is hardly peculiar to them. It characterized Napoleonic family law across Latin America and continental Europe in the nineteenth century. Forbidding people to file paternity suits to identify their fathers did not so much defer to a biological fact as ensure that this fact could not be discovered. The law suppressed exactly the kinds of social knowledge, such as witness testimony, that had historically defined paternity in courts and communities. It was thus not nature that made paternity uncertain but law that did so.

A similar logic prevailed in colonial societies, where Europeans treated paternity as ontologically distinct from paternity in the metropole. As an Italian long resident in East Africa put it, "the certainty of paternity, in the case of indigenous women, is quite different, if not impossible" because African women were supposedly more promiscuous than European ones.[23] Such beliefs justified individual colonists' reluctance to assume responsibility for their biracial progeny. In France a 1912 reform of the century-old Napoleonic prohibition of paternity suits allowed mothers to pursue legal action to identify the fathers of their children. But the reform was not applied in France's colonies because it would permit *métis* children born of French men and native women to identify their fathers and thereby acquire French citizenship.[24] Paternity was deemed unknowable when it involved higher-status men and lower-status, nonwhite women; conversely, the father was often considered obvious when a white woman birthed a nonwhite infant. If paternal uncertainty is a historical idea, the fact that some fathers have been more uncertain than others suggests it is also a political one. Paternal uncertainty is not just an idea but an ideology, one that has been strategically mobilized in the service of slavery, patriarchy, and empire.

The idea of the unknown father has persisted across millennia, but it is not timeless or immutable. It has a history, meaning that it has changed over time in response to political, economic, cultural, and legal circumstances. If the nineteenth century was the high-water mark of paternal uncertainty, it is not because the natural facts of human procreation changed circa 1800. Nor did those facts differ in colony and metropole. If the legal regulation of paternity differs today in Brazil and France, it is not because Brazilians are biologically distinct from the French. What has changed over time and what differs across place is the way people think about those facts. By recognizing transformation and variation, we can wrest paternity from biology and restore it to where it belongs, to history.[25]

IN THE FIRST DECADES of the twentieth century, this history took a dramatic new turn. Alongside the older orthodoxy of paternal uncertainty, a new set of ideas—modern paternity—began to take hold across the transatlantic world. In this new conception, paternity was a physical condition, and the father was the biological progenitor. What is more, his identity could be made certain thanks to new scientific methods. If previously the father was considered not just unknown but potentially unknowable, this moment heralded the birth of paternity as an empirical fact. If in the past the father was sometimes better left unknown, modern paternity advanced the imperative to reveal. Whereas in the nineteenth century, law was the chief arbiter of paternity, increasingly scientific authority was. Paternity was always embedded in relations of power, shaped by colonialism and slavery, but modern paternity's racial inflections were especially marked. Finally, this new conception of paternity was resolutely public: promoted by states, it took shape in courts but also in the press. Modern paternity thus had three parents: science, the state, and mass media.

If modern paternity was a set of ideas, the scientific paternity test was the method by which it was actualized. The rapid proliferation of new methods for assessing parentage beginning in the 1920s did not necessarily make paternity knowable, however. The ensuing decades of scientific progress made it possible to establish the identity of the father only in a limited number of cases. Rather, the transformation of how people thought about paternity—from an intractable mystery of nature to an empirical fact that could be discovered—made it possible to imagine a test to detect it.

The idea of a test is likewise significant. People had long been aware of the physical resemblances between parents and children and sometimes invoked them as evidence of paternity. Family likeness was often considered casually evident to a layperson. A paternity test, in contrast, implied a formal method based on specialized knowledge and promising an objective conclusion. Such a method required expertise, although at first it was not necessarily scientific or medical in nature. As late as the 1930s, Anglo-American courts sometimes deposed artists to perform paternity analyses, that is, to appraise bodies for similarities that indicated kinship. Increasingly, however, doctors and scientists became the undisputed arbiters of paternity.

Modern paternity augured a shift from legal epistemologies to scientific ones. While the courtroom was still the mise-en-scène for paternity determinations, what in the nineteenth century was couched as a legal puzzle was increasingly understood as a biomedical one. Language itself reflected this shift. From the 1830s to the 1880s, use of the phrase "law of paternity" increased, only to decline and bottom out around 1900. Beginning in the 1910s, a new phrase signaling a method of physical detection, the "paternity test," surged. An older legal framing was being displaced by an emerging science of paternity.[26]

This new science grew out of two taproots, the first of which was individual identification. The late nineteenth century witnessed an "identification revolution," as the emergence of nations and borders and unprecedented flows of global migration prompted states to monitor the movement and identity of citizens and foreigners.[27] Such surveillance required new technologies of identification, including those based on documents, such as the passport, and others that established identity by way of the body. The latter included the anthropometric techniques of the famous nineteenth-century French police official Alphonse Bertillon, photography, and fingerprinting.[28] Parentage testing followed a parallel logic, treating the body as a source of truth about family identity rather than individual identity.

Paternity science's other taproot was hereditarianism, the belief that heredity decisively shaped human nature and difference. The word "heredity" first appeared in the nineteenth century, and Charles Darwin's *On the Origin of Species* (1859) helped make it a central problem of biology. Already in the 1880s, the "father of eugenics" (and cousin of Darwin) Francis Galton mused, "It is not improbable, and worth taking pains to inquire, whether each person may not carry visibly about his body undeniable evidence of his parentage and near kinships."[29] Not coincidentally, the birth of kinship testing in the 1920s coincided with burgeoning scientific, political, and popular interest in eugenics and racial science. Paternity testing was a product of the same ethos and developed in close association with these fields.

But even as it drew on the common well of hereditarian thinking, paternity science differed from eugenics and racial hygiene in several important ways. Eugenicists were concerned with the classification and improvement of human hereditary stock. They postulated that heredity was either good or bad and that good heredity should be encouraged and bad heredity discouraged. As such, they operated at the level of the population, seeking to shape the fitness of races or nations through policies vis-à-vis marriage, reproduction, and immigration.

Paternity scientists, in contrast, sought not to shape reproductive outcomes but merely to discover them. They recognized no normative distinction between good and bad heredity. They simply used knowledge of inheritance to uncover biological kinship. Theirs was an eminently practical pursuit: they sought to know if this man was the father of that child. The scale of study was therefore different, directed at individual bodies—typically those of the mother, child, and putative father—rather than at populations. Compared with the grandiose and grotesque ambitions of eugenics and racial science, parentage testing appears modest and innocuous. It made no claims to inherent human inferiority or superiority. It was practical, not utopian (or dystopian).

Yet the scientific enterprise to know the father was hardly benign. Racial thinking is a defining characteristic of modern paternity, and its technologies were deeply entangled with the eugenic and racial projects of the first half of the twentieth century. Paternity science helped legitimate racial research and became one of hereditarianism's most prosaic and enduring applications. In German-speaking Europe, parentage tests developed in the 1920s morphed into racial ancestry tests in the 1930s and 1940s. After the war, when former race scientists and Nazi collaborators found themselves barred from their earlier professional activities, many quietly earned a living conducting paternity assessments for the courts.

If paternity science became a discrete pursuit—and for some a lucrative one—most researchers did not set out to discover a practical method of parentage assessment. Techniques of kinship testing were frequently accidental by-products of other research. When scientists discovered that human blood groups followed the laws of Mendelian inheritance around 1910, they were hardly looking for a paternity test, but this discovery suggested a practical method for resolving cases of disputed parentage. Twentieth-century paternity science relied heavily on techniques repurposed from nineteenth-century race science, identification methods, and criminology. Some paternity testers used Bertillon's anthropometric methods into the 1940s—a half-century after they were discarded by their original adherents in police departments. Paternity testing was an incidental pursuit of many fields but the exclusive focus of none. Medical doctors, physical anthropologists, eugenicists, biologists, and medicolegal experts all dabbled in kinship assessment as part of broader interests in race, heredity, bacteriology, serology, forensic science, and identification. Thanks to the diversity of its practitioners, its techniques were likewise eclectic—and often contested.

The eclectic and often exotic early methods purporting to establish physical kinship raise an obvious question: Did they really work? Was it (is it) possible to find a father in the teeth, the fingerprints, earwax? The answer is generally no. For much of the twentieth century, paternity science was not just rudimentary and experimental; sometimes it was patently absurd. A San Francisco doctor claimed he could reveal paternal origin with the oscillophore, a machine that measured electronic vibrations of the blood. German race scientists believed they had solved the problem of identifying both paternal and racial origin thanks to elaborate analyses of noses, ears, and other traits.

But paternity science also included the hereditary analysis of ABO blood types, a line of inquiry that eventually lead to modern genetics and garnered a Nobel Prize for its pioneer, Karl Landsteiner. Even dubious ideas drew on scientific symbols and were frequently peddled by well-respected, well-credentialed experts. If some of those ideas remained absurd (such as the os-cillophore), other ideas, no less implausible at the time they were proposed, ultimately panned out (the idea that paternity could be revealed in a drop of blood in the first place). Only hindsight can distinguish the preposterous from the prophetic.

Instead, paternity science should be taken as a field of knowledge on its own terms, significant not because of the validity of its truths but because people perceived a need for it in the first place.[30] Rather than asking whether paternity tests developed over the course of the twentieth century worked, a more fruitful question is, what work did they do? This shifts the focus from scientific veracity to social function. Considering DNA fingerprinting and the oscillophore in the same frame invites a new question: not whether they worked, but how they did so.

MODERN PATERNITY REFLECTED the belief that knowing the father was a social necessity as well as a scientific possibility. Dramatic changes in gender roles in the years following World War I inspired growing interest in kin-ship science. In Europe and the Americas, women's demands for and some-times acquisition of civil rights, their changing relationships to work and to consumption, and newfound social and sexual freedoms prompted cri-tiques of "the tyranny of modern women."[31] This "tyranny" inspired some men's rights groups to call for the reform of paternity laws perceived as bi-ased against them and the expansion of scientific testing to fend off spurious paternity claims. Women's rights advocates also lobbied for paternity reform, albeit for very different reasons. In the first decades of the twentieth century, the Napoleonic laws and Victorian morality that had long condemned the unwed mother and child appeared increasingly archaic. Once upon a time it had seemed appropriate that mothers who bore children out of wedlock should shoulder the burden of their transgression, but now activists forcefully challenged that double standard. From France to Argentina, they successfully campaigned to reform laws that had relieved men of responsibility for their illegitimate children.[32] In this context, too, paternity science acquired appeal. Would these new technologies protect men from female cunning or women from male irresponsibility? The gender politics of modern paternity were hardly a foregone conclusion.

Modern ideologies surrounding childhood also galvanized the scientific search for the father. In the late nineteenth century, new fields of knowledge—demography, public health, social hygiene, eugenics, sexology—subjected the family to novel forms of state power.[33] Public authorities imbued children with biopolitical significance, linking their welfare to society's eugenic, moral, and economic fortunes. They regarded paternity science as a tool of reproductive fitness because fatherless children were assumed to be more vulnerable to disease and death. Meanwhile, where once children worked in factories and fields, increasingly they were sent to school and to play. The child's dwindling economic value raised the stakes of parental responsibility. Who would shoulder the increased burden of this new, nonproductive life stage?

Modern ideologies of childhood also reimagined the young waif, long an object of private charity, as a citizen who bore rights. Children now enjoyed an expanded right to paternal support as well as to an entitlement that had not previously existed at all: the right to an identity, which appeared in contexts ranging from new protocols for birth registration to new legislation surrounding paternity investigation. These rights trumped the man's right to privacy, the legitimate family's right to patrimony, and the dictates of public morality, ideologies that in the nineteenth century had justified concealing the father.[34]

Modern paternity is also in part a creation of the modern welfare state. Public authorities were increasingly concerned with finding fathers lest poor children overwhelm public rolls. As one Norwegian statesman put it, "anonymous paternity is an offense against the child and against the State."[35] In the twentieth century, large welfare bureaucracies were designed to track down biological fathers, establish their paternity, and hold them economically responsible for their children—that is, to make paternity both known and actualized.[36] States were also concerned with making family relations legible because it was through family that they conferred modern resources ranging from social security to military pensions to nationality. "Civilized legislative systems and contemporary social structures," asserted one Brazilian observer, made it "necessary to embed the citizen in a network of blood ties."[37]

If twentieth-century transformations of gender, sexuality, childhood, and the family shaped modern paternity and its technologies, so too did transatlantic ideas and practices of race. Paternity science grew out of eugenics and race science, but modern paternity's racialization is most apparent in its application. Like paternity, race was understood to be an innate physical quality, an essential truth that could be hidden, ambiguous, or unknown. With striking frequency, racial truth and paternal truth were treated as mutually referential and revealing, as when an individual's race was considered uncertain because his father was unknown, or when a child's racial characteristics supposedly revealed her paternity. The trope of the white woman who births a brown baby—present in classical sources, endlessly recycled in the context of modern paternity, and still with us today—captures this conceptual association.[38] It is a story about interracial sex but also about the reciprocal ambiguity of race and paternity. Paternal uncertainty seemed to enable racial mixture, pollution, and indeterminacy. In the transatlantic imaginary, the abiding belief that paternity is inherently uncertain provided fertile ground for racial anxieties to bloom. It also shaped racial governance. From Nazi Germany to postwar immigration policy, paternity has figured centrally in the efforts of twentieth-century states to create and police boundaries of race, nation, and citizenship.

A more intangible but no less powerful impetus for modern paternity was the intense popular fascination with genealogical origins. "In all ages questions of identity have excited the interest of men," anthropometrist Alphonse Bertillon once mused. "Is it not at bottom a problem of this sort that forms the basis of the everlasting popular melodrama about lost, exchanged, and recovered children?"[39] The scrappy orphan who discovers he is the son of a gentleman may belong to the nineteenth-century novel, but stories about identity retained their salience in the twentieth century.

Ordinary people could easily imagine the uses of the new kinship science because it spoke to familiar tropes: the child swapped at birth or kidnapped by gypsies; the abandoned mother, her innocent babe, and the unscrupulous Don Juan; the adulterous wife and her cuckolded husband; the white mother and her racially ambiguous child.

Beginning in the 1920s, newspapers evinced a keen interest in the role of science in addressing such scenarios. Often they did so in the context of lurid scandals involving the rich and famous, as in Charlie Chaplin's paternity tribulations. Stories of sex and scandal, of identities lost and found, and of the sensational power of science to find the father had broad cross-cultural appeal. They circulated along the proliferating networks of the international wire services. The press explained new scientific developments to eager publics (often with dubious accuracy), and in some countries, such as the United States, it was the principal medium through which judges, lawyers, and state officials learned of advances they then put into practice. The press, then, was not merely a passive conduit of information; it actively helped produce new ways of thinking about paternity and spreading the technologies associated with them.

MODERN PATERNITY AROSE in the decades after World War I in a transatlantic milieu encompassing North America, Latin America, and Europe. Its history across this far-flung terrain is best told from a transnational perspective, to account for the circulation of these ideas, and a comparative one, to capture similarities and differences in how societies responded to them. Stories from certain countries—Argentina, Brazil, Germany, Italy, and the United States—while by no means exhaustive, are illustrative, capturing both global patterns and local variations.

In the first half of the twentieth century, dense networks of circulation and exchange traversed the transatlantic scientific community. This was especially true for the fields of expertise most relevant to paternity science, namely, forensic science, eugenics, and hereditary and racial science. Through multilingual journals, international conferences, personal collaborations, and friendships, scientists shared the knowledge, laboratory techniques, and even material supplies (such as blood serum) that underwrote the science of kinship. To be sure, filaments of this network extended to other parts of the globe, such as Japan and Egypt. But the transatlantic grid was especially dense.

The production and circulation of scientific knowledge was not distributed evenly across this geographic space, however. German-speaking Europe and the United States were the undisputed centers of early-twentieth-century scientific modernity, and certain key techniques, like blood group tests, emerged from their labs. Latin Americans were perennially more attuned to European and North American developments than vice versa. But spatial hierarchies blurred as scientists traveled across them: a number of key paternity scientists were Jews who fled Europe for the United States and Latin America, bringing their expertise to adopted countries where the discipline was less established. Moreover, paternity science was never a single set of practices or ideas; it had local variants as researchers in different locations developed their own methods to know the father. In this sense, paternity science had no clear centers and peripheries.

Modern paternity's transatlantic cradle also reflects a shared Euro-American tradition of family law. This tradition transcends the obvious distinctions between the civil law of continental Europe and Latin America and the Anglo-American common law. It cuts across important variations and peculiarities in national laws. Family law across Euro-American societies rested on Roman antecedents and drew on similar constructs of maternity and paternity, filiation and kinship, legitimacy and illegitimacy. At the beginning of the twentieth century, these included the basic assumption that the identity of the mother was inherently certain while that of the father was inherently uncertain. It included the presumption of marital legitimacy, which held that a married woman's husband was always and automatically the legal father of her children. It included the idea that illegitimacy was a morally and legally inferior status that justified discrimination against the unmarried mother and her child, including restrictions on the right of that child to discover its father. Not only did these countries begin the century with a shared legal tradition, but in succeeding decades, as social and political change gradually eroded time-honored legal orthodoxies, trajectories of reform tended to move in parallel fashion.[40]

Finally, modern paternity arose out of a shared set of cultural beliefs about kinship. Euro-American kinship rests on a series of oppositions, between nature and culture, the biological and the social, truth and fiction, which motherhood and fatherhood themselves epitomize. Where maternity is associated with nature, biology, and physical certainty, paternity is associated with culture, law, and intellectual presumption. In this framing, the true facts of kinship exist in nature, and knowledge consists of the revelation of these facts.[41] A scientific paternity test—a procedure that promised to rescue natural truth from the fictions of law and culture—made sense in this part of the world because it was consonant with prevailing cultural understandings of how kinship worked.

To be sure, Euro-American kinship was and is characterized by marked national and regional, as well as ethnic and class, variations. This is especially the case for Latin America, where historical patterns of marriage and filiation have contrasted markedly with those of Europe and North America. At the turn of the twentieth century, Latin America had the lowest rates of marriage and highest rates of illegitimacy in the world. By the 1920s and early 1930s, the years marking modern paternity's genesis, illegitimacy rates in many countries in the region were ten times higher than in Europe.[42] Little wonder that for many Latin American observers, the unknown father seemed an especially pressing problem. The category of Euro-American kinship thus signals not sameness but comparability, that cross-national and cross-regional differences can be fruitfully compared and contrasted.

MODERN PATERNITY IMPLIED a new set of beliefs and practices, a novel twist on the old quest for the father. The idea of modernity and progress has in fact long been central to the way Euro-Americans have thought about paternity. Victorian theorists looked backward, fascinated by notions of descent in the distant human past and how they differed from the present. Later ethnologists and anthropologists looked sideways, fixating on the "primitive paternity" of contemporary savages that supposedly contrasted with their own civilized understandings.[43] The medical and forensic experts who worked to develop a science of paternity were captivated by these histories and ethnographies. They wrote frequently about the colorful beliefs about fatherhood among ancients and aborigines.[44] Primitive paternity captivated them because it was the foil against which they defined their own scientific quest. If paternity's past was one of ignorance and doubt, its present was distinguished by comprehension, and its future, perhaps, by certainty. It was not just the scientists who understood things in these terms.

The press too lauded scientific experts as "modern Solomons," who had ushered in a solution to the problem of "obscured paternity" characterizing "life and art . . . since the earliest recorded history of man."[45] Part of what defined modern paternity was its insistence on its own modernity.

Yet this insistence obscures strong elements of continuity. Many of the "new" scientific techniques for finding the father were in fact recycled, and they were also frankly limited in their power to find him. Above all, the continuities were social. Modern paternity heralded new ways of thinking about paternity, identity, and ancestry, but in practice it did not displace the old ways. The idea that science could expose paternal truth provoked fascination but also deep ambivalence, for it could challenge established orthodoxies surrounding gender, sexuality, and the family. Older visions of paternity lived on in popular beliefs and were defended in a variety of contexts by political, legal, religious, and military authorities. Most surprisingly, doctors and scientists themselves—the consummate avatars of modern paternity—sometimes privileged traditional ideas over modern scientific verities.

The quest for paternity in the twentieth century exposed long-standing tensions between the social and the biological; the scientific and the legal; the imperatives of truth and those of justice, morality, and social order. Rather than resolving those tensions, it reified them. It is thus not a story about how a new set of beliefs, practices, and technologies eradicated the old. It is a story about the contested paternities that persist into the present.

In his miraculous paternity test, Saint Anthony asked the baby, "Who is your father?" and the child turned to the man and declared, "that is my father." The baby may have given the right answer, but the saint had asked the wrong question. As the history of modern paternity shows, the truly difficult question is not who the father is, but what do we want him to be?

2

THE CHARLATAN AND THE OSCILLOPHORE

> The work of a great scientist may be full of error, and conversely, the
> work of a charlatan full of surprising significance.
>
> —Abraham Goldfarb, Biology Department, City College of
> New York, 1930

IN 1921, MRS. ROSA VITTORI filed charges in a San Francisco
court against her former husband. Paul Vittori refused to pay child support
for her two-month-old daughter Virginia because, he insisted, the baby was
not his. Paul was an Italian immigrant and streetcar conductor. Rosa was
from Spain. The couple had been married two tumultuous years, and this
was merely the latest twist in the catastrophic demise of their union. By the
time they wound up in the courtroom, the couple had been granted two
separate divorces on grounds of cruelty, and Rosa had been briefly jailed
after an altercation involving a revolver. It was a fairly conventional, if heated,
story of domestic misery, but Paul's refusal to recognize the baby would soon
land the Vittoris in the pages of newspapers around the world.

The couple appeared in the courtroom of Judge Thomas F. Graham of the
Superior Court of San Francisco. Judge Graham was a celebrated local figure
who in his twenty years on the bench had become known as the "Prince of
Peacemakers" for his ministrations to estranged couples. The *San Francisco
Chronicle* regularly reported on his courtroom conciliations ("Reconciled by

Judge Graham: Quarreling Couples Make Up," "Judge Effects Another Reconciliation"). The judge supported the right of couples to end their marriages but lamented the social scourge of easy divorce, a message he disseminated both in court and public lectures. For Rosa and Paul Vittori, however, marital reconciliation was no longer a possibility, nor even at issue in the legal proceeding. At issue was the question, who was baby Virginia's father?

In addressing this question, the law was squarely on Rosa Vittori's side. If divorce was increasingly common, challenging a child's legitimacy was quite a different matter. California law followed the deep-rooted legal tradition, dating back to Roman law, which declared a married man to be the father of his wife's children. The law made it very difficult, and sometimes impossible, for a husband to challenge this "presumption of paternity," a restriction intended to protect the rights of legitimate children and the integrity of marriage in society.

Yet rather than rest on the presumption of the law, the young lawyer representing Mrs. Vittori opted for a different, and highly unusual, approach: a blood test. Attorney Stanley F. Nolan was twenty-four years old and just a few years out of law school. He had read of new scientific experiments conducted in Paris and at Johns Hopkins that purported to demonstrate parentage through blood, and inspired by these experiments, he proposed a scientific test to prove baby Virginia's paternity. Nolan approached Central Emergency Hospital, a large public facility, but was told they had "no equipment for such a test." A doctor at the hospital advised him to consult an expert in gynecology and obstetrics. Perhaps a specialist at one of the local universities could help.[1]

The young attorney's quest was unconventional, but it was also prescient. No blood test to determine parentage was yet in routine use anywhere in the world, and it is unclear what tests, in Paris or at Johns Hopkins, he had read about. But his inquiries suggest that the idea of such a test was already in circulation. What is more, it appeared credible enough that a lawyer could assume that such a procedure, though perhaps not yet widely known or routinely used, not only existed but that it could help his client.

An article on the curious case appeared in the local newspaper. It ended inconclusively, noting that the lawyer would "endeavor to interest [scientific experts] in the case, in the belief that both medicine and law will benefit thereby." In the meantime, two obvious questions remained unresolved: Who was baby Virginia's father? And could science shed any light on this question? Little could the lawyer, the judge, or the unhappy couple foresee the extraordinary interest these questions would generate.

PROBABLY THE FIRST CASE of its kind in the United States, the Vittori suit appeared on the cusp of a veritable explosion of scientific, legal, and popular interest in paternity testing. Soon academic journals, professional conferences, and the international press would bubble with talk of blood, bodies, and parentage. "Deep down in those minute eddies of the human blood, Nature has placed the hallmark of every man's heredity," mused the *Atlanta Constitution.* "In his blood cells is bound up the unmistakable record of his fatherhood."[2] In the 1920s, the "unmistakable record" of the father, so carefully concealed by Mother Nature, seemed poised to be unmasked. An anthropologist in Buenos Aires developed a technique based on the principles of Mendelian heredity to solve an inheritance dispute. In Russia, Austria, and Scandinavia, scientists designed paternity analyses based on physical similarity. In Berlin, a court accepted a test of paternity based on ABO blood typing. Shortly thereafter two forensic doctors in São Paulo became the first in the Western hemisphere to perform a blood group test. Hospital officials from Cleveland to Havana attempted to solve sensational cases of baby mix-ups using scientific methods. By the mid-1920s, courts in Germany and Austria began to routinely accept biological evidence in paternity disputes, and within a few years, at least 5,000 such tests had been performed. Methods based on hereditary blood groups jostled with more dubious, but often more captivating, claims that parentage could be determined through blood crystals, electronic vibrations, and light particles.

Through the newspapers, global publics followed the exciting developments occurring in laboratories and courtrooms. In the United States, readers learned about techniques of parentage determination in magazines like *Popular Science Monthly* and *Popular Mechanics* and even in detective stories. Argentines followed them in the celebrated society magazine *Caras y Caretas.* Readers of the *Times of India* learned how scientific evidence had allayed the suspicions of a doubting father in rural Nebraska. Some of the claims of fantastic scientific achievement were bona fide, others less so. The press rarely made such distinctions, and slowly the idea that science could establish a kin relation through an examination of blood and body took hold.

Such developments signaled the birth of modern paternity. In the nineteenth century, the biological identity of the father was considered not just unknown but unknowable, concealed by an "impenetrable veil." By the 1920s, such perspectives looked increasingly quaint. Science appeared poised to pull back the veil, to reveal the ineffable mystery of kinship in general and paternity in particular. When a group of forensic doctors in New York City gathered to discuss recent developments in paternity science, they were willing to entertain the possibility of even far-fetched new claims: "things which seemed impossible," observed one participant, had blossomed into promising new advances. For the lawyer Stanley Nolan, as for many observers, the question was less whether such a test was possible than where a doubting father, abandoned mother, or conscientious young attorney might obtain one.

Two weeks after the initial hearing in Judge Graham's court, the Vittori case took a dramatic turn. Nolan had located a specialist willing to perform the necessary test, a local doctor by the name of Albert Abrams. Dr. Abrams conducted an analysis of the blood of the three individuals, Rosa, Paul, and Virginia Vittori and announced his verdict, which he considered "absolutely conclusive." Despite his adamant protestations, Paul Vittori was Virginia's father.[3] Judge Graham ordered the errant father to pay $25 a month in child support to his ex-wife and declared Abrams's test "one of the biggest things established by medical science in years."[4]

THE SAVANT RESPONSIBLE for the startling breakthrough, Dr. Albert Abrams, was no ordinary doctor, and indeed some would say that he was not a doctor at all. To be sure, the San Francisco native had obtained a medical degree in the early 1880s at the prestigious University of Heidelberg and had enjoyed a successful career in medical research and treatment. At the time of the Vittori case, he was affiliated with the recently established medical school of Stanford University. But over the previous decade, Abrams had moved increasingly beyond the fold of medical orthodoxy. He had developed a theory of the body as an electrical system, which he called ERA, the Electronic Reactions of Abrams. According to this theory, diseased and healthy parts of the body gave off electric vibrations that could be measured by a special machine and then interpreted by a trained ERA diagnostician. Through a series of astonishing gadgets of his own invention, Abrams claimed to be able to diagnose and cure a range of ailments, from tuberculosis to syphilis to cancer. By the time of the Vittori case, he had founded a journal, a laboratory, and a special school devoted to electronic medicine. He also had a growing cadre of disciples, both medical practitioners who trained in his methods and grateful patients who benefited from them. But it was Virginia Vittori's paternity that first vaulted Abrams into the national and international limelight.[5]

It is unclear how Abrams came to lend his services to the court in the Vittori case, but the most likely scenario is that he read about Nolan's search for a blood test in the newspaper and came forward to offer his services.[6] If this was the case, it presaged the synergies between science, the courts, and the media that would promote the dissemination of paternity testing.

However Abrams first became involved, the Vittori case was an excellent opportunity to showcase his scientific repertoire. Up until then, his circle of followers knew him for his ability to cure illnesses, not identify wayward fathers. But among Abrams's inventions was a machine known as the oscillophore (also referred to as an oscillospore, oscillophone, or electroradiometer), which purported to measure the vibrations of electrons in a drop of blood. Rates of blood vibration, according to Abrams, varied across age, sex, race, and other characteristics. He had worked out various rates at which blood vibrated according to the ethnic ancestry of the individual (Jewish blood, 7 ohms; Irish, 15 ohms; German, 13 ohms; and so on). What is more, because "racial rates of vibration are transmitted to the progeny" and "a child through generations has the same vibratory rate as its parents," a comparison of the blood of different individuals could reveal whether they were related.[7]

This is how Abrams had determined that Paul Vittori was Virginia's father. Because the case happened to involve parents of two different "races" (a concept that at this time encompassed what we would now call nationality), it also demonstrated the oscillophore's remarkable powers of racial identification. Baby Virginia's blood revealed that "on the father's side she was Italian and on the mother's side 16–25ths of an ohm Spanish and 3–25ths of an ohm French, measured electrically."[8] The oscillophore creatively fused two popular-scientific obsessions of the era: electricity and heredity. It also reflected an enduring characteristic of modern paternity and its science: whether based on blood groups, physical traits, or electronic vibrations, it was inextricably bound up with the idea of biological race.

"TO THE AVERAGE INDIVIDUAL hampered by tradition it may appear incredible for science to have so far progressed to determine parentage by the examination of a few drops of blood," Abrams would observe several months after the stunning Vittori verdict.[9] Indeed, some skeptics would dismiss his test as "preposterous," and the medical establishment would later denounce Abrams as the "dean of twentieth century charlatans."[10]

Yet for every adamant skeptic, there was an expert open to the possibility that maybe, just maybe, Abrams was on to something. At the September 1921 meeting of the Society of Forensic Medicine in New York City, members engaged in "quite a discussion" about the new paternity test and decided to appoint a committee to investigate it. If its efficacy could be confirmed, Albert Abrams "deserved to have his name inscribed in perpetuity in the annals of Forensic Medicine."[11] That a reputable group of scientists was willing to entertain the possibility that a machine could reveal the secret of parentage through electronic vibrations suggests that there was nothing intrinsically ridiculous about this claim. It was extraordinary, perhaps, but not impossible. After all, "extensive progress has been made along serological lines during the past few years," the assistant medical examiner of New York reminded his colleagues. "Things which seemed impossible then, have since developed into tests of real diagnostic value."[12]

If Abrams's paternity test was credible, it was because it located identity precisely where, in the 1920s, most people would have expected to find it: the veins. The idea that blood carried the essence of selfhood was a deeply compelling one. Blood is perhaps the most culturally ubiquitous idiom for talking about race, identity, and family, "the major symbol of our kinship system," and "a liquid rich in allegorical meaning."[13] The idea of a blood test of paternity combined the ancient cultural association of blood and ancestry with the modern preoccupation with heredity. If in conventional wisdom the veins were a metaphorical locus of ancestry, Abrams's oscillo-phore made them a literal one.

ALBERT ABRAMS WAS not the first investigator to distill ancestry from a drop of blood. Several years earlier, Edward Tyson Reichert, a professor of physiology at the University of Pennsylvania, made a similar claim based on a technique called crystallography. Reichert and a colleague amassed blood samples of more than a hundred vertebrate species and painstakingly measured the angles of the crystallization patterns of their hemoglobin. They surmised that the patterns could serve to differentiate between species and elucidate the evolutionary relationships between them. Older methods of comparing species relied on morphological similarity, but such methods were increasingly criticized as subjective. After all, similarity and difference could rest in the eye of the beholder. Reichert's efforts were part of a broad current of research that sought to discover a more objective method of comparison based on some measurable biochemical property.[14]

Today, Reichert's crystallography is remembered as an important contri-bution to molecular biology and the understanding of evolutionary relation-ships. Forgotten is his attempt to use it to find similarity and difference among humans. For if crystal patterns could distinguish between species, perhaps they could also distinguish human races, or even individuals. Per-haps each person's hemoglobin had a characteristic crystal pattern, akin to a fingerprint. The practical applications of crystallography, from racial an-thropology to forensic science, seemed endless. And then, of course, there were kin relations. Reichert's studies of the blood of children and parents revealed that "there are certain phenomena present in the blood of a child which are also apparent in the blood of the father." "Within the next year," Reichert told a journalist in 1913, "I will have established proof that he-redity can be traced through the blood crystals."[15]

Like the oscillophore a few years later, crystallography fired the popular imagination. Reichert's work was reported in the press and even appeared in detective fiction. The celebrated writer Arthur B. Reeve narrated the ad-ventures of a "scientific detective" known as Craig Kennedy (also dubbed the American Sherlock Holmes) for *Cosmopolitan* in the 1910s. In one ad-venture, the artful sleuth uses crystallography to reveal that a blonde woman is in fact a disguised "negress." "I wonder if you have ever heard of the Reichert blood test?" Detective Kennedy asks his amazed associate. "Already they can actually distinguish among the races of men, whether a certain sample of blood, by its crystals, is from a Chinaman, a Caucasian, or a negro." Tests of race inevitably suggested tests of kinship: "They even hope soon to be able to tell the difference between individuals so closely that they can trace parentage by these tests."[16]

Despite the optimistic predictions of the fictional detective and of Reichert himself, a crystallographic proof of paternity never came to fruition. But his quest was not futile, for it primed the public for Albert Abrams and his oscillophore several years later. Rosa Vittori's lawyer explained that he first decided to seek out a scientific test of paternity because he had read about such a method. Two years before the Vittori case, an article on Reichert's crystallography had appeared in the *San Francisco Chronicle.* That article was the likely source of the lawyer's inspiration. In a sense, then, thanks to the newspapers, Reichert's test eventually did enter the courtroom—even though it did not actually exist.[17]

HAVING BEGUN AS JUST another colorful vignette from Judge Graham's courtroom, reports of Abrams's paternity test soon traveled across the country and then the world. Wire stories reached England, France, and Italy. Several Australian papers speculated on how the "New Yankee Theory" could apply to a divorce case in Perth. "Seducers . . . watch out for the hematogram!" warned a French observer. In a small municipality in northeast Brazil, the local newspaper predicted that Abrams would be "overwhelmed by requests from the four corners of the world" given the invention's promise to "unveil profound mysteries, innumerable surprises, and deceptions!!!" An Argentine society magazine declared "the vulgarization of the use of this apparatus . . . of transcendental importance for the world."[18]

Back in the United States, Albert Abrams went from a colorful local character to a national celebrity. A few months after the Vittori verdict, the 800,000 readers of *Popular Mechanics* could learn about hot air balloons, electric watches, suspension bridges, and the oscillophore.[19] Attorney Nolan and Judge Graham were deluged with inquiries about the case. And as the

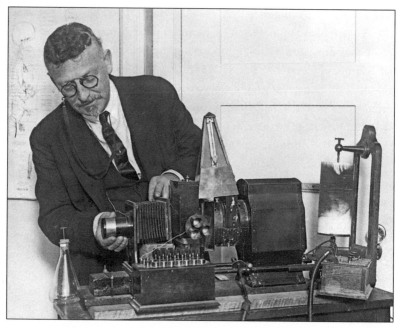

Dr. Albert Abrams and his sensational oscillophore, 1921.
Bettmann / Getty Images.

Brazilian newspaper predicted, a steady stream of requests began to arrive on Abrams's doorstep.

One request came from Atlanta. For more than a year, the city had been riveted by an alleged hospital baby mix-up, in which two families claimed a single infant. One family's lawyer read about Abrams's test in the newspaper and sent an inquiry to Judge Graham, asking if he considered the test "reliable." Abrams sent him materials to collect blood samples from his clients.[20] In Boston, the mother of a child "stolen by gypsies" appealed to a hospital for a blood test to determine whether a young boy recovered by police was her son. In Chicago, the technique promised to resolve a conflict between two Polish women both claiming maternity of a child. "I realize now the true wisdom of Solomon in his decision," the judge mused, pledging that "blood tests and expert medical testimony" would identify the correct mother and send the perjurer to prison.[21] Meanwhile, enterprising observers found entirely new applications for Abrams's method. In Illinois a farmer disputing ownership of a young bull requested "a blood test similar to that used recently to determine a child's parentage in California" to demonstrate the animal's provenance. (A veterinarian testified that the test was not applicable to cow blood.)[22]

If the oscillophore promised to restore swapped infants, stolen children, and the occasional lost bull, its most widespread public association was with cases like that of the Vittoris, involving disputed paternity. Within weeks of the Vittori verdict, Abrams was summoned back to Judge Graham's court, this time by a woman named Mrs. Mamie Del Secco, who wanted to prove that her ex-husband was not the father of the eight-year-old son of whom she sought sole custody. Again, the oscillophore offered a dramatic resolution to the case when it affirmed the ex-husband's paternity. Throughout much of the year (and for a decade thereafter), the papers followed the noisy divorce suit of James A. Stillman, the fabulously wealthy president of the National City Bank of New York. Stillman charged that the fourth child of his wife of twenty years, the former Fifi Potter, had been fathered by a "Canadian Indian" employed on their estate. In another riveting case of alleged adultery, John Tiernan, a law professor at the University of Notre Dame, attributed paternity of his wife Augusta's fourth child to a South Bend haberdasher. The lurid twists and turns of the Stillman and Tiernan cases captivated the public for months, and Abrams's test was a central feature in the coverage. (Eventually, both children were found to have been fathered by their mothers' husbands.) The papers also explored its applicability to the "Russell baby case," a divorce and paternity dispute that so scandalized English society in the early 1920s, it led to legal restrictions on how the press covered divorces.[23]

Scandalous stories of divorce, adultery, and illicit affairs were a mainstay of the press in the 1920s, and questions of paternity became a standard part of these narratives. Litigants and lawyers began to inquire about, and sometimes formally request, blood tests as evidence of paternity. Press coverage of high-profile cases now routinely referred to such evidence. Which party wanted a test, whether it would be performed, and what it might show became integral parts of the public narratives of scandal. The cases that received the most attention featured wealthy men—a "clubman" in St. Louis; a coal merchant in Bethesda; a corset manufacturer in New Haven; Cornelius Vanderbilt Whitney, scion of not one but two dynasties. The mothers, the estranged wives and paramours of these wealthy and powerful men, were often dancers, "chorus girls," and onetime "stage beauties." Meanwhile, Abrams took his oscillophore on the road, "called from coast to coast to use his tests as evidence in sex trials." In Chicago, he presided over conventions on electronic medicine and met with Mr. and Mrs. Tiernan. In New York, he gave a lecture at Carnegie Hall, with admission proceeds going to fund a free Abrams clinic in the city. In Boston, he spoke to an enthusiastic crowd of more than 1,000 people packed into the dining hall and balconies of the Copley Plaza.[24] Modern paternity was resolutely public, intimately intertwined with the culture of sensation and celebrity.

But ordinary men and women were also drawn in. They read about Abrams's test and saw in it a panacea for their own personal dramas. One Everett Campbell of Los Angeles accepted his fatherhood of an infant girl after a doctor showed him the results of the oscillophore.[25] From Lamont, Oklahoma (population 585), a Mrs. C. W. Womack wrote Dr. Abrams in April 1921: "I see by the papers that you can test blood and wondered if you would help me any in my case." She was the mother of three small children, but her estranged husband refused to recognize the youngest one, a three-month-old daughter. "I am so sorry to think a father would do such a thing and I want to show to the world and also to him and his lawyers [and] also mine that it is realy [sic] and truly his."[26]

These dramas of the seduced and the cuckolded, the deceived and the abandoned, could have been plucked from a nineteenth-century novel. What was new was the idea that science could resolve these most intimate of disputes. "With . . . a scientifically accurate determination of paternity," observed a Baltimore paper, "a world of misery, suspicion, lies and faithlessness would be abolished from the earth."[27] "In the course of time," predicted the *Atlanta Constitution*, "extended controversy over [parentage] will not be possible. . . . The court will simply say 'The oscillophore declares that John Doe is the father of the child—and what the oscillophore says—is law.'"[28]

The technology was particularly well suited to the 1920s, for in the sultry hothouse of sexual modernity, such dramas appeared to be proliferating. The flapper, Freud, and Fitzgerald created the impression, if perhaps not always the reality, that sexual mores were coming undone. Gender roles were also in transition. During the war, women entered the labor force in unprecedented numbers, demonstrating their aptitude for jobs for which they were once considered incapable. Only a few months before the Vittori case broke, the Nineteenth Amendment established women's suffrage and, with it, their political citizenship. Some contemporaries detected in these transformations "a slackening of husbandly and parental authority."[29] And yet in spite of changing sexual norms, or perhaps because of them, a central trope of Victorian sexuality lingered: the uncertain father.

Enter Albert Abrams and his miraculous contrivance. "A simple set of blood tests," he told a journalist, "will settle the matter beyond question."[30] "The matter" in this particular reference was the paternity of two-year-old Guy, whose paternity was disputed by James Stillman, the New York banker. More generally, "the matter" was the roiling tempest of modern gender roles.

One interpretation of Abrams's electronic gadget is that it registered a moral panic about female adultery. The most notorious cases through which Abrams achieved front-page status—Vittori, Del Secco, Tiernan, and Stillman—all involved married couples. The cuckolded husbands hailed from a cross-section of (white) America: they included an immigrant street car conductor, a law professor, and a blue-blood banker, their dramas set in San Francisco and New York as well as small-town Indiana. Perhaps the idea of a scientific paternity test gained credence because it reflected a widespread fantasy of anxious husbands in the face of quickening social change.

But if the oscillophore could expose the perfidy of women, it could also expose the iniquity of men. After all, it was not Mr. Womack of Lamont, Oklahoma, who felt moved to write Dr. Abrams for assistance. It was his wife, who appealed to the doctor to absolve her reputation before her husband and her community: "I want to show them I am a true woman and they don't half [*sic*] to take my word for it."[31] Like many future paternity testers, Abrams emphasized how his technology would serve as a tool of justice by vindicating wronged women. "I am convinced that this invention will effect a revolution in this and similar types of cases," he declared in an interview about the Vittori verdict. "It will prevent an injustice being done in the case of a girl who has been wronged by some scoundrel who tries to evade his legal responsibilities." Thanks to the oscillophore, "I can imagine many [scoundrels] who must be shaking in their shoes today."[32]

Yet the cases did not always conform to a simple narrative of female vindication either. In two of Abrams's high-profile cases, wives solicited the oscillophore to prove not their innocence but rather their "guilt." In order to gain custody (Mrs. Del Secco) or to punish a faithless former lover (Mrs. Tiernan), wives asked Abrams to show that the disputed child was not their husband's. Abrams's scientific method of paternity determination and others like it would settle a range of conflicts and abet a variety of interests.

The sexual politics of the oscillophore were thus fluid and volatile, and in this Abrams's gadget presaged the sexual politics of paternity testing generally. Genetic technologies could serve the interests of men or women, and what those interests were depended on context. Technology might promise to restore an older sociosexual order or to usher in a new one. And rather than settling long-standing problems, science could create new ones. Mrs. Womack saw Abrams and his test as her ally, but perhaps the media sensation about adultery had helped plant the seeds of doubt in Mr. Womack's mind in the first place. A central paradox of modern paternity and its technologies is that the very promise of scientific certainty often sowed doubt.

FOR MORE THAN TWO YEARS, the oscillophore and its inventor continued to make headlines. But as blood testing became a central part of the public narrative of American paternity suits, its status in legal practice was deeply contested. In the Stillman case, skeptical lawyers rejected Abrams's analysis; in the Atlanta hospital baby swap case, opposing counsel mused that the oscillophore could no more identify the correct parents than could a Ouija board. The test was endlessly discussed in relation to both cases but was ultimately performed in neither. Meanwhile, Judge Graham accepted Abrams's blood results (twice), but another San Francisco judge refused to permit his findings into evidence. In the Tiernan case, Abrams met with Mr. and Mrs. Tiernan, who pled with him to perform the test. But citing his desire to remain an impartial expert, he refused to do so without a formal request from the court. No such request was forthcoming.

Abrams's bid for legal legitimacy is telling. For all the excitement it generated, the oscillophore does not appear to have determined a verdict in any courtroom besides Judge Graham's. And even there its influence is unclear: while the papers widely reported that Abrams's test had been the deciding factor that convinced the judge of Paul Vittori's paternity, the verdict may simply have followed the traditional legal presumption of marital legitimacy in California law.[33]

The *Los Angeles Evening Herald* reports the verdict in the Vittori case, 1921.

"Parentage of Girl Determined by Blood Test," *Los Angeles Evening Herald*, February 15, 1921, 1.

Abrams's oscillophore clearly did not gain universal scientific or legal acceptance. What his machine did do was establish the idea of a biological parentage test. Initially this idea was closely associated with Abrams and his invention, but not for long. Within a few months of the Vittori case, newspaper reports on cases of disputed paternity began referencing generic scientific "blood tests" without mentioning Abrams. Who performed such tests? What did they consist of? These details were usually left unspecified. It was thus not the oscillophore per se that gained legitimacy, although

Abrams proved a particularly effective spokesman for his method, but the abiding idea of modern paternity: that scientists could reveal the physical link between parent and child.

The press helped bring forth this idea. The 1920s was the era of "banners and spice," as American newspapers eagerly embraced stories of sex and crime to attract readership.[34] Modern paternity rode in on the coattails of media scandal. But the Abrams story was not just about sex; it was also about the miracles of modern science. Although tabloid journalism was a hallmark of this period, the oscillophore's exploits were actually reported in older, conventional newspapers. The press framed it as a story of scandal but also of scientific breakthrough. It adopted a didactic stance, explaining to readers the science of paternity, even if the lessons it proffered were frequently incomplete or garbled.

The press did not simply transmit these developments but helped drive them. Newspapers promoted new paternity technologies by linking scientists to lawyers and the public. In the Vittori case, Abrams probably learned of Attorney Nolan's search for a paternity test for baby Virginia through the *San Francisco Chronicle*'s reporting. Mrs. Womack in Lamont, Oklahoma, wrote to Dr. Abrams after reading about him in the paper. A few months after the Vittori verdict, the *Atlanta Constitution* published a long piece that described the miraculous electric paternity test, its uses in the Vittori and Stillman cases, and its possible application to Atlanta's still unresolved baby swapping dispute.[35] The lawyer for one set of the Atlanta parents read the article and within days contacted Judge Graham, wondering if the blood test might aid his own clients. He was at once cautious and intrigued by the headlines: "Although I fear that the newspapers have used the theory as a basis for news stories, I feel that the subject is worthy of investigation."[36] The article from the Atlanta paper, including its nine photographs and two illustrations, was then reproduced in the *San Francisco Chronicle*. Three days later, the San Francisco paper followed with a story about the Atlanta lawyer's inquiry. Its gleeful banner read: "Chronicle Story May Aid Georgian." By linking the lawyers and parents in Georgia with the scientist and judge in California, the newspaper itself had become a protagonist in the drama.[37]

The press further stimulated paternity science by acting as an arbiter of the new method. Judge Graham's finding that baby Virginia was Paul Vittori's daughter may have rested less on the determination of the oscillophore than on the presumption of marital legitimacy in California law, but the headlines told a different story: "Court Accepts 'Blood Test' of Parentage," "Settles Paternity by Test of Blood," "Court Establishes Parentage of Baby by Electric Blood Test." In reporting, perhaps incorrectly, that the court had embraced the oscillophore's validity and established a groundbreaking legal precedent, the press brought that precedent into being. Newspapers did not passively communicate verdicts, in other words; they actively helped create them in the court of public opinion. The press would continue to play this role as new technologies emerged in succeeding decades.

ABRAMS ATTRACTED ARDENT supporters who, though probably always a minority, were vocal and remarkably high profile. The past president of the British Medical Association enthusiastically endorsed the Electronic Reaction of Abrams. The chief surgeon of the Chinese navy, visiting the United States as a delegate to the American Medical Association at the moment the Vittori case hit the press, described Abrams's paternity test as "infallible" and ordered an oscillophore himself.[38] In a lively meeting, members of New York's Society of Forensic Medicine expressed enthusiastic, if cautious, curiosity about the Abrams test. Their chief reservation was with Abrams's claim that his test was infallible, a categorical assertion that vexed scientists used to grappling with probabilistic claims. Still, they expressed great interest in the method: "no doubt the test must have some value," opined one member.[39]

Abrams did not just attract a following; he developed a franchise. For $200, "reputable physicians" could take courses in electronic medicine. Six months after the Vittori case, several dozen doctors traveled to San Francisco each month to visit his laboratory and study his methods. For those who could not make it to California, there were correspondence courses. Abrams leased one of his inventions, the oscilloclast, at $200 down plus $5 a month. By late 1921, more than 130 practitioners had taken out a lease, some with multiple machines, and the number soon doubled.[40] Meanwhile, annual subscriptions to Abrams's journal, *Physico-Clinical Medicine*, were $2 ($2.50 outside the United States, Canada, and Mexico), and Abrams's books ran $5 a copy.[41] By October 1923, according to one estimate, the number of physicians using his methods had ballooned to 3,000.[42] While most practitioners were probably interested in his diagnostic and curative techniques, they also performed parentage analyses. One American doctor and follower of Abrams resident in London, a Dr. McCouney, recounted to a journalist how he had assisted a wealthy patient in identifying the father of his maid's child (contrary to the man's suspicions, his son had not fathered the child, but by coincidence, another patient of Dr. McCouney's had).[43]

Meanwhile, Abrams's unconventional methods attracted high-profile adherents beyond the field of medicine, including the writers Sir Arthur Conan Doyle and Upton Sinclair. Sent by *Pearson's Magazine* to investigate Abrams's sensational claims, the muckraking Sinclair was captivated by what he witnessed in Abrams's San Francisco clinic. Sinclair's laudatory report, "The House of Wonder," probably did as much to further Abrams's public standing as the legions of grateful patients he had restored to health. The article circulated widely in the United States and was also published in France and Brazil.[44]

Indeed, Abrams's devotees extended well beyond the United States. The lessees of his gadgets hailed from five foreign countries. A correspondence course in Abrams's methods was available in Rio de Janeiro.[45] He boasted a particularly lively group of disciples in Britain, probably thanks to the very public, and very controversial, support of the former head of the British medical guild. He also cultivated a devoted following among the Mexican political elite. The personal doctor of former dictator Porfirio Díaz (1877–1910) was a follower of ERA. When President Álvaro Obregon (1920–1924) invited Abrams to advise him on health matters in 1923, he sent a private car to pick up the healer in El Paso. Obregon's successor, Plutarco Calles (1924–1928), was another follower. Plagued by chronic ailments resulting from his military service during the Mexican Revolution, he consulted two North American authorities in search of relief: the Mayo Clinic and Albert Abrams.[46]

But Abrams's methods also attracted widespread skepticism. Among some, and perhaps most, orthodox medical practitioners, news of a scientific test of paternity was met not just with skepticism but with disdain. The dean of the University of Michigan medical school declared it "preposterous"; the head of the Pasteur Institute in Paris pronounced it "nonsense."[47] When the mother of a child "stolen by gypsies" appeared at a Boston hospital asking for a blood test, a hospital physician did not mince words. "This blood test scheme originated in San Francisco, then came to Chicago and was heard of again in Iowa [*sic*: Indiana was meant here, a reference to the Tiernan case]. Next we heard of it in the Stillman case. . . . And at last it has reached staid old Boston. You may be sure, however, that such a test will not be made at this hospital." The mother went home "much disappointed." The official's acerbic statement, which came a week after Abrams's warm reception by throngs of admirers at the Copley Plaza, made front-page news.[48]

But by far Abrams's most vocal, and powerful, critic was the American Medical Association (AMA). The AMA, which considered the exposure of medical quackery to be one of its central missions, orchestrated a rancorous, single-minded, but only partially successful campaign to discredit the eccentric healer. The organization's Propaganda Department, which investigated charlatanism, was inundated by inquiries about ERA from doctors and members of the public alike. Its august mouthpiece, the *Journal of the American Medical Association (JAMA),* published multiple articles lambasting Abrams and then distributed thousands of reprints in pamphlet form for doctors and the public.[49] The journal's editor Morris Fishbein, who for decades was the public face of the medical profession, wrote best-selling books on quackery that prominently featured Abrams. In the late 1930s, an AMA publication for the lay public declared "the name of Albert Abrams . . . leads all the rest in the history of medical charlatanry in the first quarter of the present century."[50]

The motives for the AMA's ire were clear enough. As ERA practitioners began to pop up in communities across the United States, conventional doctors found themselves hard-pressed to compete with the method's miraculous claims. Abrams's technique for curing dreaded diseases "virtually ridicules established medical science by putting diagnosis and treatment upon just as positive a basis as the measuring of an electric generator's output or the location of trouble in an electric circuit."[51] If Abrams's gadgets did what he claimed, diagnosis and cure were but a dial's turn away and traditional medicine obsolete. The AMA dismissed ERA as a crass scheme to enrich its founder, but economic interests clearly motivated the guild as well: it feared Abrams's impact on conventional doctors' legitimacy and livelihood.

Significantly, however, while the AMA spared no opprobrium for Abrams's therapies, it took no interest in his paternity test. The national newspapers had been broadcasting the oscillophore's paternity verdicts for more than a year before *JAMA* first mentioned Abrams. When it finally did, it was to condemn his medical cures, not his parentage assessments. At the height of the furor over Abrams's scientific test of paternity, *JAMA* published an editorial entitled, "Is Parentage Determinable by Blood Tests?" The editorial (which concluded that it was not) was actually an overview of the nascent research on the heredity of ABO blood groups (a parallel field of paternity research at this time).[52] It made no mention of Albert Abrams, despite the fact that the newspapers in those very weeks were replete with coverage of the expert "famed in his exploits in determining parentage."[53] The AMA's anti-Abrams campaign targeted his extravagant therapies but entirely ignored his equally astonishing parentage test.[54]

This selective attention no doubt reflects the threat that ERA therapies posed to orthodox medical practitioners. By extension it also tells us something about the relationship of paternity testing to medical science in the early 1920s. Most doctors probably regarded Abrams's paternity test as laughable—the experts quoted in the press suggest as much—but they did not regard it as threatening. This was because the American medical establishment had not yet claimed parentage testing as an arena of expertise. In the early 1920s, biological testing was born, but it was born first as a popular idea and only later as a scientific practice. In the United States, that idea was pioneered not by the medical profession but in the first instance by the news media.

A "SHORT, WELL-BUILT, active, bald-headed man" nearing sixty, Albert Abrams cut a gentlemanly figure. Having begun his career as a physician trained at and affiliated with the most prestigious institutions of his day, Abrams had by the time of the Vittori affair been a respected doctor for more than two decades. Twice widowed by wealthy wives, he was well off, widely traveled, and harbored literary aspirations (his book *Scattered Leaves from a Physician's Diary* contained airy reminiscences of his early medical career).

Over the years, the respected scientist and genteel man of letters moved radically beyond the bounds of medical orthodoxy. He made the most astonishing of claims in the most phlegmatic of terms but treated his critics with gentlemanly patience. When a University of California biochemist expressed skepticism about the oscillophore, Abrams responded with equanimity: "Dr. Bloor is perfectly right in doubting my blood test theory." "He admits that he has never heard of my method, so he should not be expected to believe in it." Abrams invited the naysayer to his laboratory for a demonstration.[55] Some were won over by this gracious self-possession. In response to *JAMA*'s vicious invective against Abrams, Upton Sinclair noted: "Dr. Abrams follows the policy of ignoring attacks on his work, taking the view that in the long run, the man who cures disease makes his way in the world in spite of opposition."[56]

It is not by chance that Abrams counted among his supporters a social critic like Sinclair as well as Sir James Barr, the Irish-born president of the British Medical Association, known for his medical ministry to prisoners. Abrams cultivated a populist image and received widespread publicity for his plan to establish free ERA clinics for "rich and poor alike." He declaimed the oscillophore's consequences for wealthy men who refused to recognize their illegitimate children. "The present inheritors of the wealth of a millionaire are his legitimate offspring," but his paternity test promised to "dispossess these proud people."[57] The well-heeled Dr. Abrams embraced not only medical heterodoxy but also his role as a scientific Robin Hood.

And then, at the height of his celebrity and his notoriety, Albert Abrams died suddenly. His death from pneumonia in January 1924 came exactly three years after the Vittori case first made the papers. It also came in the midst of a major investigation of ERA by *Scientific American* magazine—a fitting end for a career that blossomed in the glare of the media. The magazine released a posthumous report concluding the "entire Abrams electronic technique" to be "at best . . . an illusion" and "at worst . . . a colossal fraud."[58] Most assessments concur that in the wake of his death, Abrams's star quickly faded. Although pockets of diehard ERA followers persisted for decades, Abrams as a popular, mass phenomenon disappeared as quickly as he had emerged. But assessments of his demise, like those of his career, are based on the fate of Abrams's medical therapies and claims to miraculous panaceas.

His paternity test, it turns out, had a much longer life. This was not because the oscillophore could actually diagnose paternity, of course, or even because most people believed it could. Rather, it was because the Abrams affair rehearsed the quest for the father as it would unfold over the rest of the century. Today Abrams is remembered for his outlandish therapeutic claims because the American Medical Association, which powerfully shaped how Abrams has been defined and remembered, was preoccupied with his medical treatments and not his paternity test. But it was Abrams's paternity test that, in the 1920s, made him a household name. The eccentric practitioner first appeared in the national newspapers thanks to the Vittori case, and his involvement in "sex trials" consistently received more press coverage than his medical cures. This skewed coverage was a source of frustration for his supporters. As Upton Sinclair lamented, "It is a curious commentary upon our journalism that it telegraphs all over the country the news that Abrams has renounced a certain child to be illegitimate, while it says not a word about the fact that there come to his clinic every day people who have been cured of all three of the dread scourges of our race, syphilis, tuberculosis and cancer."[59] The newspapers may have unfairly shortchanged Abrams's therapeutic achievements, but the whole episode portended the central role of the media in disseminating and legitimating paternity testing. Long before daytime television's love affair with paternity scandals, the Abrams episode vividly captured the public nature of modern paternity and how science, law, and the media produced it.

Abrams was also remarkably prescient as a businessman: he not only asserted that a drop of blood could reveal parentage, he recognized that there was profit to be made from this fact. What made Abrams the "dean of twentieth-century charlatans" was not his outlandish claims or madcap gadgets—plenty of visionaries and swindlers trolled the margins of medical orthodoxy in the early twentieth century—but the commercial acumen he brought to the enterprise. Abrams attracted a following and also created a franchise. At the time of his death, his detractors claimed the ERA empire was valued at at least $2,000,000 (Abrams always claimed that his profits supported the construction of free clinics and training facilities). There is no better evidence for the value of his brand than the fact that it inspired numerous imitators and copycats—some of whom Abrams sued for fraud.[60]

Abrams was also a pioneer of mail-order therapeutics. For $10, a patient or doctor could send a drop of blood on white blotting paper and receive a diagnosis. As the *Scientific American* report noted, "He made it possible for the patient in New York to be examined by the doctor in San Francisco."[61] The prospect of diagnosing dread diseases by mail was as ludicrous to many people then as it is now. But consider Abrams's mail-order paternity test: it foreshadowed the method used by commercial DNA testing today. An individual takes a blood sample (or snip of hair, cheek swab, or saliva sample) and sends it in a special envelope to a far-away lab, which remits the results. In 1993, the first biotech company began to market paternity tests directly to U.S. consumers.[62] Abrams's oscillophore anticipated, some seventy years before, the commercial strategies associated with the modern genetic testing industry.

THESE DEVELOPMENTS WOULD EMERGE only many decades later. More immediately, Albert Abrams cast a lingering shadow over the whole enterprise of parentage testing. For years after, testing in U.S. public discourse retained the distinct odor of quackery. When another hospital baby swapping scandal occurred a few years after Abrams's death, a commentator urged caution in the use of scientific assessments: "Tests of human parentage by blood, by facial appearance, by the bumps on the head, by a hundred other irrelevancies, have always been plentiful along the quack-infested fringes of true anthropology."[63] Perhaps the initial credulity of Judge Graham, the press, and a certain portion of the American public explain the skepticism that crept in later on. Future blood tests of paternity, including those unambiguously endorsed by scientists, would have difficulty gaining traction in U.S. courts.

It is tempting to see Albert Abrams as a setback for the progress of modern paternity and its technologies. His charlatanry cast a shadow over future, more legitimate methods of kinship assessment. Yet Abrams's most significant and longest lasting legacy was arguably just the opposite: to introduce into American public consciousness the idea that paternity could be known and that modern medical science was the way to know it. Abrams was not the first to make such a claim, but for several years in the 1920s, he offered a particularly compelling version of this story. It is one of the many ironies of this consummate snake oil salesman that his chief legacy was to create a closer association between the quest for the father and medical science.

By 1924, Dr. Albert Abrams was dead, and Judge Graham, still on the bench, had begun offering marital advice through a new, modern medium: a radio show. The young lawyer Stanley Nolan had returned to quiet professional anonymity, and Rosa, Paul, and Virginia Vittori disappeared from the public record altogether. But in parts of Europe and Latin America, other methods of scientific paternity determination were just getting started. There, Nolan's suggestion that was so unusual in 1921—that science might be able to find the father of a fatherless baby—would by the end of the decade become both conventional wisdom and established legal practice.

3

BLOOD WORK

> Scientific research should not obey immediate practical intentions
> because later on, applications will spontaneously emerge from the
> granitic pedestal of newly conquered truths.
>
> —Leone Lattes, Italian serologist, 1927

LESS THAN TWO WEEKS after Albert Abrams's death in January 1924, a group of forensic experts gathered in Berlin to hear a lecture. The speaker, Dr. Fritz Schiff, was a bespectacled bacteriologist at the municipal hospital, and the subject of his lecture was the use of blood tests to assess paternity. But Schiff's blood test had little in common with the electronic methods Abrams had pioneered. He explained how the four human blood groups were passed from parent to child according to predictable patterns and how these patterns might be used in a case of disputed parentage.[1]

His audience was probably already familiar with the substance of his talk. Systematic investigation into blood group heredity had begun more than a decade earlier, and Germany was a center of this research. But although scientists had speculated on the application of blood groups for finding the father, it remained a purely hypothetical proposition. "Sooner or later," the British medical journal *The Lancet* had mused some eighteen months before, "the test is sure to be used [as judicial] evidence."[2] Now Fritz Schiff proposed to do just that. And so, a world away from San Francisco and less than a month after Albert Abrams's death, another blood test of paternity was in the process of being born.

To be sure, this was a very different test. Unlike Abrams's sensational but short-lived oscillophore, it would garner widespread legitimacy in scientific and then judicial circles. Within months, the new method was used in courts around Germany. Soon it would spread to other countries in central and northern Europe, to Latin America, to Australia and North America. And as the method became the first widely applied method of paternity testing, Schiff himself would become the world's foremost expert in the new field of kinship serology. A decade later, however, his career was in shambles, and Schiff, who was Jewish, was forced to flee Germany. The international reputation and professional networks he had developed through his work would save his life and that of his family. But the persecutors who forced him out of Germany would appropriate his research on hereditary serology for grotesque ends.

Modern paternity advanced the idea that the secret of parentage lay in the body and promised new technologies for finding it. Schiff's blood typing and Abrams's oscillophore were just two of a panoply of "blood tests" to emerge in the 1920s that sought to do this. Only Schiff's found systematic practical application, but many methods received widespread publicity and shaped how courts and publics regarded the scientific quest for the father. "The scientific facts give no reason for a special blood mystique," Schiff once cautioned, yet that mystique proved irresistible. Blood in these years was an increasingly politicized metaphor and one with ever more ominous racial overtones. The search for a serological proof of paternity was intimately shaped by this context, and the quest for biological paternity was inextricable from the simultaneous search for biological race.[3]

The test Schiff proposed was not, strictly speaking, a test of paternity at all. Rather, it was a test of *non*paternity. Blood type analysis could exclude impossible fathers, or at least some of them, but it could not identify the progenitor of a given child. This probably explains why it initially inspired much less interest among global publics than the miraculous oscillophore. The introduction and rapid expansion of blood testing for paternity in Germany and beyond barely registered in the international press. Even the international scientific community was slow to perceive the implications of these developments. But the method's less-than-dramatic debut belied its eventual importance. Schiff's test was based on the four ABO blood groups, but over time the discovery of new serological properties—MN, Rh, P, and an alphabet soup of others—meant that blood typing would become increasingly powerful, capable of excluding more impossible fathers, and therefore more useful for forensic purposes. It was the most widely used method of assessing biological parentage into the 1970s.

Over time serological science became more powerful, but its very limitations shaped its uses and meanings. In fact, early blood group science proved forensically useful as a test of parentage not in spite of its limitations but sometimes because of them. Even the rudimentary methods of the 1920s would find surprisingly widespread, varied, and creative applications. For while the new method exploited universal laws of heredity in the service of incontrovertible biological truths, the truths it produced were always shaped by local circumstance. Pioneered in Berlin, the hereditary blood typing technique soon circulated widely across the transatlantic, but as it did so, it was put to work solving very different problems. The technology of modern paternity gave rise not to global convergence but to quite the opposite: remarkable local variation in terms of how courts used it and the meanings societies gave it.

THE DISCOVERY OF HUMAN blood groups dated back a quarter century. Around 1900, Austrian researcher Karl Landsteiner observed the patterns by which the blood of some individuals reacted to the blood of others by clumping. Out of these patterns of clumping, or agglutination, he discerned three blood groups (a fourth, rarer one would be identified later). The groups came to be referred to as A, B, O, and AB.

The discovery of human blood types would have momentous consequences for human health and open vast new fields of scientific inquiry. Knowledge of blood group incompatibility made safe blood transfusions possible and led to therapies for hemolytic disease, in which the incompatibility of blood type with the mother caused the death of the fetus. It had applications in forensic criminology, since the characteristics of a bloodstain allowed it to be traced to certain individuals and not others. Research suggesting that the frequency of blood groups varied across ethnoracial groups led some scientists to believe that they had located a biochemical marker of race. In 1930, Karl Landsteiner received the Nobel Prize for his work on blood groups.[4]

Although Landsteiner had anticipated a number of these applications, he did not intuit another crucial characteristic of blood groups: their hereditary character. This discovery came in a Heidelberg laboratory a decade later. A young Polish scientist named Ludwik Hirszfeld and his mentor, the German Emil von Dungern, recruited their colleagues and the colleagues' wives and children as research subjects to test the hypothesis that blood groups were heritable. In all, they studied 348 individuals belonging to seventy-two families. The results were summarized in a chart organized by the colleagues' names—Prof. J, Prof. K, Dr. W, Dr. H, and so on—recording their blood groups and those of their wives and children. Hirszfeld later quipped, "for many years, people talked about an odd professor [von Dungern] and his assistant [Hirszfeld] who discretely inquired about the marital happiness of professorial families lest a cuckoo's egg might overthrow the scientific law they had established."[5]

The charts of these familial groupings revealed a remarkable pattern: human blood groups were not only transmitted from parents to children but followed the fixed, predictable pattern of Mendelian inheritance. Gregor Mendel's nineteenth-century studies of pea plants had recently been rediscovered, and blood groups were the first trait identified in humans that followed Mendelian laws.[6] These laws were simple and yet extremely powerful precisely because of their simplicity. Types A and B were dominant while O was recessive. From Hirszfeld and von Dungern's investigation and follow-up studies by others, there emerged a series of axioms of possible and impossible inheritance. Based on the blood group of a child and one of its parents (typically the mother), one could predict the possible blood group or groups of the other parent (the father). If in a case of disputed paternity, a man did not have one of the possible blood groups, he could be categorically excluded as the progenitor. The rules of transmission—often presented as a chart of "possible" and "impossible" relationships—were reproduced in countless medical, legal, and popular texts.

Certain characteristics of blood groups made them especially well suited for forensic application. They were stable and unambiguous and their inheritance predictable and universal. A person's blood group did not change over time in relation to disease, environmental influence, or aging, and all human beings had a blood type that could be revealed by a simple test. These

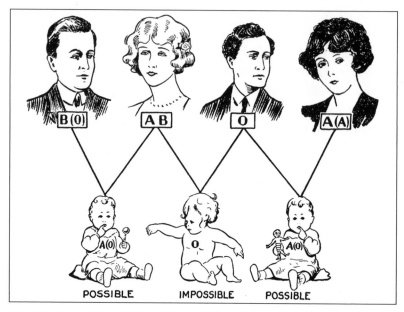

Countless charts of "possible" and "impossible" combinations of inherited blood groups appeared in medical and popular literature.

Reproduced from Edward Podolsky, *Sex Today in Wedded Life* (New York: Simon Publishing, 1942), 99.

characteristics distinguished blood groups from other physical traits. Certain traits such as eye color were known to be inherited, but they obeyed complex laws and were impossible to predict. Likewise, the outward, or phenotypic, manifestation of an individual's genetic makeup, or genotype, was often difficult or impossible to evaluate. Where there were four blood types, there were infinite gradations of eye, hair, and skin color. Meanwhile, many hereditary traits were subject to environmental influences. An individual's height was a matter of inheritance but also of nutrition; hair color was inherited but could be altered by the sun or artificial coloring. Even the rarest blood types occurred frequently enough to make them suitable for practical application. Blood groups were thus superior to certain traits, such as extra fingers or a shock of white hair that followed predictable hereditary laws but were too rare to be routinely useful.

For all these reasons, the ABO blood groups were a hereditarian holy grail, "the ideal human genetic marker, nature untouched by nurture." They opened up a world of research questions and laid the basis for modern genetic science.[7] They also had immediate, practical applications. Hirszfeld's quip about the cuckoo's egg—that blood groups could reveal spurious paternity—anticipates one of the most obvious uses to which they would soon be put.

Yet blood groups did not provide a definitive test of paternity. With just four types to test for, the probability of excluding an incorrect father was about 16 percent. If the mother and child were the same blood type, for example, no paternal blood group, and therefore no possible father, could be excluded. Gradually, new blood characteristics were discovered, and the power of exclusion increased. What did not change, however, was the basic logic of exclusion itself: blood group testing had the power to eliminate "impossible fathers," and thereby to exonerate men falsely accused of paternity, at least in some cases. But it could not positively identify an actual father.

IF HIRSZFELD AND VON DUNGERN foresaw the forensic application of blood groups to establish kinship as early as 1910, their discoveries were not immediately taken up by either scientists or jurists. It would be another decade before researchers embarked on the large-scale family studies necessary to confirm the universality of the patterns first discerned among the professorial families of Heidelberg.[8] Perhaps the interruption of World War I explains this lag. But ironically, it was a serendipitous experiment created by the war itself that sparked renewed interest in the hereditary dynamics of blood groups.

The year 1919 found Ludwik Hirszfeld, together with his wife, Hannah, a pediatrician, stationed in Salonika, providing medical care to the Allied troops at the Macedonian front. The agglomeration of soldiers of different nationalities and ethnicities provided a fortuitous scientific experiment. The Hirszfelds began testing soldiers' blood types and noticed they were not distributed randomly among the sixteen ethnonational groups analyzed, which included local Greeks, Turks, and Jews and troops from central, western, and eastern Europe and various Asian and African colonies. Certain blood types tended to predominate in certain groups. What is more, they seemed to exhibit a distinct geographical distribution, in which moving eastward out of Western Europe, populations were characterized by an increasing proportion of type B blood and a declining proportion of type A. Scientists had searched in vain for a biological marker of race, exploring everything from skull shape to fingerprints. The Hirszfelds' finding raised the tantalizing possibility that blood groups were that long-sought marker.[9]

Their study triggered an avalanche of global research on the ethnonational distribution of blood types. A new scientific field, racial serology, was born. Over the next decade, researchers from dozens of countries conducted hundreds of studies that sought to link blood type to race as well as to what were considered hereditary forms of degeneracy (for example, epilepsy, criminality, and "feeblemindedness"). The initial excitement proved premature. While the proportions of blood groups varied somewhat across some groups as the Salonika study suggested, research failed to find what many sought: a serological proxy for race. This failure did not convince the more committed race scientists of the chimeric quality of their object of study, however; it merely spurred them to keep looking.

In an attempt to better understand mechanisms of ABO inheritance, scientists also revisited the question of parent–child transmission. "In order that the distribution of the groups may be used in studies of racial relationships," noted one well-known American researcher in 1926, "it must be shown that they are hereditary, and the mode of their inheritance must be demonstrated."[10] Scientists now took up where Hirszfeld and von Dungern's study of faculty families in Heidelberg had left off a decade earlier. A series of investigations around the world recorded the blood types of parents and children. This research refined scientists' understanding of the genetic mechanisms underlying blood groups but confirmed the central insight that they were passed from parent to child according to predictable Mendelian laws. The fact that these family studies examined not "generic" parents and children but families defined by a specific ethnicity or nationality—English, American, Italian, German, Norwegian, Japanese, Korean, aboriginal Australian—reflects how prevailing concepts of heredity imagined race and parentage as somehow connected.

Because blood groups proved imperfect tracers of ancestry, they spurred scientists to look for better ones. Among them were Russian biochemists E. O. Manoiloff and his protégé Anna Poliakowa, both of the State Institute of Public Health Commissariat in Leningrad. Manoiloff's earlier work had sought to identify biochemical markers distinguishing the blood of men and women. "By analogy to the presence of hormones characterizing this or that sex," he wrote, "there must be something correspondingly specific of race in the blood of different races of mankind. This specific substance gives the seal of the given race and serves to distinguish one race from another."[11] In the 1920s, he began to test the blood of two "racial" groups—Jews and Russians—in an effort to locate this substance.

Within a few years, Manoiloff had stumbled upon a remarkable discovery: a chemical reaction that he believed revealed an individual's racial identity. He described an experiment so simple it evoked the concoction a child might create with the contents of a kitchen cabinet. He mixed an emulsion of red blood corpuscles and salt, stirring with a glass stick "so as to obtain a rather thick emulsion." Then he added a sequence of five ordinary dyes and chemicals commonly found in any chemist's toolkit: "add 1 drop of the first reagent, shake; 5 drops of second reagent, shake again; 3 drops of third reagent, shake; 1 drop of fourth reagent, and, lastly, 3 to 5 drops of the fifth reagent."[12] The resulting solution supposedly turned different colors according to the "race" of the blood: Jewish blood was blue-greenish, while Russian blood was blue-red. This simple technique, Manoiloff claimed, could correctly distinguish between Jews and Russians more than 91 percent of the time. Additional research by Anna Poliakowa revealed additional "race reactions": Korean blood turned reddish-violet; Estonian blood, reddish-brownish; that of Poles, reddish-greenish; and so on.

Manoiloff and Poliakowa's experiments in chromatic alchemy, initiated in 1922 and first published in 1925, coincided with the flowering of interest in the racial and familial inheritance of blood groups and was in explicit dialogue with this research.[13] "Manoiloff's race reaction," as it came to be known, appeared to furnish a simple, elegant, and unambiguous technique for differentiating races, one much more powerful than blood groups. His findings were cited by other scientists and inspired attempts at replication.[14] The American press was fascinated, announcing "Gentile and Jewish Blood Unlike" and marveling at the "startling accuracy" of Manoiloff's technique.[15] The African American and Jewish press was more skeptical. Interviewed by the Jewish News Service, anthropologist Ruth Benedict judged Manoiloff's claims to be "at best . . . dubious." A medical columnist for the Associated Negro Press voiced a different critique. He did not question the validity of the method but suggested it might become "unpopular with whites" when it revealed the fiction of white racial purity.[16]

As Manoiloff noted, his discovery had forensic applications to paternity testing. A doctor who had heard of his experiment presented him with three vials of blood, two from adult subjects and one from a child, and asked him "to determine, by the blood, to whom the child belonged." Manoiloff determined that the first vial belonged to a Jew, a result that "made a strong impression" on the doctor, who revealed that in fact one of the two possible fathers was Jewish. Next Manoiloff tested the blood of the other man and the child and placed all three vials alongside each other, backlit by an electric lamp of "50–100 candles." The child's blood looked more like that of the first vial. The biochemist triumphantly concluded the Jewish man was the father.

The doctor was "greatly pleased with my investigation," Manoiloff recounted, which turned out to confirm what he had suspected all along. As the doctor explained, the Jewish man was married to a Russian woman, but after the birth of her child, he had come to suspect she had had an affair with a Russian paramour. In revealing the child's blood as Jewish, the experiment laid the husband's fears to rest.

Manoiloff's "race reaction" thus doubled as a paternity test, an application that his associate Anna Poliakowa developed more systematically. Where Manoiloff's original experiment focused on Jews and Russians and relied on subjects whose ancestry could be traced back three generations to find the "purest" blood and induce the strongest reaction, Poliakowa compared the blood of children born of "mixed" marriages (with parents of different "nationalities") and "pure" marriages (with parents of the same "nationality"). She concluded that while the blood of children of "pure" marriages showed the same reaction as their parents, that of "mixed" children reacted according to the specific combination of racial parentage. Children with a Russian father and Jewish, Armenian, or Polish mother, for example, exhibited a blood reaction more like the mother, but in those with a Russian father and a German, Finnish, or Tartar mother, the blood reacted more like that of the father. Poliakowa concluded that Manoiloff's race reaction "has apparently a great practical importance" for determining paternity. If the blood of a mother and child provoked different reactions, for example, it could be surmised that the father had to be a man of a different "nationality."[17]

Manoiloff and Poliakowa were hardly the first researchers to parlay a race test into a paternity test. Edward Tyson Reichert's crystallography and Albert Abrams's oscillophore purported to do much the same thing. For these investigators and many others, race and paternity were at once homologous and mutually revealing. Both were regarded as essential biological truths located on the body. But they were also potentially hidden and elusive, and scientific experts doggedly pursued physical markers that would reveal them. Manoiloff and Poliakowa used racial knowledge to discover paternity; in other contexts, notably Germany after 1933, knowledge of paternity would be used to determine racial identity. A defining feature of modern paternity was its constant conceptual and practical reference to race.

Blood group research also reproduced this association. The Hirszfelds' Salonika study of racial variation inspired family studies that made the forensic application of blood groups to parentage possible. Early manuals on the science of blood groups discussed both their racial significance and their application to parentage, and scientists tended to study both racial and familial transmission.[18] At the very moment Fritz Schiff was advocating for blood group testing in paternity cases, he was conducting a major study of ABO distribution among Berlin's Jews.[19] German-speaking Europe was the global birthplace of racial serology in the 1920s. That it was also a center of early paternity testing was not accidental. The two quests were born together.

SCIENTISTS WERE INITIALLY cautious about applying blood groups to actual paternity cases; they wondered if the rules of hereditary transmission were truly fixed and whether there might be exceptions. "Given the state of things, we would not feel comfortable deciding a question of filiation solely on the basis of hematological investigation," wrote Italian serologist Leone Lattes, author of the first textbook on blood groups, in 1923. *JAMA* concurred: "Science knows of no blood test by which parentage can be determined."[20]

Within a few years, however, such reservations collapsed under the burgeoning accumulation of evidence. By the mid-1920s, researchers had amassed data on 1,900 families with a combined total of almost 4,500 children in eight countries. The largest of these studies was one led by Fritz Schiff involving some 500 families.[21] The regular pattern of ABO inheritance discerned in the Heidelberg study appeared incontrovertible. Four years after expressing caution about the forensic use of blood groups, Lattes had changed his mind. "The data acquired in the meantime have confirmed the suspicion" that any exceptions to the Mendelian laws were due not to biology but to technical errors or adultery—Hirszfeld's cuckoo eggs. If blood tests were not yet used in legal proceedings, it was not because their scientific value was in question but because, as *JAMA* now asserted, "much misinformation has been disseminated as to the nature of the test"—perhaps a reference to the lingering fallout from the Abrams affair?[22]

The international scientific community had come around: blood group science was ready for a courtroom debut. But as it turned out, such pronouncements were tardy. In Germany, paternity testing based on human blood groups was already well under way.

If Germany was the cradle of forensic paternity testing, Fritz Schiff was its midwife. Born into a professional Jewish family in 1889, Schiff grew up with his mother and younger sister after the early death of his father. He studied medicine and then served as a hygienist in World War I, traveling widely in the Levant and Turkey, where he was briefly a prisoner of war. Schiff was distinguished by his singularly eclectic intellectual curiosity. Early in his career he published articles on the anthropology of Crete, sanitation in Jerusalem, the craniology of the Czechs, typhoid, and tuberculosis.

By the early 1920s, Schiff had become head of the bacteriology laboratory in a Berlin hospital. It is unclear when and why he first developed an interest in blood group heredity, but by 1923, he was conducting both racial and family studies. Around this time, he also became acquainted with Leone Lattes's *Individuality of the Blood in Biology and in Clinical and Forensic Medicine* (1923), the first text in the field of forensic serology. Ever the inquisitive mind, Schiff translated the book from Italian to German and struck up a lifelong friendship with its author.

Throughout the 1920s and early 1930s, Schiff presided over a thriving lab and directed numerous students. He also brought his scientific curiosity home with him. Schiff circulated at family gatherings collecting blood samples from the attendees, probably as part of his studies of blood group heredity.[23] He was a productive and prolific researcher, publishing some 150 works in less than three decades. His opus included dozens of works on the medicolegal application of blood group inheritance and paternity testing in particular. Among them was a how-to manual, *The Technique of Blood Group Research* (first published in 1926), which became the go-to resource for European forensic lab experts. *The Blood Groups and Their Areas of Application* (1933) would also become a widely cited classic. Reflecting both Schiff's eclectic interests and the broader ideology of modern paternity, the book not only instructed readers in the theory and technique of forensic blood group testing but also explored cultural beliefs about paternity from China to Babylon.

Above all, Schiff became known for shepherding scientific knowledge to practical application. Perhaps his intellectual eclecticism primed him to venture from the lab into the courtroom. Beginning with the 1924 lecture to the forensic society in Berlin, he became a tireless advocate for blood testing in paternity disputes. It was Schiff who performed the first tests for the courts. And it was Schiff who would become the world's most experienced paternity tester, enjoying, according to his friend Leone Lattes, "great celebrity the world over as a medico-legal expert in questions relating to the inheritance of blood properties." That this assessment was pronounced by an Italian colleague in an Argentine journal is testament to its veracity.[24]

WITHIN MONTHS OF HIS Berlin lecture, Schiff had teamed up with a pathologist named Georg Strassmann to perform the first blood tests of paternity for the local courts. Strassmann was director of the Institute for Forensic Medicine at the University of Berlin and thus an invaluable liaison with the courts and legal community, arenas with which Schiff himself had no previous experience.[25] Schiff instructed Strassmann in the testing technique and provided him with blood serum. Soon the two colleagues were performing tests for the court and advocating for the new method before scientific and legal audiences.[26]

Schiff also turned his attention outward, toward the public. Although the international press did not register these developments, the German press and public were immediately intrigued. In fact, Schiff credited the newspapers with popularizing the new method. At the same time, he found the publicity a potential liability. The biggest challenge was not defending the veracity of the test or promoting its usefulness but quite the opposite: tamping down inflated public expectations. The newspapers hailed the new technique under misleading headlines that implied that a definitive method to identify the father was at hand. In Germany as in the United States, the press was an avid participant in the making of modern paternity. Perhaps too avid: Schiff worried that the method could lose credibility in the face of these expectations.[27]

Perhaps fortunately in this regard, the results of the technique were initially anticlimactic. For two years, blood group analysis had no impact on an actual paternity case because not a single exclusion was obtained. The delay reflected the limited powers of a test based on just four blood groups. Of course, it could also reflect the fact that a preponderance of the paternity claims filed in court by German mothers were actually true. After all, the only cases that blood group analysis could clarify were those involving false paternal identifications.

Still, word of the method spread and additional practitioners around German-speaking Europe began adopting it. The procedure traveled rapidly not least because it was quick, simple, and inexpensive. No special technical skills or equipment were required to ascertain blood groups other than a steady hand, a careful eye, and good quality serum. While different scientists developed slight variations, the basic technique was straightforward. The doctor drew a blood sample from the finger and mixed it on a glass slide or in a test tube with serum for blood groups A or B. After a few minutes, four possible patterns of agglutination could be observed: only the sample mixed with A clumped, only the sample mixed with B clumped, neither sample clumped, or both did, thus revealing the blood to correspond to one of the four groups, A, B, AB, or O. To be sure, the investigator had to be well versed in the delicate technique; novices tempted by its simplicity could produce erroneous results. But this was nothing that instruction could not remedy. This was why a serologist like Schiff could easily teach the technique to a forensic doctor like Strassmann.

In 1926, the blood group test finally resulted in a conclusive finding. By this time, the technique had been performed in around one hundred cases in the German courts and had also spread to Austria. In a child support case in Vienna, a putative father, an engineer, was found to have a blood type incompatible with the mother and child, thereby demonstrating that he could not be the progenitor.[28] Having achieved the first successful result, testing now expanded rapidly. Initially requests for testing had originated with the parties' lawyers, but soon courts and child welfare authorities began to ask for it.

Its expansion was not uncontested. In 1927, Schiff became embroiled in a noisy dispute with a Prussian Court of Appeals judge who questioned the validity of the method. The controversy sparked a national debate about blood testing and thrust Schiff into the limelight as the method's most prom-

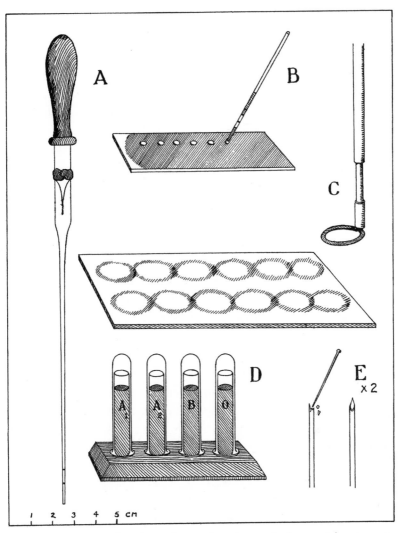

Blood grouping was a relatively simple procedure that required no special equipment.

Reproduced from David Harley, *Medico-Legal Blood Group Determination: Theory, Technique, Practice* (London: William Heinemann Books, 1943), 37.

inent public advocate. He worked with his brother-in-law, a lawyer, to successfully beat back the challenge.[29] Scientists and jurists alike rallied in defense of testing, and it continued to expand.[30]

What began as a trickle soon swelled to a flood. By 1929, blood tests had been used in some 5,000 paternity cases in German courts. Schiff, who collected the data as special advisor to the Reich health office (Reichsgesund-heitsamt), commented that the numbers were "surprising even for those acquainted with German conditions." In Austria, some 700 cases were recorded in the three years after testing was introduced in Vienna in 1926. By the early 1930s, the Ministries of Justice of larger German states and Austria had issued official endorsements of blood group analysis, directives used by courts to guide procedure.[31]

The method also spread beyond German-speaking Europe. In 1926, a Soviet medical council authorized blood group testing for paternity establishment. The method made inroads in Czechoslovakia and Poland as well as Scandinavia. By 1930, Danish courts had heard more than 500 cases involving blood tests of parentage, and over the next few years the University of Copenhagen's Institute of Legal Medicine would perform more than 3,000 such tests. In smaller numbers, courts in Norway and Sweden also began to accept blood group evidence in cases of disputed paternity.[32]

The method traveled beyond Europe. The first instance of exoneration by blood test, the 1926 case of the Viennese engineer, was picked up by the Associated Press wire service and reported in newspapers from Rio de Janeiro to Boston. It is probably not by chance that shortly thereafter, a judge in São Paulo began requesting tests at a local medicolegal institute, in apparently the first systematic application of the new method in the Western Hemisphere. A year later, the method was put to use in a marriage annulment case in the Children's Court in Perth, Australia, the first use of blood group testing in the British Commonwealth (it was inconclusive). In 1929, the test revealed a case of baby swapping at the maternity hospital in Havana. Within a few years, courts had requested the test in paternity cases in Colombia and Peru.[33]

Still, it was not taken up everywhere. Apart from isolated cases reported in the press, into the 1930s group testing in the United States remained largely absent from paternity proceedings.[34] By 1940, only seven states had passed laws admitting serological evidence in these cases, a decade after many German states and Austria had done so. Of course, the most famous example of the limitations of blood testing in U.S. courts was the Chaplin suit. In France, the courts first admitted blood group tests only in 1937 and thereafter pronounced contradictory decisions on their admissibility. In England before 1939, the test had been introduced into just twenty-one affiliation suits.[35] In all these places, the introduction of blood group evidence into paternity proceedings was not just slow and uneven but often deeply contested.

ONE EXPLANATION FOR THESE divergent patterns and, most conspicuously, German-speaking Europe's rapid and enthusiastic embrace of testing is the advanced state of heredity science in the region. Most of the important milestones in the scientific quest for the father were associated with researchers trained in or affiliated with German-speaking institutions, including Landsteiner's discovery of blood groups, von Dungern and Hirszfeld's work on their hereditary transmission, and the Salonika study by the Hirszfelds.[36] Many of the earliest and largest family studies of blood group inheritance were likewise carried out in Germany, most notably by Fritz Schiff.

Yet Germany and Austria could hardly claim a monopoly on paternity science. In 1923, Landsteiner emigrated to the United States and together with American collaborators continued to make important contributions to blood group science. These included the 1927 discovery of the M, N, and P antigens, which doubled the odds of exclusion in parentage analysis, and later, in 1940, the Rh blood group system.[37] Landsteiner's collaborators in this research became the foremost advocates of testing in the United States. In Italy, Leone Lattes was an international spokesman for forensic serology, and a number of family and racial-regional studies of blood groups were conducted there in the 1920s. Yet neither the United States nor Italy embraced paternity testing with anything like the zeal of Austria or Germany. The mere existence of cutting-edge paternity science did not guarantee its use in local courts or its acceptance by the public.

Instead, legal and social factors were determinant. The structural relationship between science and the court could facilitate or inhibit the uptake of scientific evidence. In the continental legal system, including the German one, forensic analysis entered the courtroom by way of a *Gerichtsarzt,* an expert certified and appointed by the court. The German Gerichtsarzt contrasted with the expert witness of the Anglo-American system, who was hired by the litigants and therefore represented their interests. The structure of expert witnessing in Anglo-American courts encouraged contending parties to question the qualifications of the other side's expert and dispute new scientific knowledge, including parentage testing. In German courts, in contrast, the Gerichtsarzt's expertise was employed by the court and was therefore above the adversarial fray. Courts received scientific information— "expert and impartial"—from "Institutes of Forensic Medicine, Medico-legal Councils, Ministries of Justice and like agencies."[38] This arrangement tended to facilitate the introduction of scientific knowledge into the legal arena.[39] The absence of such institutions in the United States may explain why the press played such an important role as an informal conduit between laboratory, courtroom, and public. It was from the newspapers that lawyers and laypeople typically learned about new scientific advances.

In the 1920s, paternity testing was added to the list of forensic tasks in the Gerichtsarzt's tool kit.[40] Schiff's collaborator Georg Strassmann was a Gerichtsarzt, which probably explains how they first introduced blood testing into the court (Schiff was not initially credentialed as a court expert). The importance of these professional collaborations went well beyond the courtroom. Schiff, Strassmann, and soon others formed a vocal cadre advocating in favor of blood testing in scientific, legal, and popular fora.

German paternity law also conditioned courts' reception of biological evidence. In Germany, blood testing was mostly used in child support cases brought by single mothers. Its application in these proceedings was facilitated by the concept in German paternity law of "obvious impossibility." According to this concept, a man who had cohabited with the mother during the period of conception would be presumed the father unless it was "obviously impossible" that he could have engendered the child. Obvious impossibility traditionally referred to physical impotence or cases where the putative father and mother were physically separated during the period of conception. An exception to this presumption was the *exceptio plurium:* if it was proven that the mother had more than one partner, paternity could not be assigned to either of the men. If, however, the mother could demonstrate the "obvious impossibility" of one partner's paternity, the other man could be held liable. Blood typing was fortuitously suited to the principle of impossibility, for impossible paternity was precisely what the method could show. Men could use blood group exclusion to prove the obvious impossibility of their paternity. More rarely, women used it to refute the *exceptio plurium,* excluding one partner and holding the other liable.

Judicial institutions and the law could facilitate forensic evidence, but for a scientific solution to be embraced, it had to address a perceived problem.

The problem that blood testing addressed—the work it was called on to do—varied widely across societies. In Germany and Austria in the 1920s, it concerned unmarried mothers and their illegitimate children. This scenario was common in German-speaking Europe, where rates of extramarital birth were high. In Germany, some 12 percent of children were born to unmarried parents, compared with just over 8 percent in France, 4.6 percent in Britain, and less than 2 percent in the Netherlands. In Austria, it reached a remarkable 27 percent—the highest, by far, in Europe.[41]

The plight of illegitimate children and their mothers inspired no small amount of public hand-wringing. The high mortality rate among illegitimate children was of particular concern, and German officials of all political stripes worried about demographic decline. In 1922, a law made illegitimate children wards of district welfare authorities, assigning them local guardians who were charged with monitoring their welfare.[42] This new guardianship system sought to secure paternal support for children.

The impact of these developments was felt in the courts. Of the 150,000 extramarital births in Germany each year, Schiff estimated that roughly half prompted legal actions against putative fathers for economic support.[43] Such suits may also have been fueled by changing cultural norms: with the rise of the so-called New Woman, less fettered by traditional notions of sexual morality, single mothers may have been more likely to pursue paternity claims in the first place.[44] What is more, courts appeared sympathetic to their petitions. While outcomes varied widely across the region, the evidence suggests that mothers won paternity suits between 55 and 86 percent of the time.[45]

Some judicial officials had become uncomfortable with this state of affairs. Paternity proceedings tended to be "he said / she said" disputes that they regarded as sordid and frustrating. "In no other field . . . do we encounter so many lies and so much perjury as in paternity suits," charged one Viennese judge.[46] The *exceptio plurium* cases were especially troubling and lengthy. Courts found themselves adjudicating cases in which as many as five men were called to testify as possible fathers.[47] In this context, a new method promising a swift and efficient resolution to unseemly disputes garnered the enthusiasm of many judges and soon child welfare authorities, the press, and the public generally.[48] As Schiff observed, "the court is often confronted with tasks that go beyond human power. In this situation, blood tests have often produced clarity."[49]

It was a clarity about some matters and not others, however. Tests based on blood group exclusion could not find fathers for the fatherless. What they could do was exclude certain impossible fathers. That is, they could ferret out some specious accusations and vindicate some innocent men. In so doing, they established not paternity but an equally important fact: that the mother, who swore that the accused was her only sexual partner and hence the only possible progenitor of her child, had lied under oath. The limited powers of exclusion of early blood testing thus generated a different sort of knowledge than its practitioners originally imagined. It revealed not paternity but perjury and, by extension, promiscuity. The method was an assessment less of certain biological facts about the father than of certain social facts about the mother.

Indeed, blood group testing did not just reveal the dishonesty of women; its very value depended on it, and its usefulness grew in proportion to the incidence of false accusations. As German and Austrian scientists entered the medicolegal trenches, performing first hundreds and then thousands of tests, the success of their labors required widespread female perjury to generate conclusive scientific results. The technique thus became squarely associated with the defense of men.

Some observers lamented the unintended sexual politics of their science. "Serology has served only to exclude paternity, more in the defense of the man than of the child or the mother," observed Ludwik Hirszfeld. He and other forensic experts periodically expressed the hope that women and children might benefit "as progress is realized, with the identification of new group properties" that would allow a positive identification of the actual father rather than a negative exclusion of impossible ones. But many other observers, scientists and jurists alike, insisted that the exclusionary test already had an important social function: protecting men. Before the introduction of blood group tests, they claimed, "hundreds of false judgments" had saddled innocent men with responsibility for strangers' children. The "capriciousness of mothers" became a constant refrain of German medicolegal writing on blood testing in the 1920s.[50]

Blood tests of paternity soon migrated from the civil suits for child support that women filed against men to a new judicial arena: criminal suits men filed against women for perjury. The relatively liberal rules of evidence in German criminal proceedings facilitated this use.[51] In 1926, a criminal court in the small town of Ellwangen (Württemberg) heard testimony from scientific experts including Fritz Schiff. On the basis of a blood test excluding the putative father, the court convicted the mother of perjury and sentenced her to six months' jail time. The prospect that a hereditary blood test could serve as a kind of lie detector (another technology being developed at this time) was provocative. The case made the international wires and was reported in papers from India to Ireland to provincial Argentina—one of the few times the wire services registered the progress of the new blood test in German-speaking Europe.[52]

Other perjury convictions followed.[53] Introduced to expeditiously resolve paternity disputes, blood groups now threatened to unleash a whole new category of litigation. Within a few years, judicial procedures were altered so that testing was used to screen out false claims before paternity suits were filed, thereby heading off any eventual perjury charges. This change had the effect of removing scientific evidence from the adversarial dynamic of the courtroom and making the laboratory the obligatory first stop for all paternity disputes. Henceforth, scientists, rather than judges, would have the first crack at finding the father.

BLOOD TESTS HAD CONSEQUENCES not only for women, men, and children but also for the scientists who performed them, not least because they proved lucrative. Around 1927, blood group analysis in Germany cost somewhere around 10–12 marks per test, for a total of 30–36 marks for the analysis of mother, child, and putative father.[54] It was not a large sum—monthly child support payments ranged from a low of 15 marks to a high of perhaps 50.[55] Then again, the test took a technician only minutes to perform. As the method spread and courts adopted the practice of pretesting incompatibility in all child support cases, it created a steady demand that could benefit individual scientists and forensic institutes alike.

One of those scientists was Fritz Schiff. For the first few years after the procedure was introduced, Schiff was the sole paternity tester for the Berlin courts.[56] By the early 1930s, he was performing blood tests and serving as expert witness for courts around Germany and elsewhere in Europe. The activity proved lucrative. Forensic analysis helped underwrite his active laboratory but also his growing economic responsibilities as a paterfamilias. In 1920, Schiff married his childhood sweetheart, Hildegard Caro. By 1930, the couple had become the parents of three young sons, Hans (born 1922), Reinhart (1927), and Hellmut (1930). The household also included Schiff's mother, Adele, and younger sister Hedwig.[57] Schiff was the household's sole breadwinner, and within a few years the income generated from his forensic activities came to constitute "the vast majority of his income."[58]

The importance of these activities can be gauged by the calamitous effect when they suddenly dried up. In the early 1930s, Schiff lost his credentials as a court expert due to the Nazis' restrictions on Jewish professionals and was "almost without money for his living," according to a colleague. As the father of the paternity blood test, Schiff's experience was no doubt exceptional. Yet it foreshadowed the growing economic importance of paternity testing for individual experts and scientific institutes, particularly when more complex, laborious, and therefore expensive methods were introduced in the 1930s.

THE GROWING USE of blood group testing in child support cases—as well as the method's evident limitations—inspired scientists to continue exploring alternatives. Increasingly they were motivated not only by general interests in heredity or serology but by the specific, practical problem of paternity. The promise of a biochemical marker that could positively identify the father continued to beckon. An obstetrician and gynecologist at Königsberg University in Eastern Prussia, Wilhelm Zangemeister, took up this problem. Delivering the illegitimate infants of single mothers at the Women's Clinic, Zangemeister developed "the desire to establish, for forensic purposes, a more dependable test for paternity than yet exists."[59]

In 1928, he claimed to have discovered such a method. When the blood sera of parents and children were mixed together and viewed through a photometer, an apparatus that dispersed light through prisms, changes in turbidity became visible. (One source described the telltale reaction as "a certain kind of small, dancing specks . . . illuminated in a special way under a powerful microscope.")[60] The sera of unrelated people reacted differently, thus allowing the test to distinguish between kin and nonkin. While Zangemeister could not explain the biochemistry behind his results, he surmised they had something to do with the development of maternal antibodies in reaction to the albumin of the father's sperm. His test could thus be used to establish not only a genetic relationship between a parent and child but also a sexual relationship between the mother and father. In this way, his paternity method tested not only kinship but sex.

It is doubtful whether Zangemeister's test—a serological chimera no less than Manoiloff's "race reaction"—was ever put into use. But his findings provoked widespread interest in German, Anglo-American, and Argentine scientific circles and were reported in *JAMA* and a variety of forensic and eugenic publications. In Brazil and the United States, Zangemeister's test was enthusiastically discussed in popular magazines into the mid-1930s.[61] The attention it attracted suggests that scientific communities and global publics alike were primed for better parentage tests.

Indeed, paternity testing galvanized sectors of the general public. They included a group calling itself the Rights for Men League, which surfaced in Vienna in 1926. The organization called for the eradication of the unjust privileges that the modern, emancipated woman enjoyed at the expense of downtrodden men. The figure of the New Woman was a bellwether of anxious modernity in the interwar years, and the League positioned itself as a bulwark against her growing domination. The organization published a weekly newsletter ("Self Defense") and boasted, probably apocryphally, of some 25,000 members. Among its international adherents, it claimed none other than Charlie Chaplin, whose sensational recent divorce (from his second wife, Lita Grey) had supposedly convinced him of "the necessity of a worldwide movement for the emancipation of the oppressed husbands."[62] In 1929, the League attempted to organize a world congress of disgruntled men. Invitations addressed to "American Men" were mailed to newspapers as far away as Spokane, Washington.[63]

Paternity was central to the organization's vision of male vindication. In addition to the reform of divorce law, it called for greater rights for unmarried fathers and the reform of marital paternity law, which automatically assigned fatherhood to husbands. The League's founder, Sigurd Höberth, was himself embroiled in a paternity action against his former wife. Echoing the allegations of widespread perjury voiced by medicolegal experts, the League alleged that "under the present Austrian law, the unmarried mother has practically her choice as to whom she wants to pick as the father of her child" and called for the expanded use of blood testing in paternity suits. In fact, the League appears intimately tied to developments in scientific paternity testing: it was formally founded in March 1926, just as a Viennese engineer became the first putative father exonerated by a blood test. The organization's platform was widely reported in the papers, and a number of politicians expressed sympathy with its goals. Medicolegal experts who advocated for expanded paternity testing pointed to the organization as evidence of the "bitter feelings" that these proceedings engendered among the falsely accused. For their part, some women's rights leaders refuted the organization's claims.[64]

The 1929 economic crash put an end to the League's ambitious plans for a world men's congress. A year later, the group's stately Viennese storefront had been turned into a women's shoe store.[65] By then the organization had splintered into two branches in a dispute over the admission of women members, but a surviving branch lasted until Höberth's death in 1938. Of course, the League's appeal to Charlie Chaplin turned out to be deeply prophetic, given that the comic would find himself at the center of perhaps the most famous dispute over paternity testing of the twentieth century. Several of its proposals would likewise prove prophetic, becoming law not in defense of the downtrodden male but in the service of racial policy under National Socialism.

MEANWHILE, THE QUEST for the father was unfolding elsewhere in the world. In May 1927, forensic scientists in Brazil became the first in the Western Hemisphere—and possibly the first anywhere outside German-speaking Europe—to perform the new blood test in the context of a judicial investigation.[66] At the request of a São Paulo criminal court, the forensic Instituto Oscar Freire (IOF) performed an analysis of the blood of Olinda de Jesus, her baby Julia, and putative father Julio Baptista da Costa. It was the first of hundreds of such tests.

At first blush, São Paulo was an unlikely cradle for the precocious adoption of paternity testing. While Brazil was a center of Latin American eugenic science, Brazilian scientists were not involved in the research that ushered blood group testing to the courts. In fact, by the late 1920s, only a handful of studies of blood groups had been conducted in Brazil, none of them concerned with questions of racial or familial heredity.[67]

The speed and enthusiasm with which Brazilian scientists embraced the new test reflects how easily the method traveled. If cutting-edge paternity science did not automatically lead to the adoption of the new method, the opposite was also true: the IOF's experts could be consumers of the new test without being producers of the research on which it was based. Indeed, Flamínio Favero and Arnaldo Ferreira, the two doctors who became Brazil's most experienced blood testers, were neither serologists nor genetic experts; they were *medico-legistas*, forensic scientists whose expertise ranged from autopsies to fingerprinting. In this they resembled Georg Strassmann, the collaborator whom Fritz Schiff had instructed in the testing technique. Favero and Ferreira had no Fritz Schiff in their midst, but with the IOF's formidably stocked library featuring the latest scientific literature in a half-dozen languages at their disposal, they quickly became autodidacts in the new method.

As it turned out, a São Paulo judge had asked the two colleagues to perform a blood group analysis a year or more earlier, coinciding with the very first tests in German-speaking courts.[68] The precocious request suggests that knowledge of the new method spread quickly not only among scientists but also among judicial authorities. Favero and Ferreira had declined the judge's request because, they said, they lacked the necessary blood serum and sufficient study of the method.[69]

Their reservations did not last long. A public laboratory near Rio de Janeiro soon began producing serum, and with the help of the IOF's library, they set about educating themselves in the theory and method of blood agglutination.[70] Reflecting the global circulation of blood group expertise, they chose a testing technique developed in France during World War I by a blood transfusion expert from Massachusetts. (Later, they would also adopt Schiff's method.)[71] The following year when a judge again approached them about performing a test for the court, Favero and Ferreira were ready.

If the new technique was gleaned from the foreign literature, the task to which they applied it was peculiarly Brazilian. The dispute that prompted the Fourth Criminal Court of São Paulo to request scientific evidence of paternity involved not child support but a criminal investigation of sexual assault. The accused, Julio Baptista da Costa, was a forty-seven-year-old doctor and teacher of applied biology (if the scientists noted this irony, they did not comment on it in their expert report). The mother, Olinda de Jesus, was his house servant. Five months earlier, she had given birth to baby Julia. Da Costa was charged with raping her, and while the sexual abuse of domestics was no doubt common, this case was especially egregious: Olinda de Jesus was just thirteen years old. The judge sought a scientific test to determine whether da Costa, who denied the charge, was the father of her baby.

On the face of it, this was an odd application for a blood group analysis. The test could not determine whether da Costa was guilty of rape because it could not positively identify him, or any man, as the father of baby Julia. Yet even if the exclusionary method could not assign criminal culpability to a man, it could assign moral culpability to a woman. For if the test could rule out da Costa as baby Julia's father, then her mother must have had another sexual partner. And a "dishonest" female—even one who was just a girl herself—would have a hard time pursuing a criminal complaint of sexual assault. If da Costa could be excluded as Julia's father, he would likely be acquitted of assaulting her mother.

São Paulo in the 1920s was a rapidly growing industrial city, home to a burgeoning population of foreign immigrants and rural migrants. The perils of urban modernization—violence, poverty, labor exploitation—kept the IOF's forensic scientists busy. Favero and Ferreira analyzed the wounds of the city's crime victims, studied the maimed limbs of its industrial workers, and performed autopsies on its dead. They also expended tremendous time and effort examining the hymens of the city's daughters.

The study of virginity was a Brazilian medicolegal specialty. Accusations of deflowering, seduction, rape, and other sexual crimes abounded among urban working-class Brazilians in the first decades of the twentieth century.[72] Brazilian medico-legistas, including those at the IOF, were experienced "hymenologists"—experts in the evaluation of anatomical virginity. Such examinations were frequent at the IOF, and Favero and Ferreira published scientific work on the topic.[73] Given Olinda de Jesus's pregnancy, the basic fact of her deflowering was not at issue; instead, the court sought to identify the author of her condition. Using genetic analysis rather than hymenology to solve sexual crimes represented a novel approach to an otherwise familiar forensic problem.

In the case against Julio Baptista da Costa as in so many others, the test was inconclusive. The accused had a compatible blood type, meaning he could not be ruled out as the father of Olinda de Jesus's baby. Presumably on the basis of other evidence, the judge found him not guilty.[74] The IOF would perform hundreds of paternity analyses for the court in succeeding decades, a significant proportion of them in cases of sexual transgressions such as deflowering, seduction, and rape. Here was a distinctly Brazilian application of blood testing, one that reflects the different kinds of work the technique could be called on to do. The limitations of the rudimentary method foreclosed some uses, but they created others.

At the IOF as in German perjury cases, those uses implicated sex as well as kinship, and women as much as men. Paternity testing tested not male parentage but female credibility and sexual honor. The genetic makeup of the child was merely a convenient way to assess a mother's moral standing and by extension whether the male defendant could be held responsible for illicit sex with her. In such cases, modern paternity reinscribed rather than challenged traditional notions of gender and sexuality.

AS BLOOD GROUP TESTING advanced in parts of Europe and Latin America, in the United States it remained more a colorful news item than a routine forensic tool. As Abrams's oscillophore disappeared from the front pages, the newspapers shifted seamlessly to discussing parentage testing based on blood typing—often with no distinction between the two methods. Throughout the 1920s, the papers reported on the use of "blood tests" in paternity disputes involving wealthy men like the Chicago clothier Modell and Cornelius Vanderbilt Whitney. But mostly the story was that the parties asked for tests, not that they were performed (eight years later, Whitney's former paramour was still trying to secure a test for her now nine-year-old son; she announced her intention to travel to Italy to consult with serologist Leone Lattes). The press also reported cases in which blood group analysis was applied to the extraordinary predicaments of ordinary people: a paternity dispute in a tiny rural community in Nebraska, a search for the kidnapped son of a New York building superintendent, a feud between two Iowa women claiming a child as their own.[75] Clearly lawyers and litigants sometimes requested the assistance of science, but throughout the 1920s, blood testing remained exotic enough to make news.

One sort of dilemma dominated the coverage: hospital baby swaps. In the sweltering Chicago summer of 1930, Mr. William Watkins, an unemployed foreman and new father, was observing his newborn son's bath when he noticed a small piece of tape on the baby's back. It said "Bamberger" in red ink. His wife, Margaret, had just returned home from the hospital, where she had shared a room with another new mother, Joanna Bamberger; Watkins had become friendly with Mr. Bamberger, an unemployed bricklayer, while paying the hospital bill. The alarmed father rushed to the Bambergers' home a few miles away, where a similar tape labeled "Watkins" was soon discovered in a wastebasket. The parents were flummoxed, and soon the press and public would be too. Which had been switched, the labels or the babies?

Throughout the 1920s, a succession of baby swap cases in the United States illustrated yet another use to which the new blood tests could be put, involving not sex, lies, and morality but the fragile enigma of identity. Serological methods were prominently featured in these stories, and scientists tried to capitalize on the public's attention to promote new advances in hereditary blood testing. Yet in the American press, these episodes served to highlight less the triumphs of science than its limitations.

The Watkins–Bamberger debacle was merely the latest version of a drama rehearsed at least twice before in the previous decade. In May 1919, both Mrs. John Garner and Mrs. David Pittman (the mothers' given names never appear in the press coverage) gave birth to a baby girl in Grady Hospital in Atlanta. Mrs. Garner became convinced that she had been given the wrong baby. The Garners filed a legal suit, and at some point the child in their custody, whom Mrs. Garner claimed was not hers, died. Two years after the birth, the case was still unresolved when the Garners' lawyer proposed a blood test. The Atlanta paper described the efforts of "local students of hereditary tendencies" to apply the theories of "George [sic] Mendel." The paper explained Mendelian theory to readers: "the odd numbered children of a family are supposed to be dominated by the parent whose sex is their own and the even numbered children by their opposite parent. Where the dominancy exists science is declared to have shown that there will be close resemblance, not only in contour and general characteristics but that the finger prints will be practically identical."[76] Clearly the press continued to take its pedagogical role seriously, even if its lessons could be garbled.

The test probably involved blood type analysis (elsewhere, the Garners' lawyer explained that scientists had "placed the blood specimens in four classes," suggesting the four blood groups).[77] If this was the case, it may well have been the first application of blood group testing to determine parentage in the United States. The case predated the large-scale family studies that, a few years later, would establish scientific consensus about the hereditary laws governing group transmission. At any rate, the analysis was inconclusive. The Garners' lawyer then wrote to Dr. Albert Abrams, whose oscillophore was at that moment all over the papers, to ask for a different sort of blood test. Ultimately, the Atlanta court declined to admit any tests at all, the parents never reached a mutual resolution, and the case faded from public view.

A second baby swap sensation emerged in 1927, this time in Cleveland. When Mrs. Sam Smith awakened after giving birth at Fairview Hospital, the nurse congratulated her on the arrival of a strapping baby boy. Mrs. Smith was thrilled; baby George joined three sisters and a brother. A week after the birth, she peeked into his diaper and received a shock: baby George was a girl. As it turned out, three infants with the last name Smith had been born the same day at Fairview; the other two were boys. Convinced that her son had been mistakenly taken by another Smith mother, she refused to leave the hospital without him, and her husband filed a writ of habeas corpus against the hospital.

Again, the press swooped in. This time, the possibility of a scientific proof appeared more quickly and was more central to the story. "Science to Decide Parentage of Baby," the headlines announced. "Blood Tests May Decide Parentage."[78] A judge assembled a team of six experts, including two anthropologists, a dermatologist, and a forensic specialist. For her part, Mrs. Smith was doubtful. "They say they will prove it by blood tests and by the baby's hands and feet that this child is mine," she told reporters. "But always there will be that doubt. How can I be sure?"[79]

The experts examined the couple, their four other children, and baby girl "George," declaring that the infant had the "same expression" as two-year-old brother Peter and the same ears as ten-year-old Angeline.[80] Meanwhile, an avalanche of letters suggesting solutions to the Smith case arrived from around the country. A Southern California palmist offered her services; a Massachusetts man proffered an apparatus that sounded suspiciously like an oscillophore. The judge forwarded all the letters to the scientists, who deemed them uniformly worthless.[81] The *New York Herald Tribune* congratulated the "physicians and judges of Cleveland" for their "admirable refusal to be stampeded into scientific quackery" as so often occurred in predicaments involving parentage. Cleveland authorities had made appropriately cautious use of reliable blood analyses, the paper observed. It then cited as examples of reliable tests those based on blood groups as well as Manoiloff's "special blood reactions."[82] (There is no evidence that Cleveland authorities actually conducted the Russian's race test.)

By the experts' own admission, however, the scientific evidence in the Smith case was "flimsy." Now the headlines changed course. "Blood Tests Fail to Solve 'Baby Shuffle,'" they announced; "Blood Will Not Tell."[83] A month after the birth, the judge called a final hearing, reported on front pages across the country. The court was packed with observers. Based on the testimony of the bereft parents, the hospital staff, and the panel of experts, the judge declared that a careless nurse had mistakenly informed the mother that her baby was a boy: "Baby George" had never existed. Pandemonium ensued. According to press reports, "fathers and mothers who had crowded the court room during the hearing broke through the railing weeping hysterically to surround the heart-broken mother."[84] The case had touched a deep nerve, and many observers perceived the outcome as unjust or unsatisfactory, but the verdict was final.

Baby mix-ups were sensational events, their feverish coverage in the press no doubt entirely out of proportion to their incidence. But they registered a real revolution in American women's reproductive lives: the rapid transition from home to hospital birth. In 1900, less than 5 percent of births occurred in hospitals. By 1921, the number had risen to at least 30 percent in many urban areas, and in some cities it surpassed half. Throughout the 1920s, hospital births continued to rise. In Cleveland, where Mrs. Smith demanded the return of Baby George, hospital deliveries jumped from 22 percent in 1920 to 55 percent a decade later. It is likely that Mrs. Smith's four older children had been born at home.[85]

"What a difference between childbirth of today and yesterday!" mused one observer before noting the obvious: "had baby Smith not been born in a hospital the confusion could not have resulted."[86] For many women, giving birth in the hospital—clean, safe, modern, scientific—was preferable to the home.[87] But the maternity ward also evoked the impersonal ethos of factory production. "In this age of efficiency, when babies come in quantity lots at hospitals, there is a hint of peril in the anonymity of large numbers." In law and culture, maternity was supposedly certain, empirically verifiable at the moment of birth. Yet in the modern hospital nursery, this most intimate and indelible of ties could be severed forever by a moment of banal carelessness. With infants arrayed as "a screaming, undifferentiated lot of bundles in a row of white bassinets," who could tell them apart?[88] Here parentage testing was charged with a new task: not finding fathers or disciplining women but repairing the fragile fabric of identity that medical modernity itself rent asunder.

Doctors tended to dismiss baby mix-ups as the "bogy of first time mothers." In a "well organized" hospital, they simply did not happen.[89] Morris Fishbein, the chief spokesman of the American Medical Association (and Albert Abrams's nemesis), suggested that the Smith case would stimulate scientists to "find in the human blood positive means of identification"—not in order to undo baby swaps but to demonstrate they had never occurred in the first place. Fishbein claimed, somewhat improbably, that the whole Smith affair had "served to allay the doubts and fears of many questioning mothers."[90]

Three years later, the Watkins–Bamberger debacle suggested that such fears might not be figments of the maternal imagination after all. This would be the loudest of the three baby-swap dramas and the one most elaborately staged by the press. It also most prominently showcased the role of science. From its debut in the papers under headlines like "Baby Mix-Up Stumps Sages" and "Baby Shuffle Still Puckers Sages' Brow; but Science Hopes to Solve Problem," the press narrated the story as a thrilling scientific riddle. As soon as the parents came forward with the suspicious labels, the public health commissioner of Chicago, Dr. Arnold H. Kegel (of pelvic floor fame), assembled a panel of eleven experts to examine the infants and parents for clues to their identities. The specialists in ophthalmology, dermatology, anatomy, obstetrics, pathology, and forensic science conducted analyses of fingerprints, birthmarks, head measurements, and blood types. When all the evidence was in, nine of the eleven scientists voted that the babies had been switched, a verdict endorsed by Commissioner Kegel.

The parents, however, remained uncertain. Mr. Bamberger in particular rejected the finding outright and at one point offered a dramatic twist when he stormed out of a meeting with officials and disappeared with the baby now deemed the Watkinses' child.[91] The Watkinses seemed more willing to accept the scientific verdict but were also ambivalent, so their lawyer suggested an alternative authority: a "jury of mothers" that would resolve the puzzle using maternal instinct "where science and law had failed."[92] A planned "mass meeting" of 2,000 Chicago mothers turned out seventy-five women at the Hotel Sheraton, where a group calling itself the Associated Mothers of Illinois called for reforms to hospital procedures to avoid future

The Watkinses (left) and Bambergers (right) pose with Chicago's health commissioner, who convened a panel of scientific experts to investigate the identity of the couples' disputed babies.

"Shuffled Babies Howl as Science Toils on Puzzle," *Chicago Daily Tribune*, July 23, 1930, 3.

mix-ups. Meanwhile, the Bambergers baptized the baby in their custody, and the Watkinses filed a $100,000 suit against the hospital for negligence. For its part, the hospital categorically denied the babies had ever been switched and claimed they had been with the correct parents all along, thus contradicting Commissioner Kegel and the scientific panel's verdict. Hospital authorities further complained that the health department's involvement in the case had "resulted in additional publicity, including photographs and statements of various experts who were asked to solve the 'mystery.'"[93] In other words, science was less a means to solve the (nonexistent) puzzle than another sensational element in a media circus.

A month after the case had first surfaced, the Bambergers abruptly changed their mind and agreed that the babies had indeed been swapped. The two couples met and exchanged the infants in an encounter carefully staged for press photographers.

The case's resolution was definitive, but two stories emerged about how it had been reached. Forever after, scientists would hail the case as a triumph of modern serological science.[94] In Chicago as in other recent instances of baby swapping, in Cuba and in France, blood grouping had solved otherwise intractable puzzles of identity.[95] Fritz Schiff noted that alleged mix-ups involving four parents and two babies were ideally suited to the technique. Alexander Wiener, a collaborator of Landsteiner's and soon one of the United States' most prominent paternity testers, calculated that the probabilities of an exclusion in such scenarios was seven in ten.[96] In other words, blood group analysis could solve a significant majority of baby swaps. At the height of the Watkins–Bamberger hoopla, Wiener capitalized on the case to write a letter to *JAMA* explaining the science behind ABO typing and the recently discovered MN blood factors.[97] In a fitting coincidence, two months after the babies were exchanged, Karl Landsteiner received the Nobel Prize for his discovery of the human blood groups.

This was how the scientific establishment narrated the case. The press, the public, and the parents themselves, however, told a very different story— starting with a crucial detail at the very beginning. As it turned out, just one week after Mr. Watkins noticed the suspicious label on his newborn, around the time the case debuted in the press, the blood of the infants and their parents had been tested and had revealed that the babies had indeed been switched. In other words, the entire Watkins–Bamberger drama as it played out in the newspapers—the panel of experts, the mothers' jury, the baptism, the legal suit, the hospital's categorical denial of a mix-up—had occurred after a blood test had conclusively demonstrated the babies had been given to the wrong parents. Conclusively, that is, from the point of view of blood group scientists.

In the popular narrative, blood group testing was just one among many scientific methods, all of which were inconclusive, none of which were infallible.[98] Press accounts described the blood test as having "failed."[99] To be sure, scientists were central protagonists in this story—sometimes literally. The experts' examination of the infants at Commissioner Kegel's office was interrupted to allow "scientists and babies to pose for taking moving pictures." The panel's fingerprint expert, an eccentric Viennese, played a starring role. He was portrayed perspiring profusely as he hunched over a baby attempting to coax its fist open in order to take charcoal prints: "Ooch, such work this is."[100]

But the sages, for all their color, did not have answers. The press stressed the thrilling but ultimately futile attempts of science to solve the riddle of the babies' identities. The director of Chicago's police identification unit, for example, characterized the babies' charcoal fingerprints as "a collection of smears."[101] The municipal court judge charged with hearing the case opined that in his experience, "human testimony is better than all this science."[102] The papers debated the authority of maternal instinct—"the oldest and, as some say, the surest of knowledge"—as compared with the wisdom of the scientists: "Though it has [as] yet received no scientific test, it has been stated often that 'a mother will unerringly know her own child,' that the call of blood is irresistible between a woman and the infant she has brought into the world."[103] And why indeed would the press have chosen to credit scientific methods? After all, a riveting puzzle that had been solved before it had even begun was not much of a puzzle at all.

For their part, the parents of the disputed babies, at least as they were portrayed in the newspapers, were no less skeptical. The Watkinses went back and forth on the question of the babies' identity, guided mostly by their shifting perceptions of the infants' physical resemblance to other family members. Mr. Bamberger discounted the conclusions of the scientific panel altogether. "I'm sick of this science business," he reportedly scoffed. "My wife knows it's our baby and I guess a mother's instinct is as good as the experts."[104] The very idea of a jury of mothers, conspicuously posed against the jury of experts, suggested that perhaps science was not the best way to discover parentage after all. To be sure, the matrons of this jury were portrayed as slightly ridiculous—but no more so than the quirky Viennese fingerprint expert and his charcoal smudges.

The idea of a scientific method of establishing parentage had emerged in the American press almost a decade earlier, during the Albert Abrams affair, and clearly popular fascination with the idea was alive and well. Yet fascination was different from faith: the press and the public played with the idea without fully accepting it. Indeed, debating the validity of parentage science was part of its allure.

A few years after the Bamberger–Watkins case, a district attorney in a paternity suit in Pennsylvania expressed skepticism about the value of blood groups: "The first time I heard about this grouping business was out in Chicago where two children got mixed in the hospital. Well, they grouped and grouped, and as far as I know they are grouping yet. . . . The mothers finally decided: This is my baby, and the other mother said, This is my baby, and they all went home happy and forgot all about this grouping business." Here the story had an entirely different moral from the one that scientists told. As one newspaper summarized it: "the mothers took matters into their own hands and swapped babies against expert say-so."[105]

By the early 1930s, "this grouping business" was widespread and uncontroversial in German and Austrian courts, and it had spread, albeit on a much more limited scale, to Brazil and other countries. In the United States, where there was a lingering skepticism about the ability of science to solve such quandaries, the method had less legitimacy and more limited application. Above all, it continued to be associated with scandal and sensation. The meanings of blood differed across these contexts.

IN 1932, THE BRAZILIAN forensic expert Afrânio Peixoto observed that the "voice of blood is now beginning to be heard."[106] Serological tests of parentage had become ever more familiar to judicial and popular audiences, although the voice was more of a cacophony—different methods of wildly varying legitimacy. Peixoto's invocation of a singular "voice" is nevertheless apt because that is how the press reported it, and that is how international publics heard it.

The American press was especially wont to conflate the many "blood tests of parentage" then circulating. It confused ABO analysis with the oscillophore and described Mendelian inheritance as a theory of birth order.[107] *Popular Science Monthly* described the Landsteiner test and Zangemeister's "entirely new method" in one breath, thereby giving equal validity to Nobel Prize–winning research and the dubious alchemy of the Prussian obstetrician.[108] The press's tendency to conflate the various tests circulating in the 1920s and into the 1930s reflected neither its ignorance nor its gullibility, however. After all, scientific experts sometimes did the same thing. Leone Lattes cited Zangemeister's paternity research in his foundational textbook on blood groups, and Manoiloff and Poliakowa published their studies in the most prestigious American journal of physical anthropology.[109] Distinctions between science and pseudoscience, query and quackery, truth and artifice, were drawn long after the fact. The various methods of finding parentage in the blood—whether based on vibrations, crystals, colors, or clumps—developed in dialogue with one another. Rather than dwelling on which of these techniques worked, the more interesting question is, what work did they do?

Blood group testing garnered the strongest scientific consensus and widest practical application of all these methods. But it was not the intrinsic power of the method that determined its value; it was the social and legal questions being asked of it. Modern paternity promised a scientific solution to the problem of the unknown father—and on occasion the unknown mother—but not only did these solutions vary, so too did the nature of the problem they aimed to solve. Fritz Schiff surely could not have anticipated the wide-ranging transatlantic reverberations of the simple technique he had helped bring to the courts. In Germany and Austria, the test combated a perceived crisis of maternal perjury in the courts. In Brazil, it served as a new tool for resolving allegations of sexual crimes. In the United States, it became associated with hospital baby swaps but was given limited credence by the press and the public. The voice of blood spoke about race and family, about women and men. It could speak about truth and justice, sex and morality, kinship and identity. But the voice was never just an utterance of scientific fact, and it resounded differently as it echoed around the world.

4

CITY OF STRANGERS

The free will of deceased fathers must always be respected.

—Judge Adolfo Casabal, Civil Court of Buenos Aires

THE PATERNITY TEST took six weeks, was valued at $150,000 pesos, and involved the examination of the bodies of sixteen individuals from four generations. Three of the subjects were no longer alive and could be examined only in photographs. One of the dead was Roque Arcardini, the putative father, whose intemperate private life had made the paternity investigation necessary in the first place.

Arcardini died in Buenos Aires in 1914 at age sixty-one, leaving behind an estate worth millions of pesos. He also left two families locked in an ugly conflict. The first one, his natal family, consisted of his ailing and elderly mother, his three sisters, and various nieces and nephews. The second family, the one he had formed in the final decade or so of his life, was composed of his consort—a woman he never married—and three young children, whose paternity was ambiguous. Now it was up to the court to determine whether the youngsters were really Roque Arcardini's children and heirs or whether his sprawling patrimony belonged to his mother and siblings.

In 1914, studying physical bodies to resolve a dispute over an inheritance was a most unusual approach, in Argentina or anywhere else. As part of their suit, the Arcardini family hired Dr. Roberto Lehmann

Nitsche, head of the anthropology department of the Museum of La Plata, the premiere natural history museum in South America. He examined the noses and ears of the sixteen individuals, studied their hairlines, and assessed their eye color and skin tone. Seven years before Albert Abrams first made global headlines with his oscillophore, ten years before Fritz Schiff's groundbreaking blood group analyses, Lehmann Nitsche's method suggested a different approach to proving paternity, one that sought proof not in the blood but on the body. While courts in Argentina and elsewhere sometimes considered physical resemblance in their deliberations on parentage, the Arcardini case was hailed as the first one anywhere in the world in which an expert applied modern "Mendelian principles" to ascertain kinship. The expert report Lehmann Nitsche submitted to the court circulated widely and would be referenced for decades by experts in Latin America and Europe.[1] In succeeding years, transatlantic scientists developed, often independently of one another, a variety of approaches for analyzing somatic traits and physical similarities to establish parentage. Such methods were referred to as somatic, morphological, anthropological, anthropomorphic, prosopographic, or comparative analysis or in the preferred German moniker that translated as "the similarity method."

Lehmann Nitsche introduced into a Belle Époque Buenos Aires courtroom a new scientific technique for investigating the old question: who is the father? But his approach also redefined the question itself. In Argentina as in many other societies, especially those following the civil law tradition, paternity was less a problem of physical proof than of social knowledge. Lehmann Nitsche's new method advanced the central conceit of modern paternity: that modern, mass society required a new understanding of who the father was and how he could be known. Whether focused on blood or body, the science of parentage determination redefined paternity itself, transforming a social quality into a physical one.

THE TENDENCY OF OFFSPRING to look like one or both parents had been noted for millennia. New knowledge of heredity promised to scientize similarity, to reveal patterns of somatic transmission that were objective and predictable. Already in the 1880s, Francis Galton, the English polymath, father of eugenics, and cousin of Charles Darwin had mused, "It is not improbable, and worth taking pains to inquire, whether each person may not carry visibly about his body undeniable evidence of his parentage and near kinships."[2] Could ears, noses, or anthropometric analysis provide such evidence? Might techniques of individual identification, such as forensic dentistry or fingerprinting, find kinship on the body?

The search for a biochemical marker of relatedness in the blood and the search for somatic or morphological markers on the body were intertwined, complementary rather than competing. Both approaches sought to parlay knowledge of human heredity into a practical method to identify family members. Both emerged at the same moment, often in the same places, and sometimes among the very same scientists. In judicial proceedings, the two methods were often used in tandem.

But blood and body also differed. Blood group analysis was simple, quick, and cheap and, thanks to the universal laws of group inheritance, objective and unambiguous. When blood typing excluded a progenitor, it did so incontrovertibly. But its practical uses were also limited: it could exclude an impossible father but never identify an actual father. Into the 1950s, blood group analysis could do no more than classify large swathes of the population as possible or impossible progenitors.

In contrast, visual appraisal of morphological similarity seemed highly subjective. No bodily traits, even if known to have a hereditary element, provided the clear-cut objectivity of blood typing. There were four types of blood, but there was an endless variety of noses and ears. How could one identify, describe, or measure similarity between traits? And what about the even more ineffable quality of resemblance between two people said to "look alike"? At best, comparing bodies could be complex, laborious, and expensive; at worst it looked suspiciously subjective. And yet for all their bewildering imprecision, bodies held out the tantalizing promise of a method for identifying the father. Drawing on an eclectic toolkit of new techniques and old, experts analyzed resemblance, described it, measured it, and even attempted to quantify it, in an effort to identify, as Galton had mused, the "undeniable evidence of parentage." This was the task Dr. Lehmann Nitsche set for himself in 1914 in the Buenos Aires court.

ROCCO ARCARDINI WAS five or six years old when he and his parents emigrated from the Piedmont region of northern Italy in the early 1850s. They left behind a rural village of fewer than 2,500 souls for a faraway city of almost 100,000, a third of whose residents were, like them, foreigners.[3] Once established in Buenos Aires, four more children were born to the family in rapid succession. The very first Argentine census, carried out in 1869, finds Rocco—now Hispanicized as Roque—an adolescent, working as a tailor alongside his father, Luigi—now Luis—on the Calle del Buen Orden. The street teemed with stores, artisans' workshops, and warehouses populated by working-class Argentines and a growing number of immigrants like themselves. But the Arcardinis would soon leave this world behind as their fortunes began to rise.[4]

In the last decades of the nineteenth century, Argentina experienced an astonishing process of economic growth as the country became integrated into global markets. Its fertile pampas produced wheat, wool, and meat that were loaded onto steamships bound for Europe. In exchange, it received foreign capital and a flood of foreign workers, the most numerous of whom, like the Arcardinis, hailed from Italy. The Argentine economy boomed, its gross domestic product growing faster than that of any country in the world. By 1914, the year World War I interrupted the transatlantic flows of commodities, capital, and immigrants and also, coincidentally, the year of Roque Arcardini's death, Argentina was among the ten richest countries in the world.

Having arrived at the cusp of this transformation, the Arcardinis were well poised to take advantage of it. Luis Arcardini began to pursue a series of savvy investments that closely tracked the rapid growth of his adoptive homeland. He bought shares in an Italian Argentine shipping company and began purchasing small properties in Buenos Aires.[5] As the city's population expanded, the Arcardinis capitalized on rising urban property values, renting out their properties to butchers, undertakers, and grocers.

They also turned their sights beyond the capital. In the 1870s, as booming exports created incentives to expand commodity production, Argentina initiated a murderous campaign to seize lands located on the frontiers of the nation-state from their indigenous inhabitants. Those lands were put up for public auction to buyers who acquired them for a pittance and then watched their value surge. Roque Arcardini, by this time a young man and, increasingly, the family's patriarch, was among those beneficiaries.[6] Meanwhile, his younger brother, the Argentine-born Antonio, married a Spanish woman and secured a high-ranking post in the Treasury, an insider position that no doubt served the family well as the state distributed lands to private buyers.

Roque's Argentine-born sisters, María Luisa and Emilia, married a French merchant and an Italian shipowner, respectively, further expanding the family's commercial networks.

By the time of Roque Arcardini's death, a list of his properties took up the better part of a newspaper column. An inventory of his landholdings in the city and suburbs of Buenos Aires alone took five years to compile and estimated their value at more than 4 million pesos.[7] The family of Piedmontese tailors had become Argentine *estancieros* par excellence—if only Roque Arcardini had been as astute in passing down that patrimony as he had been in accumulating it.

Unlike his siblings, Roque never married. Known for his womanizing, around the turn of the century he began a public relationship with a French immigrant named Celestina Larroudé. She was described in court records as his "concubine" and "a woman of inferior condition," whom Roque had "saved from misery and an uncertain and perilous fate."[8] It is little wonder that his family was "tenaciously" opposed to their marriage. Instead, the two lived together informally, a common if not entirely respectable arrangement in Belle Époque Buenos Aires. Soon Larroudé had three children, María Mafalda (born 1903), Roque Humberto (born 1904), and María Carmen (born 1905). Roque recognized Roque Humberto and María Carmen as his natural children before the civil registrar. He did not formally recognize the oldest child, María Mafalda, but her paternity was signaled in other ways both civil and sacred. She was born in a newly renovated property owned by the Arcardinis, where her mother, Celestina, resided.[9] Her baptismal certificate identified her parents as unknown but noted she had been "adopted by" Roque Arcardini, though adoption did not formally exist in Argentine law. Finally, María Luisa Arcardini, Roque's sister, was godmother to the girl and also, a few years later, to her younger sister. Roque's longtime lawyer was godfather to the boy whose name, Roque Humberto, further signaled his paternity.

Together for more than a decade, Roque and Celestina were a quarrelsome couple. They fought about Celestina's tendency to drink too much. They argued over care of the children. Roque criticized Celestina for giving the children milk instead of food.[10] But by all accounts, he was also profoundly solicitous of her and the children. A cache of notes that became part of the court records after his death chronicled everyday life in the household. "Celestina, I'm not coming. It's really cold," wrote Roque in one note. "Put the children to bed early and send me a message with what you need from the store." "Tell me how Roquito is doing," he implored in another note when the boy was sick. "Don't let him get up and keep him covered."[11]

The degree of Arcardini's paternal solicitude was exemplary and unusual. And yet out of respect for the family that he came from, Roque Arcardini chose not to formalize it. He did not marry Celestina, although he promised he would do so when his mother died (María Trischetti de Arcardini was eighty-seven years old and bedridden when her son predeceased her). He legally recognized two of the three children born in his relationship with Celestina Larroudé, but for whatever reason—preference, oversight?—never recognized the eldest, María Mafalda. And in 1908, Roque did a strange thing: he drew up a document stating that the three children were neither his nor Celestina's, that they had been purchased as newborns and surreptitiously brought into the household. He asked Celestina to sign it, and she did.

The couple continued to live together, and the children grew. Around March of 1914, Roque placed ten-year-old Roque Humberto in a local boarding school, paying the month in advance. In the paperwork, he identified himself, ambiguously, as the new pupil's "adoptive father." Once, he returned to the school and took the boy out for a visit. Then a few days later, the solicitous father, age sixty-one, died suddenly of causes unknown.

With the death of Roque Arcardini, two families lost their patriarch. In the days that followed his untimely passing, the grieving Arcardinis did two things. First, they sent money to Celestina and the children. And second, they filed a legal suit challenging the youngsters' status as heirs to his estate.

The strange document Roque had drawn up about the fake births resurfaced. The Arcardinis claimed Celestina had simulated the three pregnancies and secretly purchased the babies to "conserve and strengthen" her relationship with Roque.[12] Roque's sister María Luisa, the girls' godmother, claimed that any care or support he had given the children reflected his "charitable and religious impulses," not his paternity.[13] Multiple witnesses corroborated the fraud, alleging in sworn testimony that Celestina had stuffed clothes under her garments to feign the pregnancies. Celestina countered that the story was patently false. She had been unaware of the document's contents and had signed it to please Roque. As for the witnesses, she said, they had all been bribed by the Arcardinis. In this case, maternity was also in doubt, but true to the special stakes of paternity, the inquest was concerned only with Roque's relationship to the children. And a lengthy inquest it was: the charge of "supposititious birth" was just the first in a long succession of suits over the identity of the three children and the fate of Roque Arcardini's estate.

ACCORDING TO THE ANTHROPOLOGIST Roberto Lehmann Nitsche, episodes like the Arcardini affair, as it would soon be known, had become all too common in Belle Époque Argentina. "As soon as an unmarried man of a certain social position, of a certain fortune, dies," putative children "spring up like mushrooms" to assert false and frivolous inheritance claims. "It would be difficult," he claimed, "to find in other countries such a large number of cases of this type."[14]

The problem, for Lehmann Nitsche, lay in the structure of Argentine law. Argentina, Latin America, and much of continental Europe followed the civil law tradition, which shared a common definition of paternity. According to civil law jurists, paternity as a physical fact was unknown and unknowable. It could therefore be established in one of two ways: either it was conferred by marriage, such that a married woman's husband was automatically considered the father of her children, or in the case of unmarried couples, the child could acquire a father through "possession of status" (posesión de estado). This figure of law referred to the man's behaviors and the public reputation that the child developed as his son or daughter. According to an old Roman trifecta nomen, tractatus, et fama, paternity came into being when a man gave the child his last name, treated it as his own, and was reputed in the eyes of the community to be the father. At the heart of this definition lay paternal will: what made a man a father was his active desire to be one. By performing his paternity toward the child or its mother before his associates or the community, a man demonstrated both his consciousness of being the father and his willingness to accept this role. As in the Arcardini case, questions of paternity often arose in disputes over inheritance, in which case the putative father was by definition already dead. Proving possession of status therefore required reconstructing the conscience of an individual who could no longer express his intentions.

Possession of status was not just a method for establishing paternity; it also defined what paternity was and what it was not. Rather than flowing

automatically from the act of procreation, paternity was social and volitional. It was made, through a man's words and deeds. The mother, conversely, was excluded from this process of paternal verification. The law in
Argentina as in many other places prohibited her from repudiating the
father of her children in court. Although this might seem illogical—who
better to know the child's father than its mother?—in fact, it was perfectly
consistent with a conception of paternity as a question not of fact but of
will, specifically the man's will, not the mother's. This explains why Celestina's extraordinary written statement that she had faked the births of the
three children was technically irrelevant to the proceedings. "What would
the principle of the stability of parentage become, subject to the caprice of
a woman?" asked the judge.[15]

Following the logic of the father's will, civil law not only discounted
mothers' testimony but also restricted the rights of children to bring paternity suits in the first place. Children could not demand a paternal relationship from a man who did not willingly recognize it. Such restrictions had the
added advantage of protecting men and legitimate families from unwanted
claims on their patrimony and good reputations. The Napoleonic Code,
which formed the basis for civil law's conception of paternity, famously prohibited paternity investigations outright, as did law in neighboring Chile and
in a number of other countries. Argentine law was relatively more liberal,
permitting some illegitimate children to bring suits against putative fathers
and to present a range of evidence to demonstrate their possession of status.
This relative openness is why we know so much about the intimate domestic
circle of Roque, Celestina, and the children: the court studied their social
and emotional lives carefully, above all the attitudes and actions of Roque
Arcardini, for it was through such evidence that his conscience could be assessed. Did he believe he was the father? Did he willingly accept that status?
Did he demonstrate it in his actions before the world?

For a growing number of critics like Lehmann Nitsche, Argentine law's
relative liberality in allowing people to ask such questions in high-stakes judicial suits had become problematic. He argued that the logic of "pos-
sessing status" made sense in an older, predominantly rural society in which
informal unions abounded. Marriage had once been inaccessible because
parish priests were few and distant. Where nonmarital relationships were
common, society had to permit a certain flexibility in proving paternity. But
in modern, urban, mass society, possession of status had become a scourge.
Now putative children cropped up like mushrooms, claiming rights that did

not belong to them. Any person could come forward and, with the help of a few corrupt witnesses, attribute to a deceased man a series of actions and deeds indicative of paternity. The proliferation of Argentine jurisprudence on the precise meanings of possession of status around the turn of the century suggests its unsettled and evolving nature.[16] Uruguayans, who in the 1910s were engrossed in debates about whether to liberalize their own more restrictive paternity laws, regarded their neighbor Argentina as a cautionary tale: "in Argentina not a rich man dies without natural children being attributed to him."[17]

There are no statistics to say whether disputes over paternity and inheritance really were increasing in Belle Époque Argentina, but it is easy to see why, amid dizzying social transformations, it might have seemed they were. Around the turn of the century, some 15 percent of children born in Buenos Aires were illegitimate, a fact Lehmann Nitsche chalked up to what he characterized as a "Neolatin" tolerance for extramarital procreation. In fact, this was a lower percentage than most other Latin American and some European cities at the time (including Lehmann Nitsche's native Berlin). Still, 15 percent was a significant minority. Moreover, thanks to Argentina's boom, there was simply greater wealth around to become the object of dispute. After the turn of the century, it was no longer possible for newcomers to achieve the kind of meteoric ascent the Arcardinis had experienced a few decades earlier: amid the flood of foreign laborers, social mobility had stalled.[18] Perhaps great wealth and limited mobility really did encourage covetous "sons and daughters" to spring up like mushrooms—or perhaps this economic panorama simply sowed fears that they would.

Regardless of whether "the cosmopolitanism of Buenos Aires" really had sparked a proliferation of fraudulent paternity suits as Lehmann Nitsche claimed, it certainly could complicate old ways of knowing paternity.[19] Transnational migration and galloping urbanization had transformed the nation and its capital. In two decades, Argentina's population nearly doubled.[20] Over three and a half decades, the population of Buenos Aires swelled from 300,000 to almost 1.6 million inhabitants, making it one of the fastest growing cities in the world. By the year of Roque Arcardini's death, 1914, half of the Argentine population resided in cities, and half of Buenos Aires's population was foreign born—a higher proportion than any city in the world at the time.[21]

Possession of status was based on intimate knowledge of the father, but in this environment, such knowledge was fallible. The court's job in a paternity dispute was to reconstruct the father's acts, attitudes, and words through the testimony of witnesses—members of the household, neighborhood, and wider community. But demographic expansion, an influx of immigrants, and urban growth undermined older forms of social knowledge, making possession of status an anachronistic, and potentially perilous, method for establishing kinship and conferring patrimony. In a world of strangers, who could know the father?

Buenos Aires was in this regard emblematic of modern paternity's urban habitat. It was not by chance that scientific tests of paternity and parentage arose in cities—Buenos Aires and Berlin, Chicago and Vienna, São Paulo and New York. This is where laboratories and courts, which provided scientific expertise and legal redress, were most likely to be located. But it is also where questions about kinship and identity were most likely to arise. The succession of hospital baby swap scandals in the United States in the 1920s reflected similar dynamics. Both the fraudulent heirs of Buenos Aires and the substituted newborns of Cleveland and Atlanta reflected how the anonymity of modern spaces and institutions—the city and the hospital— threatened relatedness. Science offered a new way to discern identity and safeguard ties of filiation. If paternity could no longer be reliably located in social behaviors and reputations, perhaps it could be found on the physical persons of parent and child.

If modernity created the need for a new scientific method of paternity assessment, in Argentina a close-knit alliance between science and statecraft was primed to provide it. In the late nineteenth century, new fields of expertise—criminology, eugenics, social hygiene, public health, identification science—offered positivist solutions to crime, disease, class conflict, and other ills associated with urban mass society. Argentina was at the forefront of these developments, a center of Latin American eugenic science and a global pioneer in the new science of fingerprinting. Statesmen were especially eager to embrace scientific expertise to solve social problems.[22] A Buenos Aires courtroom was thus a fitting birthplace for the first Mendelian paternity test.

MODERNITY HEIGHTENED the perceived need for a somatic test of parentage, but the idea of the body as a cipher of kin identity was an ancient one. Philosophers and doctors had commented on the physical resemblance between parents and children since antiquity; Hippocrates supposedly defended a woman accused of adultery when her child did not resemble its father.[23] Early modern observers debated the value of physical or temperamental likeness as evidence of parentage. Alongside social and legal presumptions based on behavior and reputation, physical resemblance might also figure in paternity disputes. But it elicited caution because, as observers noted, the vagaries of resemblance made it a potentially misleading indicator of parentage. After all, strangers might bear an uncanny likeness to each other, and children might not resemble their parents at all.

The use of similarity as evidence for kinship was further complicated by the problem of maternal imagination. A widespread early modern idea, which survived as folk belief in many places into the twentieth century, held that children were physically marked by the workings of the mother's mind. The pregnant woman's unsatisfied yearning for cherries could result in a red birthmark on her child. A mother who witnessed a public execution could produce a fetus with physical deformities. Following this logic, it was widely believed a child could resemble the man whom its mother visualized at the moment of conception, rather than the physical author of that conception. In other words, resemblance did not necessarily indicate physical parentage.[24]

Some early modern authorities, including the celebrated Italian forensic doctor Paolo Zacchia, took issue with such claims. Discounting the more fantastic powers attributed to the maternal mind, Zacchia argued that physical similarities were indeed evidence of parentage. He himself offered expert testimony on resemblance in paternity suits heard by ecclesiastical tribunals in sixteenth-century Rome.[25] Early-twentieth-century paternity testers often invoked Zacchia as the forefather of their own forensic endeavors.

Filial likeness featured in other medicolegal traditions as well. Islamic law recognized the assessment of resemblance as a discrete expertise: the *qiyāfa* was an evaluation of physical clues for the purpose of establishing kinship, performed by the *qā-if*, a physiognomic expert able to detect signs invisible to the unschooled eye. The practice, which originated in pre-Islamic Arab societies, was probably used mostly in the case of children born to slave women. As in early modern legal medicine in the West, the practice was controversial, but it enjoyed the approval of the Prophet, the Companions, and the Four Righteous Caliphs.[26]

Filial resemblance raised questions about the nature of reproduction, generation, and heredity, and as such it spilled out of medical circles into literary and philosophical debates. The child who does not look like its father, the suspicious husband who searches the visage of his wife's child for confirmation of his paternity, are recurring plot devices, from ancient myths to Shakespeare. Eighteenth-century British observers were taken with the phenomenon of resemblance as part of broader debates about the workings of nature and nurture.[27] Courts, too, worked through these ideas. A famous Scottish suit of a couple accused of faking the birth of twin sons to secure transmission of the family estate elicited long disquisitions about the wonders of family likeness in the House of Lords. Noting "the most perfect resemblance" between the second twin and his putative mother, the Lord Chancellor declared, "It is an impression stamped by God himself to prove the legitimacy of the child."[28]

In the nineteenth century and into the twentieth, such "impressions," though controversial, were routinely admitted as evidence in Anglo-American courts.[29] This practice contrasted with civil law jurisdictions. In Latin American paternity suits, witnesses sometimes commented on a child's resemblance to a man as commonsense evidence of his paternity, but such observations were seldom a topic of forensic deliberation. Where paternity was an expression of the father's will rather than a physical fact, evidence of similarity was legally irrelevant.[30] The contrasting judicial treatment of resemblance reflects distinct understandings of paternity; where civil law understood it as primarily social, in the Anglo-American tradition its physical dimension was more central.

THE ANGLO-AMERICAN CONCEPTION of paternity as a physical fact did not necessarily imply a scientific mode of assessing it, however. The very same year as the Arcardini case, another family drama played out, whose mise-en-scène stretched from rural England to California. Lieutenant Charles Slingsby and his wife were the parents of four-year-old Teddy, heir to the family estate in Yorkshire. Relatives challenged the boy's identity, claiming that the couple's "real" child, born during a sojourn in San Francisco, had died at birth and that Mrs. Slingsby had obtained another infant through an advertisement in a local newspaper.

The case was heard by an English court in 1915 and revolved around evidence of this alleged fraud. As the judge listened to the arguments of the parties, he began to study Teddy and his purported parents, who were present in the courtroom. Struck by the resemblance between them and above all by the similarity between the jaws of the boy and Lieutenant Slingsby, he decided to call in an expert to assess the matter. Sir George Frampton, a personal friend of the judge, examined the parties. Frampton not only confirmed the resemblance in paternal and filial jaws but also found a striking likeness between the left ears of Teddy and Mrs. Slingsby. The judge ruled that the boy was the true son of the Slingsbys, citing among other evidence this physical assessment as "absolutely conclusive."[31]

A sizable estate; allegations of suppositious birth; somatic likeness; an expert's assessment: the Slingsby case recapitulated the Arcardini suit but with a crucial difference. Unlike Dr. Roberto Lehmann Nitsche, Sir George Frampton was not a scientist but a sculptor (among his more noted public works are the lions guarding the British Museum). When the judge sought an expert to assess the Slingsby jaws, he decided the question was most appropriately addressed not to "a surgeon or a medical man," but to an artist.[32] Artists had been consulted as forensic experts of family resemblance for hundreds of years, and the practice persisted into the twentieth century, especially in American and British courts.[33]

The press dubbed young Teddy "the boy with the $500,000 left ear." The British medical journal *The Lancet* expounded on the "Slingsby ear," which referred to an infolding of the ear margin. It noted that the trait was actually quite common and that an anthropologist, had the court solicited one, would not have attributed much significance to it. But the journal did not question the sculptor's expertise or the magistrate's judgment in soliciting it. On the contrary, it noted, "Sir George Frampton's method of observation we do not know, but we must admit that sculptors have certainly a right to express an opinion."

The comment was not intended to be facetious. As evidence for the legitimacy of the sculptor's opinion, the editorial went on to recount an anecdote about no less a scientist than Charles Darwin. In *The Descent of Man*, Darwin had identified a certain ear nodule as a vestigial trait revealing the common origin of primates and humans. He first noticed the nodule, a feature of his own ear, because the sculptor making his bust had drawn his attention to it. (In fact, the nodule was first named the "Woolnerian tip" after the sculptor; only later did it became "Darwin's tubercle.") If a sculptor had identified a somatic trait of such transcendental scientific significance, mused *The Lancet,* surely a sculptor could also appraise traits in a parentage dispute.[34] So saying, the premiere British medical journal endorsed the court's reliance on an artist's testimony.

Several high-profile cases of disputed parentage in the first three decades of the twentieth century featured similar testimony. In a celebrated 1903 case in Germany, a Polish countess, Kwilecki, was accused, like Celestina Larraudé in Buenos Aires and Mrs. Slingsby in Yorkshire, of substituting a "supposititious" child for patrimonial purposes. Two doctors, a specialist from the police department's anthropometric unit, and a portrait painter were called in to assess resemblance, which, again like the Slingsby affair, zeroed in on a telltale ear. (One of the doctors was Fritz Strassmann of the University of Berlin, whose son Georg collaborated with Fritz Schiff to introduce blood group tests into German courts.)[35] Medicolegal experts would frequently cite the case as a precedent for the scientific analysis of physiognomy in parentage disputes.

Probably inspired by the Slingsby case, a few years later a San Francisco judge deposed a well-known sculptor to assess likeness in a paternity dispute. The sculptor himself was skeptical about this charge ("I honestly do not believe such a thing possible," he testified). The judge who deposed him was none other than Thomas Graham, of Albert Abrams fame; in fact, both Abrams and the sculptor served as experts in the case, reflecting how the scientific and the artistic were, as in the Kwilecki case, regarded as complementary modes of assessing physical parentage.[36] As late as 1932, a Scottish housepainter presented a New York City courtroom with a specially commissioned bronze bust of the deceased real estate mogul he claimed was his father. Called to the witness stand, the sculptor of the bust, which featured a removable derby, mustache, and spectacles, used it to demonstrate the striking resemblances between the putative father it portrayed and the man claiming to be his son.[37]

To be sure, artists were probably sporadic witnesses in paternity disputes. The fact that newspapers and medicolegal authors avidly followed their testimony suggests it must have been somewhat unusual, and artistic authority in such matters was hardly undisputed. When the Scottish housepainter sidled up to the bronze bust to demonstrate his resemblance, the courtroom erupted in laughter. By the early 1930s, such assessments were still occasionally introduced, but they were clearly taken with a grain of salt.

Yet the fact that artists were invited into such deliberations at all suggests that, in the first decades of the century, the status of resemblance was up for grabs. Galton had mused on the "undeniable evidence of parentage," but the question of how one might discern this evidence, and who was equipped to do so, remained unresolved. Anglo-American courts did not automatically entrust the assessment of physical similarity to medicolegal experts, anthropologists, or doctors. Some jurists and even some medical and scientific experts (as in the *Lancet* article) treated parentage assessment as an exercise not in biological investigation, still less in hereditary analysis, but rather in visual appraisal. The body might harbor perceptible clues to ancestry, but evaluating these clues, indeed, the very practice of seeing, was as much an art as a science.

The most frequent appraisers of filial resemblance in Anglo-American parentage disputes were neither artists nor scientists, however. In fact, they were not experts of any kind. They were juries and judges—laymen who could claim no special skill in seeing at all.[38] Several years after the Slingsby case, Britain was rocked by the Russell divorce case, which revealed lurid details about the sex lives of the English aristocracy when the Honorable John Russell, son and heir to a barony, denied paternity of his wife's child. The judge ordered the baby displayed in his chambers, noting, "some of the jurymen were parents and knew the sort of discussion that took place as to whom an infant resembled."[39] (Ultimately, the House of Lords ruled that evidence bastardizing a child born in marriage was inadmissible, and the baby kept his legitimacy and his barony.) American courts routinely displayed children for the jury's inspection—the "child itself as an exhibit"—and witnesses might also testify about similarities of appearance, gait, or gesture, particularly if the alleged father was deceased.[40] The Charlie Chaplin paternity suit reached a moment of high drama when mother, child, and putative father were instructed to line up before the jury box and, for forty-five seconds, stood motionless as the hushed courtroom examined their facial features. The appraisal of resemblance by judges and juries lasted in some American courtrooms into the 1960s.[41]

Physical resemblance was not only a form of evidence; it was one non-experts were equipped to evaluate. As an English jurist opined in 1923, "tribunals sufficiently familiar with the course of nature to be able to take judicial notice that a woman of seventy will have no more children, and that a husband's non-access for a year disproves the legitimacy of his wife's child, will need no aid from scientific witnesses to assure them that 'Like is apt to beget like.'"[42] This truism cast resemblance as a self-evident fact. Everyone knew that parents and children looked like each other; discerning resemblance required no more skill than the juryman's ability to observe.

Such truisms did not go uncontested. Critics charged that presenting jurors with an adorable baby played more to their heartstrings than to their critical faculties. An observer in the Russell divorce suit pronounced the spectacle of a helpless infant cooing through a trial that could brand it with the stigma of illegitimacy "as revolting as it is futile."[43] And then there were basic questions of veracity. Did resemblance between two individuals necessarily prove their kinship? And how did one know resemblance when one saw it? Surely similarity sometimes rested more in the eye of the beholder than in the faces or bodies of the beholden. The word most often used to critique likeness in English-language writings was that it was "fanciful."[44]

Was filial resemblance obvious and transparent, or was it fanciful and therefore deceptive? These two positions, which vied for dominance among Anglo-American forensic experts, seem antithetical but in fact coincided on a key point: they both imagined resemblance analysis as beyond the purview of experts, especially scientific ones. If resemblance was inherently unpredictable, then science could not read it. If it was a matter of simple common sense, then science was superfluous, for even a layperson could see it. In the first decades of the twentieth century, both positions lost ground to the idea that similarity between parent and child was neither obvious nor fanciful. Rather, it followed predictable patterns of inheritance. Expert knowledge of these patterns allowed one to discover kinship between two individuals. Slowly, artistic and vernacular modes of discerning parentage gave way to the self-styled scientific ones associated with modern paternity.

NO ONE ARTICULATED this new idea more clearly than Lehmann Nitsche himself. "In terms of inheritance, nature is not at all *'capricious'* as the general public and those unschooled in biological matters suppose, *and it is very much possible to establish scientific rules* as did Mendel and the modern school of biology."[45] The analysis of physical resemblance required scientific expertise, Lehmann Nitsche argued. "The public, in general, does not know how to observe with enough detail the color of the hair, the iris, the skin, etc . . . not even you, benevolent reader!—knows the color of your own eyes."[46]

Galton had once mused on the possibility of discerning kinship on the body. Now his hereditarian fellow travelers—physical anthropologists like Lehmann Nitsche, eugenicists, and forensic experts—began to take up this challenge. At around the time an English probate court was deposing a sculptor to comment on the shape of Teddy Slingsby's ear, in Buenos Aires a judge was reviewing the first Mendelian analysis of parentage.

Born and educated in Germany, Roberto Lehmann Nitsche had arrived in Argentina in 1897 at age twenty-five to assume a position as head of anthropology at the natural history museum in La Plata. Over the next thirty years, he researched Argentine popular culture, indigenous ethnology, material culture, linguistics, and archaeology. He was also a Victorian race scientist par excellence, for whom the human body was a trove of data that could elucidate human evolutionary origins, racial difference, and social pathology. One thread of his life's work was therefore devoted to examining the bodies of the living and the dead—the latter in a collection of ancient remains at the Museum of la Plata, the former in mental asylums, criminal institutions, and indigenous communities, where he performed research and also served as forensic consultant to criminal investigations.[47]

Lehmann Nitsche brought this wealth of somatic expertise to bear in the Arcardini case, but it broke new ground in a key respect. The usual objects of forensic anthropology were inferior and atavistic bodies: the indigenous, the working class, the pathological, the criminal, the insane. Lehmann Nitsche trained his somatic analysis on civilized bodies—the members of a prosperous and respectable family of European origins. And he did so for a new end: not to reform or control inferior populations but to protect the patrimony of superior ones.

The anthropologist's somatic method reflected his transnational scientific milieu. Lehmann Nitsche's professional ties stretched up to North America, across the Atlantic, and around Latin America. He corresponded with U.S. eugenicist Charles Davenport, German racial anthropologists Otto Reche and Eugen Fischer, the Brazilian anthropologist and racial taxonomist Edgard Roquette Pinto, and Argentine fingerprint expert Juan Vucetich, among many others.[48] Reflecting this milieu, his paternity analysis homed in on familiar racial traits. He methodically described each subject—the three children, their mother Celestina Larraude, and twelve individuals from four generations of the Arcardini family—assessing their ears, noses, facial shape, and hair, skin, and eye color. Pigmentation, according to Lehmann Nitsche, was especially useful for his analysis because it followed predictable Mendelian laws of dominance and recessivity such that one could predict a child's coloration based on that of the parents. Here he cited Charles and Gertrude Davenport's well-known research on the hereditary bases of eye color. As for family members he could not examine because they were dead—Roque Arcardini, his brother Antonio, and their father Luis—he drew on photographs and the memories of other family members. He produced a genealogical tree of the Arcardinis that, he claimed, demonstrated perfectly the Mendelian transmission of somatic traits.[49]

Having spent almost twenty years researching race and culture in Argentina, Lehmann Nitsche's approach also incorporated local knowledge. His studies of indigenous groups in northern Argentina and folklore about the mixed-race gauchos of the nineteenth-century pampas led him to consider a novel somatic trait: hair growth. In human fetuses, covered in the downy hair known as the lanugo, the hairline is indistinct, with the hair of the scalp extending to the eyebrows. In individuals of the white race, Lehmann Nitsche claimed, the scalp became differentiated from the hairless forehead during the first year of life. In nonwhite races, this differentiation did not occur. For this reason, according to the anthropologist, gauchos applied depilatory balms to their foreheads in a vain attempt to erase this mark of racial inferiority.[50]

Drawing on this racial knowledge, Lehmann Nitsche proceeded to describe his subjects. The majority of the Arcardini family members shared "notable family traits"—similar profiles, nose and ear shape, and coloring. Their blue eyes corresponded "perfectly" to the "type of iris observed in 40 percent of the population" of the Italian Piedmont, their ancestral home. Likewise, Celestina Larroude's fair physiognomy corresponded "completely to a person of central Europe" as befit her French origins. The three children, however, were dissimilar both from each other as well as from their putative parents. Nine-year-old María Carmen was fair and covered in freckles, clearly of "pure European blood" with a "physiognomy reminiscent of northern Italy or France." But eleven-year-old María Mafalda was darker, displaying the physiognomy of southern Spain. Most strikingly, ten-year-old Roque Humberto had "dark, olive skin," dark hair, and dark eyes, pigmentation the anthropologist described as "negroid." What is more, the boy exhibited the telltale hairline of the gauchos. His "inferior" somatic traits, concluded the anthropologist, "leave no doubt about the *mestizaje of European blood with a primitive colored race.*" It was a crucial clue given that Celestina and Roque were European.[51]

Lehmann Nitsche as well as contemporary and later observers claimed that his method was innovative because it was "Mendelian." By this they meant that the method traced what were known or assumed to be dominant and recessive traits. In other words, his assessment was based on the premise that certain traits presented themselves in parents and children not as a "random mix" but as a "pattern that theoretically can be predicted."[52] Characterizing his method as Mendelian highlighted what was modern, scientific, and original about it. Like blood group testing, Lehmann Nitsche's somatic analysis applied knowledge of human heredity to a new end, the practical determination of paternity.

Yet for all its self-styled scientism, his method also incorporated older ideas about resemblance. Lehmann Nitsche argued that family groups evinced an "average type" or "family resemblance" *(aire de familia)*. "The individual is not an isolated, unique object, but rather part of a more or less numerous group (the family, the lineage) whose typical characteristics reflect with greater or lesser intensity, whether closer or farther, from an ideal type." As an observer contemplated a series of painted portraits, photographs, or the living members of a given family, this type "presents itself in our imagination."[53] Here Mendelian analysis fell away in favor of the sorts of visual observations an artist or even a lay observer might make. Lehmann Nitsche concluded that even without its "*material* representation," the scientist's "*mental* conception of that average [family] type is enough."[54] If he could see the family likeness, it was objectively there.

Having studied the faces and bodies of the Arcardinis, Celestina Larraude, and the three children, the anthropologist now announced his conclusion. In his expert opinion, neither of the two oldest children, Roque Humberto and María Mafalda, were the biological offspring of Roque Arcardini and Celestina Larraude. As for the youngest child, María Carmen, the scientific evidence was inconclusive. A month later, the judge issued his verdict: he rejected Lehmann Nitsche's assessment and dismissed the Arcardinis' criminal suit for suppositious birth. But the litigation was far from over. Having lost their criminal complaint, the family now filed a civil suit challenging the filiation of the children. Lehmann Nitsche received his first payment of 2,000 pesos from the Arcardini family just in time for Christmas.[55]

AS THE SUIT ENTERED a new phase, scientists elsewhere were beginning to develop their own methods of analyzing bodies. In the United States, eugenicists became interested in parentage assessment, proposing methods very similar to Lehmann Nitsche's, although there is no evidence they knew about his work. Roswell Johnson, professor of eugenics at the University of Pittsburgh, suggested that traits "inherited in a more or less Mendelian fashion" could be useful in paternity proceedings. One hundred measurements of the child and putative parent—the ear lobe, head shape, facial features, and skeletal measurements—would produce an "index of correlation" that could reveal kinship or nonkinship. "The method of attack is well understood by competent students of heredity," Johnson asserted, with a certain blithe optimism. He presented his ideas to the American Association for the Study and Prevention of Infant Mortality, which promptly passed a resolution petitioning the federal Children's Bureau to sponsor additional research into paternity establishment.[56]

While there is no evidence the Children's Bureau ever signed on to the proposal, a handful of eugenicists continued to work on the problem. A few years later, Charles Davenport, founding father of U.S. eugenics and Johnson's mentor, gave an opening address to the Second International Congress of Eugenics, in which he judged "our knowledge of the inheritance of [certain] physical traits . . . sufficiently precise to be applied practically in cases of doubtful parentage." In a hypothetical case involving a mother and two possible fathers in which the "family stock" of the three adults could be investigated, he asserted that paternity could be established "with a high degree of certainty ranging from 75 to 99 per cent." The Eugenics Record Office, a major center of U.S. eugenic research founded by Davenport a decade before, had already been queried in a dispute over the estate of a wealthy man.[57] Eugenicists believed that the science of heredity held the key to improving human society, and a method of assessing biological paternity contributed to that goal. The real value of such a method lay not in the resolution of inheritance disputes, according to Johnson, but in the defense of child welfare. By identifying the father, science would help secure child support and reduce infant mortality.

The prospect of "eugenic tests of parentage," like Abrams's oscillophore and scientific analyses of baby mix-ups, which appeared around the same time, elicited enthusiastic interest in the American press and public. Davenport's comment on parentage assessment accounted for a single paragraph in his nine-page speech to the eugenics congress, but press coverage of the meeting featured the issue in the headlines.[58] Yet despite public interest and the optimistic predictions of leading eugenicists, there were few systematic attempts to develop somatic or morphological methods of paternity investigation in the United States. *JAMA* published dozens of articles on blood testing for parentage beginning in the 1920s, but in the first six decades of the twentieth century, not a single article on somatic analysis appeared in its pages. Its English counterpart *The Lancet* was only slightly more likely to treat the topic. After a brief uptick of interest in resemblance in the 1920s, forensic and eugenics journals abandoned the topic.

Anglo-American scientists' relative lack of interest in developing an expert method of analyzing resemblance is noteworthy given that resemblance was a long-established form of evidence in Anglo-American parentage disputes. Perhaps this is not a contradiction. Where resemblance was the speculative practice of judges, juries, and sometimes artists, it may have held less appeal to scientists: a familiar folk method was not readily "scientized." The proposals of eugenicists like Johnson and Davenport thus remained exceptional. Their colleagues never embraced Galton's idea to seek kinship on the body and remained focused on blood. Anglo-American paternity assessment hewed largely to serology.

In continental Europe and Latin America, in contrast, the scientific study of kinship likeness took off. As early as 1915, the year after Roque Arcardini's death and Roberto Lehmann Nitsche's initial foray into the field, a well-known medicolegal expert in Rio de Janeiro examined the bodies of two women and a fostered child to resolve a dispute over who was the mother. Brazilian forensic experts began systematically using somatic methods in São Paulo a decade and a half later. Meanwhile, in Norway, a 1921 paternity case was decided on the basis of somatic likeness. In the Soviet Union, one Dr. Poliakow developed a technique of paternity assessment involving 125 points of comparison. Somatic methods were taken up in Poland and Hungary, and Portuguese and Spanish medicolegal experts also debated their use.[59]

It was in Austria and Germany, however, that the analysis of the body, like that of blood, would systematically and precociously enter judicial practice. In the mid-1920s, an ambitious racial anthropologist introduced the first somatic examinations into a court in Vienna. Otto Reche was director of the Vienna Institute of Anthropology and Ethnography, and his method relied on an analysis of nineteen physical traits, including ears, nose, hair, and eye color. He developed the method at the request of a judge who had become frustrated by what he believed was rampant perjury in paternity proceedings. Within a few years, the higher courts in Germany began accepting his anthropological analysis, and in 1931 the Supreme Court of Austria began to require such analyses in cases of disputed paternity.

The multiple quests to find parentage on the body germinated in part due to the transatlantic circulation of ideas and knowledge among scientists but also through spontaneous generation. Otto Reche and Lehmann Nitsche lived on opposite sides of the Atlantic but shared much in common—both were German anthropologists of the same generation, trained in a similar intellectual milieu, and they were also professional acquaintances.[60] Both became interested in the problem of somatic assessment, but there is no evidence that Reche was aware of Lehmann Nitsche's paternity work in the Arcardini case, which preceded his own by a decade.[61] Experiments in paternity science often materialized independently, primed by a transatlantic milieu fixated on heredity and its practical applications.

BACK IN BUENOS AIRES, the Arcardini case was in full swing in civil court. Having failed to prove suppositious birth in a criminal court, the Arcardinis now tried to nullify Roque's legal recognition of the children. He had recognized the two younger siblings, Roque Humberto and María Carmen, in their baptismal registries.[62] The suit contended that the birth hoax invalidated this recognition, for Roque could not recognize children whom he had not actually fathered. The court collected testimony from dozens of witnesses—Celestina Larraude's neighbors and friends, the babysitter, the midwife. It also received scientific reports from Robert Lehmann Nitsche as well as from two additional experts, both medical doctors. The three had discussed the case and exchanged bibliographic references, but when the time came to write their expert report, they found it impossible to reach a consensus and submitted separate conclusions.[63]

Now it was time for the *fiscal,* or court prosecutor, to review the copious evidence and summarize it for the judge. Prosecutor Ernesto Quesada was a well-known public figure, a jurist and polymath with interests ranging from Roman history to Goethe to Tunisian real estate legislation. He was immediately taken with Lehmann Nitsche's expert report (the other two doctors do not seem to have produced reports to explain their conclusions). Recognizing the method's potential to revolutionize the high-stakes filiation suits common in Argentine courts, he scoured the international jurisprudence for other instances of Mendelian paternity analysis but came up empty handed: Lehmann Nitsche's method apparently had no global precedents. And so, as befit his voracious intellectual proclivities, Quesada decided to evaluate its scientific merits himself. The circle of intellectual elites in Belle Époque Buenos Aires was small, and Quesada and Lehmann Nitsche were friendly acquaintances. The prosecutor asked the anthropologist if he could consult his scientific library.[64] He dove into the literature and proceeded to write a report more than twice as long as Lehmann Nitsche's own.[65]

Quesada found the idea of a scientific method to solve riddles of kinship to be deeply seductive, but it also raised a series of questions. How could Lehmann Nitsche be so certain that there were no exceptions to Mendel's famous rules? What about the presence of hereditary atavism? Perhaps Roque Humberto was like the black lamb born of white sheep, a throwback to some tawny ancestor. Or maybe environment could change physiognomy. Take the light-skinned Goths, who invaded Italy and in just a few generations had darkened to a Mediterranean hue; perhaps the racial appearance of the children had likewise been altered by their environment. Mendelian laws were hardly "absolute or exclusive." And even if they were, Lehmann Nitsche's application of them struck Quesada as suspect because it lacked any analysis of the maternal line. While he had examined a dozen members of the paternal family, beyond a quick description of Celestina Larraude (whom Lehmann Nitsche characterized as having a "restless, false look"), an assessment of her pedigree was entirely absent.[66]

Quesada was further troubled by the fact that the three scientific experts in the case had reached contradictory conclusions. Lehmann Nitsche declared Roque Humberto and María Mafalda nonchildren and the evidence on María Carmen inconclusive. One of the doctors, however, thought that none of the three could be the offspring of Larraude and Arcardini, and the other said he could not determine the parentage of any of them. If Mendelian law was so clear, predictable, and well understood, why had the three distinguished experts failed to reach a consensus?[67]

Quesada's conclusion was thus diplomatic but emphatic: the Mendelian method could not prove paternity. While he found Lehmann Nitsche's conclusions unconvincing, however, he accepted the prospect of a scientific test itself. Quesada threw himself into the scientific literature and ultimately concluded that "according to the current state of our knowledge," science "is unable to . . . say whether a person is or is not the child of another."[68] But he invited the possibility that in the future it might do so.

IN 1918, ALMOST THREE and a half years after Roque Arcardini's death, it was time for the judge himself to pronounce on the fate of Arcardini's estate. He carefully reviewed the testimony of the witnesses, the expert reports of Lehmann Nitsche and the two doctors, and the assessment of prosecutor Quesada before declaring that the children's filiation had been proven. Roque Arcardini was the father of the two younger children, whose paternity was at stake in the civil suit. The judge was no more convinced by Lehmann Nitsche's assessment than Quesada. "If there is any practical result that can be extracted from the expert report," he concluded, "it is the impotence of science, at the present time, to resolve with the certainty that the law requires, questions of kinship by comparing physiognomic traits and family types."[69]

The judge's dismissal was as brief as it was categorical. His verdict exhaustively surveyed the witness testimony, Celestina's strange confession, and the baptismal registries, but he spent little time discussing the lengthy and complex anthropological report, or for that matter the prosecutor's even lengthier critique of it. This is because the questions that Lehmann Nitsche and Quesada addressed, namely, whether Arcardini had sired the children and how one might go about proving this fact, were different from the questions the judge himself sought to answer. For the judge, the main question was whether Roque Arcardini had legally recognized the two children as his own, and whether he had ever behaved at variance with this conviction. Given Celestina's confession and the recurring rumors of fraudulent birth that had circulated in Roque's lifetime, the judge sought to determine whether Roque himself had ever demonstrated doubt or a disinclination to fatherhood.

The magistrate concluded that he had not. Roque Arcardini's behavior was that of a man who had unequivocally embraced his status as the father. This mattered because paternity was an act of will rather than a question of fact. "As long as the letter and spirit of the civil code remain in force," he declared, "it will not be me who . . . violates [the man's] will, which to me is worthy of the highest respect, to overrule a paternity that the deceased embraced in public records and through unequivocal and repeated acts confirming it."[70] The anthropologist's august analysis of kinship was not terribly persuasive, but above all it was irrelevant to the question of will that the judge sought to answer.

The Arcardinis appealed the verdict. Their lawyers published large sections of the court transcript in a book entitled *Hijos Artificiales* (False Children), perhaps in an attempt to sway public opinion in their favor. On appeal they lost again. In the eyes of the law, Roque Humberto and María Carmen were Roque Arcardini's children and heirs. In another suit several years later, their elder sister, María Mafalda, also proved her filial identity. Even absent Arcardini's formal recognition in her baptismal registry, she was able to demonstrate her "possession of status" as his daughter.

Following Argentine inheritance law, the vast estate was divided between the deceased man's forebears (Arcardini's mother María Trinchetti, who died in the middle of the judicial proceedings) and his descendants (the three natural children). Still, the litigation continued. In the early 1940s, almost thirty years after Roque's death, the "legitimate" branch (heirs to the portion inherited by Roque's mother) and the "natural" branch (heirs to the three children) were still mired in suits related to the division.[71] By then, the legitimate branch was represented by Roque's grandnieces and -nephews. One of them was a well-known lawyer who had been a year old when his great-uncle died and was the youngest of the sixteen family members examined by Lehmann Nitsche.[72] As for the natural branch, while the youngest daughter María Carmen's fate is unknown, her siblings María Mafalda and Roque Humberto both died as young adults, and their portions of the estate passed into the hands of spouses and other heirs.[73]

For their part, the anthropologist and the jurist continued their spirited debate about the Mendelian paternity test, publishing multiple rejoinders over the next few years. Undaunted by the outcome of the Arcardini case, Lehmann Nitsche went on to contribute expert reports in other cases of disputed filiation. Other scientists adopted his somatic method. Forensic paternity assessments appeared from time to time in Argentina's medicolegal publications, but the fact they were worthy of publication suggests they continued to be a novelty. Two decades later, a new generation of scientists would remember the Arcardini affair, the thoroughness and erudition of Lehmann Nitsche's and Quesada's studies, and the interest the case had generated beyond the courtroom. As for the failure of the Mendelian method, they chalked it up to the skepticism that greets all new things.[74] As scientific assessments of paternity gradually made inroads, this would become a common refrain among scientists and some jurists, in Argentina and elsewhere: if courts rejected the science of paternity, it was because they did not understand it or felt threatened by a competing form of authority.

But the problem in the Arcardini case was not the novelty of Lehmann Nitsche's method nor lingering questions about its scientific validity. The problem was what it purported to discover in the first place. Lehmann Nitsche sought to determine Roque Arcardini's biological paternity of the three children, but the traditional object of Argentine filiation disputes was not biological paternity. It was possession of status—the man's consciousness and acceptance of being the father. Lehmann Nitsche's method challenged these older social and volitional conceptions, but it did not easily unseat them. Three decades after the Arcardini case, another frustrated scientist found Argentine judicial practice still wedded to the old paternity. "Possession of status is purely and exclusively subjective," he declared. "The biologist cannot share the conviction of the judge that . . . it constitutes a categorical and decisive proof of filiation."[75]

That exasperated scientist was none other than Leone Lattes, the Italian serologist, who had fled Mussolini's racial laws, emigrated to Argentina, and now served as an expert in filiation cases. By this time—the late 1940s—blood typing was widely known and broadly accepted as a method for proving nonpaternity, but to Lattes's chagrin this did not mean that Argentine judges embraced it. He recounted the case of a wealthy gentleman and his much younger consort who had given birth to a baby girl. For several years, the man treated the child as his daughter and maintained mother and child economically, but as the relationship cooled he began to have doubts. A blood test found the man could not be the biological father, but his actions in the first years of the child's life constituted a "precise and complete" legal proof of her possession of status. The court assigned the man paternity.[76] Thirty years after Lehmann Nitsche's failed debut, Lattes encountered the same obstacle: biological tests could not overcome the possession of status.

Against the backdrop of modernization's dislocations, as old ways of knowing were in flux and in a courtroom primed to embrace scientific authority, the Arcardini family staked their claim to Roque's patrimony not just on a new method of establishing paternity but on a new, biological way of defining it. Their story reflected social, economic, and political circumstances specific to Belle Époque Buenos Aires, but the outcome of the suit was hardly peculiar to this time and place. The new science of paternity was in frank tension with the traditional idea of possession of status, a figure of all civil law regimes. In 1935, a French court rejected blood evidence as "contrary to the general system of French law" in which paternity was "not susceptible to direct proof" and only to social "presumptions." The tensions between biological and social paternity remain a marked characteristic of French law to this day.[77] But it was not only in countries with civil law that biological evidence ran aground on the shoals of social paternity. As Charlie Chaplin discovered, American courts too sometimes favored nonbiological definitions of fatherhood. This was the nature of modern paternity: in introducing new scientific methods to discover the father, it heightened tensions between different ways of knowing him.

Ultimately, the Arcardinis lost their gamble, and with it half of Roque Arcardini's estate. Who was the father? Even as paternity science gained increasing authority, that question was not necessarily entrusted to the expertise of scientists like Roberto Lehmann Nitsche, Leone Lattes, or the doctors in the Chaplin suit. It continued to be addressed to communities, neighbors, and sometimes to putative fathers themselves, men like Roque Arcardini.

5

BODIES OF EVIDENCE

The portrait of the family . . . was truly beautiful, yet it was marked by the most disquieting of shadows: *that of doubt.*

—Argentine forensic specialist Luís Reyna Almandos, 1934

IN 1926, THE POLICE in a small town in northern Italy apprehended a man wandering in a cemetery, confused, distraught, and apparently suffering from amnesia. Interned in a mental institution, he soon stabilized, but the staff were perplexed by the new arrival. Rather than a demented vagabond, the man's demeanor suggested he was educated and refined. The director circulated a photograph of the so-called *Sconosciuto di Collegno,* the unknown man of Collegno, in the newspapers. "Does anyone know this man?" asked the caption.

A well-heeled Veronese matron, Giulia Canella, saw the picture and was struck by the resemblance to her long-lost husband. Giulio Canella had disappeared in combat on the Macedonian front some eleven years earlier. A scholar and teacher of philosophy, devout Catholic, and family man, he had been a well-respected member of local society. His body had never been found, but his wife never lost hope that he might one day return. She contacted the authorities.

From her first surreptitious glimpse of the mysterious bearded man, through a crack in the door of the mental asylum, she became convinced that he was Giulio: fragile, bereft of his past, but alive. In their first face-to-face meeting, deliberately staged by the staff as a casual encounter in the asylum's courtyard, he caught sight of her face, stopped, and stared, as if in recognition. Overcome by emotion, Giulia took out a rosary and turned her eyes to heaven. Exclamations, tears, and an embrace followed.[1] The grieving widow had found her husband.

Days later, Giulio Canella was released to his wife's care and taken home to be reunited with the two children who no longer remembered their father. The press and the public were transfixed. A full decade after the devastating war, one family's tragic loss had suddenly and improbably acquired a happy ending.

The story might have ended there were it not for the inconvenient appearance of another bereft wife. Days after Giulio Canella's release, authorities received an anonymous letter informing them that the mysterious man was not Giulio Canella at all but a typographer and anarchist named Mario Bruneri. The man's wife, Rosa Bruneri, recognized the Sconosciuto as the husband who had abandoned her and their son several years before. Mario Bruneri was a notorious con artist wanted for a series of petty crimes. It was just the sort of ruse he would pull.

Was the mysterious amnesiac a respectable professor and war veteran, or was he an anarchist crook? The attempt to answer this question sparked a five-year legal case. As the press followed the inquest, Italian public opinion split into two camps, the "bruneriani" and "canelliani." The case became increasingly politicized, with fascist officialdom and sectors of the Catholic Church drawn into the conflict. Charges swirled of police and judicial incompetence, shadowy interests, dark conspiracies. The fascist regime's informers carefully monitored the vicissitudes of public opinion, concerned that the lurid case was undermining public morality and respect for the police and the courts. At one point, Mussolini ordered the papers to cease reporting on it.[2]

The case was also reported in the international press, but it was of special interest to Brazilians. Giulia Canella's father, who underwrote the family's lengthy and expensive legal action to prove the Sconosciuto's identity as Giulio Canella, was a well-known resident of Rio de Janeiro, where he had lived for some forty years. A merchant with lucrative business interests, he was regularly referred to in the Brazilian newspapers by the honorific "Marquês." Giulia had lived in Verona since her marriage to Giulio, who was actually her father's cousin, but she had been born in Rio, a Brazilian citizen. Carioca high society rallied around her during her tribulations half a world away.[3]

The attempt to determine the Sconosciuto's identity tested new technologies of identification and new understandings of neuropsychiatry. Medicolegal experts performed fingerprinting and handwriting analyses and assessed the case in light of emerging psychological theories of trauma. They compared the Sconosciuto with the physical, mental, intellectual, and cultural profiles of Giulio Canella and Mario Bruneri to determine whom the strange man most resembled.

The legal case came to a dramatic climax in 1931. After four years of investigation and multiple lower court decisions, a divided high court reached a final verdict. The mysterious amnesiac who had taken up residence with Giulia Canella was not her long-lost husband, Giulio; he was Mario Bruneri. The tribunal sentenced the impostor to four years of jail time for earlier crimes committed by the fugitive Bruneri. Even a last-minute appeal to Il Duce himself failed.

For the Canella family, the verdict was perhaps even more disastrous than Giulio Canella's original disappearance in the war. The court not only consigned the man they claimed as their husband, father, and son-in-law to legal nonexistence; it implicitly found Giulia Canella to be living in sin with a lowly criminal. Worse still, it branded the two (soon to be three) children born to the couple since their (re?)union with the ignominious stigma of illegitimacy. The verdict destroyed the civil identity of a man but also the honor of an upstanding bourgeois family. The Canellas flatly refused to accept the court's decision.

In 1933, the Sconosciuto, now legally Mario Bruneri, was released from jail, his sentence reduced by a general amnesty commemorating the tenth anniversary of fascism. He rejoined Giulia, who had never wavered in her insistence that he was her husband. But they lived in the shadow of scandal and the lingering possibility that Rosa Bruneri could bring criminal charges against them for adultery. A few months later, the Canellas and the five children sailed for Rio de Janeiro, where they were greeted by waiting journalists eager for a photo of the famous family. There, the man who had left Italy with a passport issued to Mario Bruneri was welcomed by a sympathetic public as Professor Giulio Canella, an innocent war hero who had suffered a gross injustice at the hands of Italian authorities. In Brazil, he would also encounter a thriving community of medicolegal experts who were as fascinated by the case as their Italian counterparts. No scientist proved more tireless, outspoken, and colorful in his attempts to elucidate the mystery than Dr. Luiz Silva.[4]

DR. SILVA WAS AN improbable answer to the Canella family's plight. For one, he was a dentist. A professor and cofounder of the School of Pharmacy and Dentistry in the port city of Santos, Silva also worked for the state of São Paulo's Identification Service and taught in its School of Police. Forensic dentistry—the use of dental records for identification purposes—had gained new visibility at the turn of the twentieth century thanks to high-profile tragedies in which victims had been successfully identified through their teeth.[5]

Silva was well versed in these new forensic methods but harbored even greater ambitions for dentistry. Drawing on an eclectic mix of forensic identification, hereditarian thinking, criminology, and psychiatry, he founded a new field called "legal odontology." The remarkable span of the field can be gleaned from the titles of the articles that, for more than thirty years, Silva contributed to Brazilian newspapers: "Teeth and Madness"; "Pacifiers as a Factor of Mental Alteration"; "Workplace Accidents and Legal Odontology"; "Identification as a Basis for National Security"; and "An Authentic Portrait of Father Diogo Antônio Feijó through Legal Odontology" (apropos designs for the bust of a famous nineteenth-century statesman).[6]

Several months after the Canellas' arrival in Brazil, Silva was in Rio for a scientific conference on identification. He and two other forensic experts were summoned by Giulia's father, the Marquês, to his residence on fashionable Glória Hill. It was probably there, in the elegant quarters with spectacular views of Rio's Guanabara Bay spread before them, that the dentist and the amnesiac engaged in the extended conversations that soon convinced Silva that the Italian courts had made a terrible error. The Sconosciuto's charm, learning, and refinement were utterly inconsistent with that of a rustic typographer, Silva concluded. The mysterious man had to be Giulio Canella. But could he prove it?[7]

Taken into the Canella family's confidence, Silva began an exhaustive forensic analysis, revisiting the evidence considered by Italian officials and identifying what he believed were serious errors of fact and interpretation. But he also proposed an entirely new way of thinking about the mystery: as a problem not of identification but of kinship. Whereas the original investigation had focused on comparing the Sconosciuto with Giulio Canella and Mario Bruneri, Silva decided to compare him with the members of the Canella family. If the Sconosciuto really was Giulio Canella, then he was also the father of the two children born to the Canellas before the war. If Silva could prove that he was the progenitor, then the mystery of his identity would be solved. What had begun as a mystery of individual identity became in Silva's hands an unusual paternity investigation.

According to Silva, facial appearance was determined largely by the mouth and the jaw—the special arena of the odontologist's expertise.[8] His singular understanding of dental, oral, and maxillofacial structures therefore gave him privileged insight into the interpretation of physical resemblance and hence kinship. Silva took molds of the mouths of the Sconosciuto; Giulia Canella; her father, the Marquês; Rita and Giuseppe, the two children born to her before the war and fathered by Giulio Canella; and Elisa and Camillo, born after the war and fathered by the Sconosciuto (a fifth child, two-year-old Maria Beatrice, was judged too young for the procedure). The fact that at the time of the examination the Sconosciuto had lost most of his teeth apparently did not impede Silva's assessment. Nor did the fact that spouses Giulio and Giulia Canella were consanguineous relatives (he was a cousin of her father)—a relationship that one imagines might have complicated the analysis of hereditary likeness within the family group. Silva analyzed the dental arches and the teeth themselves. He studied the palatal ridges, the characteristic crests and furrows on the roof of the mouth. He searched the lines of the jaw and measured the angles of the face. Newspapers would dub the marvelous and peculiar study a "paternity test of the teeth." The Canella case provided an extraordinary opportunity to showcase the application of legal odontology to paternity determination. Soon Silva would apply his new method to find other fathers.

In recasting the Canella affair as a paternity investigation, Silva not only suggested a new scientific approach to the conundrum but also intuited the episode's deeper meanings. The mystery of the Sconosciuto was not just the calamity of a man who had lost his identity; for publics on both sides of the Atlantic, it was also the social and moral drama of a family. The Italian press dubbed it "the most terrible family tragedy of the century." Argentine

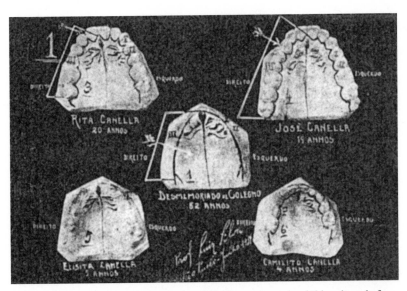

Dr. Luiz Silva took molds of the mouths of the Sconosciuto, the children born before the war, and those born after, in search of similarities that would reveal the man's paternity.

Reproduced from R. Babini, "Identificazione odonto-legale dello 'sconosciuto di Collegno,'" Estratto da *La Stomatologia* 33, no. 4 (1935): 5.

forensic expert Luís Reyna Almandos, who was also in Rio for the identification conference and accompanied Silva to the Marquês's home, agreed: the doubt surrounding Giulio Canella's identity was a "perpetual offense that mercilessly ravages the honor of a whole family."[9] As he observed Giulio, Giulia, her father, and the children, it was as much the question of kinship as that of identity that riveted him. Here was an honorable, distinguished, and prosperous group ensconced in their sumptuous abode, but what were the true relationships, marital, filial, and fraternal, among them? Was it possible to discover the truth behind this beguiling family portrait? Giulia Canella had taken the Sconosciuto into her home and her bed. They had produced three children together. Either those youngsters were the lawful offspring of an honorable wife and her husband, or they were the bastards of an adulterous relation between Giulia and a common criminal, as the Italian courts had effectively declared.

For Silva, this family portrait shadowed by doubt suggested not only the depth of the moral cataclysm but also its possible solution. If he could clarify those relationships, demonstrating that Giulia Canella's consort was the father of all her children and therefore her lawful husband Giulio Canella, he would rescue the wife, the children, and indeed the whole clan from moral ignominy. In this spirit, he titled one of his lengthy scientific disquisitions on the case "Defending an Italian and Safeguarding the Honor of a Brazilian Woman." Silva's study of the Canellas was distinct from most paternity investigations in that he sought to draw a link not from children to father but from father to children. But as in more conventional scenarios of disputed paternity, such as those involving unmarried mothers and adulterous wives, this case too was about the sexual honor of a woman and the good name of her children. The quest for the father always implicated the mother and the child. And even as modern paternity brought new technologies to bear in this effort, the quest itself was premised on traditional notions of family honor, sexual morality, and social propriety.

ALMOST A YEAR after the Canella family's arrival in Brazil and following months of odontological study, Silva reached his conclusion. He chose to reveal it not in a scientific journal but in a newspaper. Over the course of more than three weeks, a well-known São Paulo journalist published an exhaustive reexamination of the entire Canella affair, nine days of which were devoted to Silva's research. The coverage included interviews with the learned dentist and grainy photographs of his dental molds. It trumpeted "the research carried out . . . by a São Paulo scientist and the stir that the results will cause."[10]

Silva announced that the mouths and faces of Giulia Canella's two children born before the war and fathered by Giulio Canella, those of the two children fathered by the Sconosciuto, and the mouth of the Sconosciuto himself all shared a number of striking similarities. This proved that the same man had fathered all four children. The mystery was thus solved. The amnesiac found wandering in the cemetery, the man whom Giulia Canella had taken in as her husband, the subject Italian authorities had declared Mario Bruneri, was in fact Giulio Canella.

Silva went on to publish his results in scientific journals in Italy, Brazil, and Argentina. He sent copies to the pope, the king of Italy, and Benito Mussolini. The Italian press does not appear to have reported his research,

Profiles of the Marquês, Giulia Canella, her older children Rita and Giuseppe, and the Sconosciuto. According to Luiz Silva, Rita's and Giuseppe's noses were different from their mother's and their maternal grandfather's and similar to the Sconosciuto's, suggesting he must be their father.

Reproduced from Luiz Silva, "Identificação odonto-legal do 'Desconhecido de collegno,' Pericia de Investigação de Paternidade," *Revista de Identificación y Ciencias Penales* 14, no. 52–54 (1936): 73.

but this was probably because of fascist censors. Italian authorities were, in fact, well aware of Silva's activities. According to internal records of the Italian police, "a Brazilian friend"—one of Mussolini's transnational network of informers—remitted Silva's scientific work to Italian officials, who dismissed it as "worthless" but groused about the "great tumult" it had inspired.[11]

Silva's sensational finding may have caused great tumult, but its legal consequences were moot. The judicial case in Italy was closed, and Brazilian authorities insisted on retaining the civil identity with which the Sconosciuto arrived in Brazil—that of Mario Bruneri. But even if his odontological verdict could not receive a hearing in court, at least it could receive one in the court of public opinion. This is probably why Silva chose to announce his findings in the press.

For the next five years, the indefatigable dentist continued to write about the case and lobby on behalf of his conclusions and the Canella family: "I was a lone voice that rebelled against the formidable judicial error that deprived Professor Giulio Canella of his . . . personhood." In 1939, the Italian consul in São Paulo finally agreed to deliver a report of Silva's work to the authorities in Italy. "I do it personally and in homage to a Brazilian woman and her children," he declared, echoing the logic of family honor.[12]

For the Canellas, this new development would come too late. As Silva's report made its way through Italian diplomatic channels, half a world away Nazi troops were amassing on the border near Danzig. Less than two months later, Hitler would invade Poland, and any possibility that the case might be reopened evaporated. But for Silva the case had already served its purpose. Having refined his scientific method and acquired a certain popular renown, he would put his unusual paternity test to the service of new people and problems.

SILVA WAS JUST ONE Brazilian scientist engaged in the quest for the father and the Canellas just one unusual family embroiled in such a quest. In the 1930s and 1940s, scientists across Latin America discussed and debated the merits of different scientific methods of paternity analysis, and in Brazil, a lively medicolegal community began to practice paternity testing. As Silva worked on the Canella case, another group of scientists in São Paulo was developing their own techniques of paternity assessment. Several years earlier the University of São Paulo's Instituto Oscar Freire (IOF), a premier center of forensic research and teaching, had performed the first blood group analysis for a court—the first not only in Brazil but possibly in the Western Hemisphere. Paternity testing soon became a fixture of the IOF's activities. Over the next three decades, the institute's director Flamínio Favero and his colleague Arnaldo Amado Ferreira examined hundreds of men, women, and children. They also trained a new generation of forensic experts, or medico-legistas, in the science of parentage determination. Alongside instruction in traditional forensic topics—toxicology, psychopathology, how to perform an autopsy—the IOF's courses in legal medicine began to cover techniques of paternity assessment. The pursuit of the father thus became a standard element of the forensic repertoire.[13]

The IOF dealt with a clientele entirely different from the well-heeled Canellas, however. Their subjects belonged to the multiracial and immigrant working class of the burgeoning industrial metropolis. The men were mechanics, carpenters, drivers, doormen, and commercial employees; the women, factory workers and domestics. Also in contrast to the Canellas, these individuals did not seek out scientific assessment voluntarily. Almost everyone who arrived at the IOF for paternity testing had been ordered there by a local court, civil or criminal, in the city proper or in a provincial town in the state of São Paulo. The IOF's clientele were the protagonists of tawdry personal dramas, but not the kind that made international newspapers. There were allegations of seduction and estrangement and charges of rape and the deflowering of a virgin, a particularly common criminal complaint in the Brazilian urban milieu. There were support cases for children born in informal unions whose fathers refused to recognize them. In some instances, the paternity dispute involved two possible fathers. A few involved husbands who accused their wives of adultery. A handful stemmed from an inheritance claim on a deceased man's estate. Although there were a few maternity cases, it was overwhelmingly paternity that the IOF was asked to clarify. Where the Arcardini case in Argentina and more recently the Canella case in Brazil were extraordinary and sensational events, the IOF reflected the increasing routinization of the modern search for the father as scientific testing expanded to include modest subjects and mundane circumstances.

If the IOF's clientele and their circumstances differed from the Canellas, so too did the tests to which they were subjected. While Silva studied mouths and faces, the IOF medico-legistas reconnoitered the body. Arriving at the IOF on the University of São Paulo's campus, mothers, children, putative fathers, and sometimes other relatives were subjected to an elaborate battery of examinations. Their fingers were pricked for a blood draw, and their bodies were measured, photographed, classified, and described, sometimes in intimate detail. Their fingertips were inked and carefully pressed on long rectangular cards for deciphering by IOF scientists. Their mouths were scrutinized, and by the mid-1940s, they were given small pieces of paper to chew to determine if they were hereditary tasters of the chemical PTC.[14]

Over time a distinct rivalry between Silva and the IOF emerged. Where the dentist touted his pioneering method for analyzing similarity, the medico-legistas focused on heredity, claiming to represent the cutting edge of modern genetic science. Where he proclaimed odontology's unmatched power to determine paternity, they urged caution. The scientific determination of paternity, IOF scientist Arnaldo Amado Ferreira would note, was "among the most delicate, interesting, and difficult of medico-legal practices."[15] Over the years, they would argue in scientific journals and conferences, in the press, and in the expert reports they produced for courts in pursuit of the father.

This was a local rivalry between experts in two distinct subfields of forensic science working in two different city institutions, but it evinced larger trends and tensions in the science of paternity. São Paulo in the 1930s and 1940s was a microcosm that reflected the growing transatlantic use of parentage tests, the intense interest in this practice, as well as the total lack of scientific consensus surrounding it. Modern paternity science was not a unitary body of knowledge or expertise. Rather, the quest for the father was a compelling practical problem to which a motley array of techniques were applied. Some of these methods were new, some recycled; some complementary, others apparently contradictory. Contrasting techniques reflected not just scientific disagreements but also social considerations. There were different reasons the father's identity might become a source of contention, ranging from the extraordinary case of an Italian amnesiac to romances gone awry, inheritance disputes, and sexual crimes. As such, men and women both affluent and modest could now be implicated in such investigations. The subjects of paternity testers' examinations were thus not just physical bodies; they were people whose social status shaped how their bodies could be scrutinized. Even as paternity testing promised biological truths, it had to be mindful of social and moral conventions.

WHILE THE INCREASING complexity of paternity science and its expansion to a wider public was new, the methods and technologies it employed rarely were. Both Silva and the IOF medico-legistas repurposed nineteenth-century techniques for a new, twentieth-century purpose. Silva's odontological paternity test was novel, but it drew on forensic dentistry as a method of identification. Paternity examinations at the IOF likewise recycled techniques originally developed by Victorian criminologists, physical anthropologists, and race scientists to identify individual bodies, distinguish between them, and discern putative racial ancestry. Now these methods were creatively adapted to kinship analysis.

One such technique was fingerprinting. In the first decades of the twentieth century, police around the Atlantic world adopted fingerprinting to track criminal suspects. Then as now fingerprinting was of primary interest as a method of individual identification. But scientists also suspected that the prints of fingers, hands, and feet contained inherited elements. In the nineteenth century, Francis Galton had systematically explored the hereditary characteristics of "finger furrows," which he hoped might serve as a marker of racial identity. But after comparing the prints of subjects identified as English, pure Welsh, Negro, Hebrew, and Basque, he concluded "there is no peculiar pattern which characterises persons of any of the above races."[16] Even if fingerprints were useless for discerning race, however, perhaps they could be used to read family. Another pioneer of fingerprinting, Henry Faulds, pondered this possibility in 1880: "the dominancy of heredity through these infinite varieties is sometimes very striking. I have found unique patterns in a parent repeated with marvelous accuracy in his child." Perhaps, he mused, families exhibited fingerprint "types"; perhaps fingerprints might be used to identify unknown family members.[17]

In the first decades of the twentieth century, fingerprints became one of the most exhaustively investigated techniques of paternity analysis. Argentina was a center of this research. The country was at the forefront of dactiloscopy, the new science of fingerprinting, thanks to the police chief of the city of La Plata, Juan Vucetich, who had developed a system for classifying prints. Around the turn of the century, Vucetich's system became one of two systems used globally.[18] Subsequently, his colleagues and protégés explored new applications of the technique in relation to heredity: "While exhausted as a system of personal identification, [fingerprinting] has not yet borne all the fruits expected of it in terms of family identification."[19] One anthropologist studied the fingers of five generations of his own family, including those of his one-year-old daughter, in search of inherited traits (his search proved inconclusive).[20] In subsequent decades, scientists across Latin America, Europe, as well as East Asia conducted countless studies on fingerprints as a marker of parentage.[21]

Their research was based on the premise that there was not necessarily a contradiction between the singularity of individual fingerprints and the heritability of certain dactiloscopic traits. Surely the infinitely complex markings of the fingers encompassed elements both unique and inherited. "A system of familial identification," in other words, could be reconciled with "systems of personal identification."[22] As with Silva's dental method, a forensic technique originally applied to individual identification now expanded to kinship determination. In 1930, two Hamburg anthropologists deplored the fact that police departments were asked to perform fingerprint analysis in kinship disputes, when "these men are not to be ranked as . . . competent in matters relating to paternity and heredity."[23] Hereditary dactiloscopy had become a discrete field of expertise, one over which the police, whose fingerprinting experience was limited to individual identification, could not claim mastery.

Despite researchers' dogged efforts, a fail-safe system of familial identification never materialized. They could detect certain similarities in the prints of family members but were perennially frustrated in their attempts to adequately systematize, much less predict, those similarities. Fingerprints were a kind of hieroglyphics, asserted the IOF scientists, and "as with all great mysteries, there is perhaps just a small key" necessary to decipher them.[24] But the key remained elusive. Still, even as Favero and Ferreira invariably concluded that the method yielded no conclusive results, fingerprinting was a standard element of the IOF's paternity work-ups through the 1950s. This was also the case in Germany, where a conference of anthropologists who performed paternity tests reluctantly agreed to abandon fingerprinting only in 1953. As late as 1959, a Hungarian court reportedly accepted fingerprints as proof of paternity.[25] Indeed, kinship analysis through fingerprinting was never fully repudiated; it was gradually abandoned as blood testing became increasingly powerful. Today fingerprints are recognized to harbor some hereditary elements, but their complexity stymies any sort of practical application.

If dactiloscopy was a modern technique of individual identification repurposed for kinship analysis, other methods on which paternity science drew, such as anthropometry, were frankly anachronistic. In the late nineteenth century, anthropometry became a standard technique of individual identification. The most widely used system was developed in the early 1880s by the Parisian police official Alphonse Bertillon, who sought to make identification more powerful and accurate through an elaborate method of measurement, description, and classification. The anthropometrist collected a series of measurements of the subject's body that could, at least in theory, distinguish one body from all others. A detailed description, the so-called spoken portrait *(portrait parlé)*, summarized the individual's somatic features in a format that could be communicated verbally. A standardized photograph captured front and side views of the subject's face—the origin of the mug shot.

Bertillonage, as the system became known, was adopted by police departments in cities across the world in the late nineteenth century, but it proved time-consuming, difficult to apply, and expensive. Above all, critics challenged whether it could truly do what it claimed—namely, to conclusively identify a person, distinguishing him or her from all others. By the turn of the century, anthropometry was already losing ground to fingerprinting, which went on to displace it as the preferred technique of individual identification around the world.

In paternity testing, anthropometry enjoyed a second, lesser-known life. Originally designed to describe an individual body, in this context it served to identify hereditary links between two or more bodies. The famous 1903 Kwilecki case, in which a Polish countess was accused of surreptitiously adopting a baby and heir, counted on its panel of experts an anthropometrist from the Berlin police. The first somatic paternity tester in Austria was a student of Bertillon's methods, and an early Russian paternity method likewise drew on the technique.[26] In the 1930s, Argentine, Brazilian, and Cuban medicolegal experts drew on the spoken portrait to establish filiation even as they debated its value.[27] Some believed its technique of measurement and description could yield "decisive conclusions" about kinship; others entirely discounted "the spoken portrait or any other portrait" for this purpose.[28] Yet in the absence of better methods, even skeptical practitioners used it. The IOF's Ferreira opined that "physiognomic traits of resemblance" were "uncertain."[29] But for decades the institute's standard paternity method included a descriptive spoken portrait "in conformity with the rules of Bertillon" as well as photographs of the subjects taken in front and profile views, following the conventions of the system.[30]

The IOF scientists complemented fingerprinting and Bertillonage with techniques culled from racial and eugenic science. One was biotypology, a system for categorizing bodies according to complex physiological, biometric, and other characteristics that had originated in fascist Italy. The IOF boasted a Biotypology Lab, which examined subjects' bodies and pronounced their biotype—normotype, braquitype, and so on—which was assumed to be hereditary and hence of interest in parentage determinations.[31] In describing the racial characteristics of their subjects, the institute's scientists employed the classification of Brazilian sociologist Edgard Roquette Pinto.[32] Subjects were described as leucoderm (white), faioderm (mixed black-white), and melanoderm (black). The IOF's paternity analyses thus drew on a cosmopolitan pastiche of Argentine dactiloscopy, French Bertillonage, Italian biotypology, and Brazilian racial taxonomy. All of these were

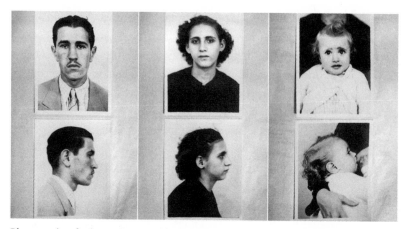

Photographs of subjects from a 1942 paternity investigation at the Instituto Oscar Freire followed the visual conventions of Bertillonage. The baby is nursing in its profile picture.

Laudos, Instituto Oscar Freire, L. 3540, 1942. Courtesy Departamento de Medicina Legal, Ética Médica e Medicina Social e do Trabalho, Faculdade de Medicina, Universidade de São Paulo.

techniques of individual identification, racial description, and eugenic classification that were creatively repurposed to determine kinship.

One striking characteristic of the IOF assessments is how they used—and did not use—racial knowledge. Scientists in Europe and the United States believed that paternity could be read through race. The classic example was that of the biracial couple: "investigation can be greatly facilitated if there are pronounced differences in the race of the father and the mother, or if the child, contrary to the mother or the supposed father, shows clear signs of a different race," noted German serologist Fritz Schiff.[33] The logic of racial difference underlay Manoiloff and Poliakowa's blood analysis of Russians and Jews, Albert Abrams's electronic blood test, and other techniques.

The IOF scientists diverged from this consensus. The subjects of their paternity examinations reflected the multiracial composition of São Paulo's working class and included a number of interracial couples—men and women whom the scientists assigned distinct racial classifications or descriptors. But they rarely considered racial difference a useful scientific fact. This was not because the scientists somehow transcended the transatlantic milieu of eugenics and scientific racism. On the contrary, techniques like biotypology and Roquette Pinto's taxonomy were products of that milieu. But when it came to paternity assessment, the IOF's practitioners argued that racial characteristics like skin color and hair type—traits invested with great significance by scientists elsewhere—had "little value" for determining parentage.[34] They made the obvious point that certain somatic characteristics might suggest a child's biracial parentage, but they could not identify a specific individual as the father any more than could a blood group test. In a case involving two women whose mother was identified as black ("preta" or melanoderm) and whose alleged father was classified as white (leucoderm), for example, they argued, "no useful conclusion can be drawn from the study of the subjects." While the two daughters were identified as faioderm (mixed race) and exhibited "racial characteristics deriving from the crossing of white and black," these were "general characteristics, the results of the crossing of any white individual with any black one and not with [the specific man] in question."[35] In notable contrast to paternity testers elsewhere, for the IOF's scientists racial difference was not transparent evidence of paternity. Modern paternity drew on racial knowledge, but what that knowledge revealed was not everywhere the same.

Everywhere, however, finding the father on the body required rethinking the nature of paternity. Nineteenth-century jurists had once characterized maternity as "a material fact, visible, subject to the domination of . . . [the] senses."[36] It was a condition that could be known at birth and could be seen on mothers' gestating, birthing, lactating bodies. The identity of the father, meanwhile, was "veiled" or "shrouded." Modern paternity quite literally stripped the male body of this mantle. Scientific assessment treated paternity like maternity, as a bodily condition that could be empirically observed. To be sure, scrutiny of the paternal body did not dwell on reproductive organs or functions as with maternal bodies. The secret of paternity could be hidden in the tips of the fingers, the curve of the nose, a telltale mole shared by child and putative parent. Still, examination of putative fathers could be intimate. In their expert reports, the IOF scientists took note of somatic features of men's bodies like the scar one man had in the subinguinal area (that is, his crotch) and the small, flat, red-purple mole on the back of the scrotum of another.[37]

AS PARENTAGE ASSESSMENT was becoming an established part of the IOF's forensic repertoire, Dr. Luiz Silva was expanding his own paternity portfolio.

With the Canella case stalled, he broadened his practice to more conventional scenarios of disputed paternity, in particular those associated with inheritance cases. Thanks to the newspapers, the Brazilian public was now familiar with his unusual paternity test. In addition to the extensive coverage of the Canella case, the press had covered a rancorous and very public dispute between Silva and a fellow dentist who had questioned the value of his odontological quest for the father. By the late 1930s, the idea of a "paternity test of the teeth" was appearing in the popular media in satirical cartoons and rhyming verse.[38] Silva's renown had grown, and with it so did requests for his expertise.

In 1940, Silva was approached regarding the case of a young man named Raul de Oliveira. Oliveira was from Baurú, a town some 200 miles from the city of São Paulo and the originating point of a great railroad that stretched a thousand miles into the interior of Brazil to the border with Bolivia. He was twenty-three years old, a man of humble upbringing, and an apprentice in the local railroad foundry. Oliveira had recently filed suit against the estate of the man he claimed was his father. Raul's mother, Maria Magdalena de Jesus, had been born in a poor village in Portugal. In 1913, in her early twenties, she met Alfredo de Oliveira, a prosperous merchant resident in Brazil, during a visit back to his native Portugal. Within months, Alfredo convinced Maria Magdalena to return to Brazil with him, paid her second-class ticket on a ship, and several years later became, according to Raul's suit, the father of her only child. Alfredo never married his "amiga," and when she died of typhoid in 1927 at age thirty-five, she left behind ten-year-old Raul. The boy was never recognized by Alfredo and therefore had no legal family in Brazil. An illiterate couple who had come from Portugal with his mother took him in. Without acknowledging his paternity, Alfredo agreed to pay them a pension until Raul was old enough to work.

By the time he died, Alfredo de Oliveira had lived in Brazil for some six decades and had done extremely well for himself. He ran a successful butcher shop, owned properties, and circulated in a network of prosperous immigrant merchants, mostly fellow Portuguese but also Spaniards and Italians. Because he was not married, had no legitimate children, and had never legally recognized Raul as his son, his estate went to his sisters. Raul de Oliveira had strong circumstantial evidence supporting his claim to be Alfredo's natural son, but his lawyers, noting the "truly astonishing advances of science" in such matters, decided to take advantage of this new form of evidence as well.[39]

Silva launched into his investigation with the same single-minded zeal he had demonstrated a few years earlier in the Canella affair. But this case presented a formidable challenge that one had not: both the parents were dead. How could he determine Raul's paternity if neither the possible father nor even the mother could be examined? Silva took dental molds from Raul and his paternal aunts and then sought recourse to another technology to solve the problem: photography. He obtained photographs of the three subjects and, elaborating on a technique he had first experimented with in the Canella case, applied his odontological method to the images. In thirty-six photographs and thirty-eight pages of accompanying analysis, he assessed everything from the shape of the subjects' nostrils to the expression of their mouths. He took minute measurements of the brow, jaw, and ear and compared them across subjects. He grafted parts of one face onto another and fused the hemi-faces of Raul and Alfredo. The result was a series of bizarre montages the effect of which, to the contemporary viewer, is oddly evocative of a Monty Python cartoon.

Where some paternity testers used family portraits to assess kinship, imbuing traditional material objects with new scientific value, Silva's method required the creation of unique images—annotated expressions, facial collages, physiognomic fusions—specially made for this purpose. He signed each image with the place and date, probably to certify its authenticity, but the effect was to make the curious portraits resemble autographed works of art.

According to Silva, the photographic exercise was conclusive. The similarities between Raul and Alfredo de Oliveira "constitute a formidable sum of identifying elements" and "convince us of a direct relation of genealogical descent between these two individuals."[40] Alfredo's sisters challenged the analysis on procedural grounds, claiming that the pictures of Alfredo had been surreptitiously removed from the dead man's belongings and could have been altered. They also questioned it on scientific grounds. Citing work by the IOF's Ferreira, they disputed whether photographs could reveal kinship between the two men. Photography's role in paternity analysis inspired heated debate as well as creative experimentation. Scientists discussed whether a two-dimensional representation, one normally limited to the face,

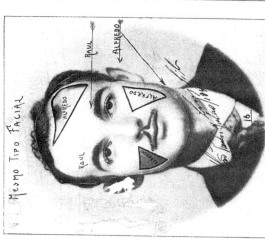

Silva found similarities in the "expression of the eyes," the left ears, and the "facial types" of Raul and Alfredo Oliveira, characteristics that he believed revealed their kinship. He autographed each image on the neck.

Reproduced from João Maringoni, Emilio Viégas, and Raul Marques Negreiros, *Um caso de investigação de paternidade* (São Paulo: Revista dos Tribunais, 1940): 53, 57, 54.

could substitute for physical bodies. They debated the value of retouched family photos and whether painted portraits could substitute for photographs.[41] And they experimented with different techniques of manipulation and interpretation.

While Silva's "prosopographic, prosopometric technique" was apparently original, the idea that photographs might be strategically manipulated to make visible an otherwise hidden ancestry dated back to the nineteenth century. In the 1870s, Francis Galton had experimented with composite photography as a way to capture "types." He took series of photographs of individuals belonging to social groups both "low" (criminals, the mentally ill, Jewish boys) and "high" (Anglican ministers, Royal Engineers) and then overlay them so as to obscure the individual and idiosyncratic and make visible some collective, physiognomic essence shared by the group. A decade later, the French photographer Arthur Batut experimented with a similar technique, exposing multiple photographs on a single plate to reveal, through a series of ghostlike visages appearing one on top of the other, "a family, tribe, or race type."[42] In the Arcardini case in Buenos Aires, Lehmann Nitsche had referenced these earlier experiments and proposed a similar method for capturing the elusive essence he referred to as "family likeness." By overlaying photographs of multiple individuals from a single family, he suggested, idiosyncratic individual features would be visually diluted and the features characteristic of the "family" would become visible. The family type presumably made for a better term of comparison with the putative child than did the potentially more idiosyncratic visage of the father.

Photography's ability to substitute for the physical body was an especially pressing issue because the body of the father was often absent. After all, it was frequently his death that catalyzed the paternity dispute in the first place. This problem was especially common in Latin America, where restrictive laws and the primacy of the man's free will in defining legal paternity made it difficult to bring filiation suits during his lifetime. The Arcardini case involved a deceased father, as did that of Oliveira and the other cases Silva worked on.[43] In the absence of the father's body, Latin American forensic specialists became particularly interested in the application of photography to paternity assessment.

The IOF's scientists also used photography in their paternity assessments, but their method diverged from Silva's, not least because most of the fathers were alive. Rather than creative manipulation, they aspired to standardization.

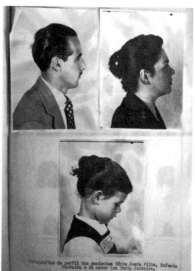

Part family portrait, part mug shot: the mother–father–child triads of the Instituto Oscar Freire's paternity examinations, 1955.

Laudos, Instituto Oscar Freire, L. 11525, 1955. Courtesy Departamento de Medicina Legal, Ética Médica e Medicina Social e do Trabalho, Faculdade de Medicina, Universidade de São Paulo.

Each mother, child, and putative father was photographed frontally and then in profile according to the specifications of Bertillon's police photography, which promised uniformity and veracity. The pictures were then pasted together in the case record. The result was an incongruous series of images, part family portrait, part mug shot.

The photographs not only reflect the intentions of the scientists who made them but also provide glimpses of how the subjects themselves might have experienced this encounter. What is most striking is the formality of their appearance: the mostly working-class subjects appear to be dressed in their Sunday best. Men sport dress shirts and ties; carefully coiffed women wear lipstick and earrings. A girl appears in a frilly dress; a boy has carefully slicked back hair. Judging by the formality of their self-presentation, going for a paternity examination at the IOF was serious business. It also suggests young children were not the most willing forensic subjects. In the photographs, some babies and toddlers are crying. Others are sleeping. Some are seated on laps, and profile shots are taken as they nursed, which kept them still and provided a convenient angle. Sometimes older women—grandmothers?—hover in the background, holding

Portrait of doubt: a mother, child, and two possible fathers, Instituto Oscar Freire, 1955.

Laudos, Instituto Oscar Freire, L. 10828, 1955. Courtesy Departamento de Medicina Legal, Ética Médica e Medicina Social e do Trabalho, Faculdade de Medicina, Universidade de São Paulo.

children's heads for the camera, suggesting there were family members present for the exam who did not appear as forensic subjects. It is hard to imagine how blurry photographs of recalcitrant toddlers, faces disfigured by howls of protest, could possibly aid in paternity assessment. Yet the IOF's scientists dutifully produced and then filed thousands of such portraits over the years.

Like the oscillophore, blood groups, and DNA, paternity photography was a genetic technology. It probed the relationship between the subjects and sought to discover the physical kinship between them. In the case of photography, the evidence was of course visual. In the IOF's Bertillonian portraits as in Silva's prosopographic confections, the subjects stare out at the viewer, posing an implicit question: is he the father? In the expert hands of the scientist, the image promised to answer that question.[44]

As for Raul Oliveira's suit, this genetic technology worked, at least partially. The judge lauded Silva's scientific analysis as "magnificent work" and "a sensational revelation of the paternity being investigated."[45] However, in deference to the defendants' objections concerning the photographs, he ordered another, even less conventional sort of examination: the exhumation of Alfredo de Oliveira's corpse, which Silva had requested in order to take cranial measurements and a plaster cast of the mouth ("if the palatal ridges are still intact").[46] The deceased man's sisters were horrified. Their brother had been laid to rest just three months before, and now he was to be disinterred and his head removed from his body? The judge was unmoved by their squeamishness. "The separation of the body from the head . . . merely anticipates the unfathomable work of nature and is never a profanation."[47]

The sisters apparently thought otherwise. They abruptly agreed to accept Raul's paternity claim in order to avoid the exhumation (one wonders, indeed, if the shocking proposal was purposefully calculated to force a settlement). In fact, it was not such an unusual request. Exhumations were discussed in relation to other paternity cases; the following year Silva requested the disinterment of a wealthy São Paulo industrialist in another inheritance dispute. In at least one Argentine case, a landowner was disinterred six years after his death to test his tissue for the MN blood type. This time the horrified party was Leone Lattes, who tried in vain to convince the judge that a blood test of the remains was futile.[48]

Alfredo de Oliveira's body remained undisturbed, but it had solved the paternity suit twice over. Silva's analysis of it won over the judge, and then its proposed desecration spurred his sisters to fold. Three months after filing suit, the modest railroad apprentice was declared Alfredo's son and heir.[49]

THE JUDGE FOUND Silva's odontological method a "sensational revelation," but scientists themselves were often much more skeptical of the myriad techniques of paternity investigation. Finding the father by comparing bodies presented a series of thorny methodological problems. Differences in age or sex between two individuals could produce variations in appearance that might mask hereditary similarities. How could one peel away such confounding factors to discover the kinship between a young girl and an old man, for example? Likewise, to interpret a particular trait as a marker of kinship, an investigator required information about the relative incidence of that trait in the general population. The fact that a child and putative parent both had red hair, for example, was a more significant piece of evidence if red hair was rare. Assessing red hair's significance thus required knowing something about its prevalence in the population. The same logic applied to blood groups: the significance of a shared blood type depended on the relative frequency of that type in the general population. But whereas the field of racial serology had generated a wealth of research on the serological characteristics of specific local, racial, regional, and national populations, there were rarely equivalent studies of other physical characteristics. Some scientists, especially in Austria and Germany, attempted to address these problems. They developed conversion indices that would allow them to discern underlying hereditary similarities between an adult and child despite differences in sex and age. They also conducted studies to determine the incidence of certain somatic traits in particular populations. One ambitious study imagined a comparative chart of physiognomic traits for individual cities—"Vienna, Rome, Paris, Rio de Janeiro."[50] But there was a potentially infinite number of somatic traits and an equally infinite number of possible populations in which to study them, and such knowledge therefore remained rudimentary.

Anomalous traits were more helpful. Features such as harelips or supernumerary nipples and fingers, when present in a child and a putative father, strongly suggested a biological relationship between the two. In the 1920s, a Norwegian court declared that a child with brachyphalangy, a hereditary condition in which joints are missing in the fingers and toes, was fathered by an accused man with the same condition after the local sheriff testified that in his twenty-five years on the job he had "come into connection with and know personally practically every person living here" and had encountered no other community member with the malformation.[51] The sheriff's intimate personal knowledge of his small community was its own kind of genetic expertise. But in a large and anonymous metropolis, such knowledge required population-level studies. In any event, such anomalies were by definition too rare to provide a systematic method of assessment. In their expert reports, the IOF's scientists touted the importance of such anomalies but rarely found them.

Another challenge was perhaps even more profound and intractable: How could something as abstract, as ineffable, as visual similarity between two faces or bodies be described, much less proven? A German anthropologist posed the problem as one of "converting form to figures."[52] Was the problem of resemblance one of the nature of that resemblance or of its degree? Were the appropriate terms of its expression therefore narrative, numerical, or visual? Should it be described with words, expressed in a mathematical index, or illustrated visually? Bertillon wrestled with a similar problem as he sought a method for describing the body in an accurate and unambiguous fashion. His solution was to combine narrative, numerical, and visual representation: the spoken portrait, anthropometric measurement, and photography. Paternity assessments likewise drew on all three modes.

Running through scientific techniques to find the father was another, even more fundamental tension, one that was not just methodological but conceptual. This was the tension between similarity and heredity, an issue over which Luiz Silva and the medico-legistas of the IOF sparred for a quarter century. Silva asserted that similarity signaled heredity. If the molars and palate of Rita and Giuseppe Canella looked like those of the Sconosciuto, this indicated a shared genetic connection. To be sure, meaningful resemblance could not be discerned by just anybody; it had to be patiently observed and objectively measured according to deliberate scientific criteria. Silva claimed that his prosopometric, prosopographic method did just that. In the hands of a competent expert, the general axiom that "like begets like" became sufficient theoretical basis for deducing kinship. The task of the scientist performing a paternity test was not to understand the mechanism behind resemblance but rather to elaborate a method for detecting it.

The IOF scientists roundly rejected this position. The visual appraisal of similarity—comparing two faces in order to determine whether they were the same person—might be sufficient for the purpose of individual identification, as in the case of a police officer who compared a recently arrested subject to a person described in police files. But for the purpose of kinship analysis, observing, measuring, or counting similarities was insufficient because "similarity is not heredity."[53] The fact two people had similar ears implied nothing about the relationship between them without an understanding of how specific traits—the infolding of the outer ear or detached earlobes, say—were passed from parent to offspring. Kinship analysis based on somatic characteristics therefore required knowledge of their dominance and recessivity. As Arnaldo Amado Ferreira repeated time and again in his expert reports to the court, "One cannot prove anything by the number of coincidences identified in two or more individuals" in the absence of an explanation of their genetic transmission.[54] The only definitive test of paternity—and of course it was really a test of nonpaternity—was blood group analysis.

The IOF doctors and others in their camp repeatedly denounced the "deceptive, dangerous" practices of certain "non-medico-legal" specialists who purported to determine kinship through prosopographic analysis of physiognomy.[55] This was, of course, a not-so-thinly veiled reference to Luiz Silva. Similarity was no better than the old ways of knowing the father through social and legal presumptions: both proofs were "pre-Mendelian" and "prescientific." The fact that Silva's techniques were "elegant" and attractive to judges did not change the fact they were based on the outmoded science of Hippocrates.[56] The two camps thus disputed what counted as "modern" paternity science. Their debate rehearsed the tendency—dating back to Engels and the other nineteenth-century theorists—to frame knowledge of paternity in terms of modernity. The stakes of this debate were not just theoretical, however; the rivals frequently found themselves arguing on opposite sides of judicial cases.[57]

If likeness versus heredity was the seemingly irreconcilable divide between the dentist and the medico-legistas, in practice the two camps diverged much less dramatically than either cared to admit. For even as the Mendelians of the IOF insisted on hereditary transmission, their analyses routinely included techniques that relied on similarity: anthropometric measurements, biotypes, the spoken portrait, photographs, and the perennially elusive fingerprint, all techniques that in their own estimation had precious little value for fixing paternity. While always cautious in wording their conclusions for the court—showing a reserve that contrasted with the blithe confidence of Silva—all these techniques relied on similarity. Sometimes they even slipped into resemblance talk themselves, as when one expert report assessing a parent and child concluded that while the forehead, eyes, nose, and upper lip of the two "could not permit a conclusive affirmation of a kin link," nevertheless "the collection of traits of similarity between the [child] and [father] is striking."[58]

The IOF medico-legistas were not exceptional: many paternity testers insisted on heredity only to fall back on similarity. In his somatic examination of the Arcardinis two decades earlier, Roberto Lehmann Nitsche characterized his method as "Mendelian" but invoked "family likeness"—an idea that sounded suspiciously like resemblance. Leone Lattes asserted that

"The similarity of the child and father are impressive." The forensic scientists of the Instituto Oscar Freire insisted that paternity analyses should be based on hereditary principles, but in practice they sometimes fell back on traditional notions of family resemblance.

Laudos, Instituto Oscar Freire, L. 4022, 1943. Courtesy Departamento de Medicina Legal, Ética Médica e Medicina Social e do Trabalho, Faculdade de Medicina, Universidade de São Paulo.

while it was preferable to rely on traits whose patterns of heredity were understood, this did not rule out the use of those whose transmission was unknown or too complicated to be predictable.[59] Fritz Schiff went so far as to suggest that "there may be similarities between father and child that obviate an analysis of the individual characteristics"—in other words, some visual truths were so self-evident they did not require Mendel to detect them. What is more, as both Schiff and the IOF's medico-legistas noted, whatever its questionable scientific merit, resemblance tended to convince judges more effectively than did technical discussions of Mendelian transmission because they could see it for themselves.[60]

In the end, what prevailed for Mendelians and non-Mendelians alike was the logic of multiple coincidences. As long as blood groups allowed only a clumsy method of exclusion rather than positive identification, as long as the hereditary elements of fingerprints remained undecipherable, as long as no other single heritable trait could illuminate kinship unequivocally, scientists were forced to adopt a kind of kitchen sink approach to paternity. Similar hair or eye color between two people, a certain resemblance of the shape of the nose or the ears, a telltale mole could not indicate a kin relation. But a pattern of such coincidences across a series of traits could be significant: "The greater the number and the degree of similarities between the child and the man in question, the more probable it is that he really is the child's father."[61]

The more-is-better approach to paternity determination was especially well suited to the field of legal medicine. As practiced in Latin America and parts of continental Europe in the first half of the twentieth century, forensic science was a multifarious field, its practitioners Renaissance men (and occasionally women) with expertise in a broad range of themes and techniques. In the 1930s, the IOF's Flamínio Favero compared legal medicine to a growing family: its vast "sphere of action" was "like a family that grows and puts out new branches."[62] The IOF's multifaceted portfolio of research and practice included forensic chemistry and hematology, workplace injuries, ballistics, embalming, fingerprinting, forensic dentistry, pregnancy determination, and virginity examinations. Paternity assessment was one more problem added to its "sphere of action," and its practitioners were well prepared to tackle it given their methodological eclecticism. Since no one trait or method could positively identify the father, their analyses drew on the wide-ranging forensic skill set they already had at their disposal: knowledge of heredity, techniques of individual identification, expertise in systems of eugenic and racial classification.

THE DISTINCT APPROACHES of the IOF and Silva reflected the lack of scientific consensus about how to find the father, but they also reflected social considerations. The body afforded evidence of biological kinship, but bodies also had social and moral meanings that determined who could be examined and how. In fact, questions of propriety had long shaped scientific examination of bodies. A recurring criticism of anthropometry was its indecorous scrutiny of the body, in particular when the subjects were women. Turn-of-the-century Brazilians charged that requiring subjects to undress to be measured and described was stigmatizing and humiliating.[63] So too was requiring subjects to pose for judicial photographs, for according to the head of Rio de Janeiro's police identification agency, "everyone is the rightful owner of his own physiognomic expression."[64] Such reservations helped usher in the use of fingerprinting, which was considered not just more accurate and efficient but also less invasive and unseemly.

Questions of propriety proved a recurring problem for paternity testers as well. In an effort to improve paternity determinations, German researchers sought to gather data on the distribution of hereditary traits in an ethnic German village in Romania in the 1930s. But when villagers refused to have their bodies examined, the investigators were forced to limit their study to subjects' heads and faces. Full-body examinations, mused one researcher, "could probably only be conducted of primitives living without clothes—in this country, on a large scale, only of athletes or in the course of clinical (constitution) studies."[65]

It was therefore not by chance that Silva's technique, used in cases of identity and inheritance among respectable, affluent, and often European subjects like the Canellas and the Oliveiras, focused exclusively on the head and mouth. Silva claimed that odontology was a more powerful method, but it was also more modest than a bodily examination. Decorum surely also shaped Roberto Lehmann Nitsche's examination of the Arcardini clan in Buenos Aires. While the anthropologist cited Bertillon as a source for his method, in fact he limited his examination to family members' heads and faces. It would have been unseemly to subject their bodies to full anthropometric examination, particularly considering that all the living Arcardinis were women. Propriety dictated that respectable people be shielded from the intimate scrutiny of scientists.

In contrast, the IOF was empowered to carry out full-body examinations because the bodies they examined were of humble and working-class, often nonwhite, and frequently immigrant subjects. Modest people, especially in the context of criminal proceedings, could be subjected to forms of scrutiny that more elite people could not. Bodily examinations were humiliating not only because they involved intimate inspection but also because they were associated with criminality. After all, anthropometry and the mug shot were not originally techniques of genetic analysis; they were borrowed from police practice. Even fingerprinting, which was less invasive, was probably still stigmatized by this association. Silva's odontological method, in contrast, not only limited exposure of the body but as a sui generis technique did not harbor the criminal associations of other methods. The contrasting photographs produced by Silva and the IOF served as visual reminders of their subjects' contrasting social positions. Whereas the mostly respectable families of Silva's assessments were represented in formal portraits, the "family portraits" of the IOF's working-class subjects were simultaneously mug shots.

THE IOF'S SCIENTISTS performed their exhaustive analyses on hundreds of people, submitting thousands of pages of carefully prepared forensic reports to the courts. Yet their examinations were almost never conclusive. The reports sometimes ventured to express a probability of paternity based on the analysis of the blood tests, but they tended to end with the same conclusion, in the 1940s as in the 1920s: "science as yet does not offer elements to be able to say, in a case of doubt, if an individual is the father of another."[66] Scientific paternity tests, even according to their most dedicated practitioners, could not find the father. Why, then, did judges keep asking for them, and why, for three decades, did the IOF's medico-legistas doggedly continue to perform them? If their paternity examinations did not work, what work did they do?

There was at least one scenario in which they did "work": when a blood test excluded an impossible progenitor. Occasionally the IOF encountered cases involving two possible fathers, in which blood could eliminate one. In the more common cases of sexual transgression—rape, deflowering, seduction—a blood test finding the defendant could not have fathered the victim's child might be useful. It could not show that the crime had not happened, but it could show that the woman had had some other partner, fatally undermining her moral character and most likely her accusation against the defendant. In such scenarios, the paternity test may have been inconclusive as to child's paternity, but it was decisive as to the woman's morality, an essential condition for her status as victim.

But these were potential scenarios based on blood groups, and they cannot explain why for decades the IOF also carried out the elaborate and time-consuming bodily inspections. Perhaps these procedures did something besides just yield scientific evidence. One Brazilian observer touted the "disciplinary power" that genetics would one day exercise when it was able to reliably identify "the genitor of every human being," but perhaps paternity tests did not need to identify the genitor to exercise disciplinary power.[67] Simply requiring people to report to a forensic institute to have their bodies measured, classified, photographed, and exposed to intimate scrutiny was itself a form of discipline. Such techniques might not expose the moral transgressions that had landed them in court, but they might themselves serve as a form of punishment.

But why these couples? After all, the scenarios encountered at the IOF were hardly unusual. Charges of deflowering and seduction were among the most common criminal complaints to Brazilian police in these years.[68] Extramarital unions and illegitimate children were likewise an endemic feature of working-class life. Why had these couples been sent to the IOF for the laborious and punitive scientific proceedings? A striking feature of many of them is the large age gap between the mothers and the alleged fathers. The average age difference between the two was twenty-five years; in two-thirds of cases, there was an eight-year or greater difference between them, and in a quarter of cases, the average age spread was thirty-six years. Meanwhile, the mothers were young, half of them sixteen or under, the youngest just eleven years old. These were not just illegitimate sexual relationships, in other words, but relationships involving very young mothers and much older men. The paternity tests at the IOF did not seek to establish fathers of generically illegitimate children but those born of a particular kind of transgressive relationship.

The forensic reports that for thirty years the scientists of the IOF typed up and bound in volume after volume, still stored in the hallway of the institute, do not record the verdicts of the cases. It is therefore not possible to know how many of the men were eventually found to be the father. But regardless of the judicial outcomes, the intimate somatic scrutiny of the IOF examination was itself punitive, both to men and to women. Ultimately, the scientists failed in their attempt to establish paternity on physical bodies, but what they achieved instead was the disciplining of social ones. In this regard, the IOF examinations shared an important commonality with the otherwise very different case of the Canellas. While the medico-legistas' tests disciplined illegitimate relations, Silva's test sought to restore legitimate ones, rehabilitating the lost honor and respectability of the Canella family. Whether it found the father or not, paternity science was a tool of social discipline and moral rehabilitation.

THE IOF CONTINUED to perform its elaborate paternity analyses through the late 1950s, at which point the scientists abruptly abandoned the finger-prints, anthropometric measurements, and photographs and focused ex-clusively on blood tests.[69] It is unclear why the analyses were radically sim-plified, but perhaps the real question, given their perennial indeterminacy, is not why they disappeared but why they lasted for so long. The persis-tence of the complex but perpetually inconclusive somatic paternity tests reinforces the impression that the work they did involved something other than finding the father.

Silva, meanwhile, continued to practice and advocate for his odontolog-ical paternity technique, which while originally a peculiar local innovation, soon gained adherents (as well as detractors) as far away as Cuba.[70] As for his most famous case of all, the Canella affair, his achievement was partial. Silva claimed that he had solved the extraordinary mystery, but his finding could not alter the erroneous legal identity assigned the Sconosciuto by an Italian court. While it may have helped to restore the family's honor in the court of public opinion, the Canellas clearly viewed their vindication as incomplete.

Born out of the destruction and sorrow of the First World War, the Canella affair was a most unusual paternity case. The fact that it became one reflects the power of modern paternity. This set of ideas foregrounded filial ties and studied the bodies of parents and children to uncover those ties, but bodies could not be stripped of their social identities. And so, in the process of determining kinship, considerations of class, status, and de-corum also came into play. Scientific techniques not only had to respect social convention and moral propriety but also became instruments to re-inforce them. Whether in the hands of an eccentric dentist or a group of forensic experts, the quest for the father was a pursuit of morality as much as of truth. Regardless of whether it "worked" or not, this is part of the work that paternity science did.

In December 1941, Giulia Canella was widowed a second time when the Sconosciuto, whom she continued to insist was her husband, died. That same year she also lost her father, the Marquês. Giulia remained in Rio with her children and soon grandchildren, but the family never stopped trying to establish the Sconosciuto's legal identity. Two years later, her stepmother, the Marquês's widow, published a book exhaustively reviewing the case. The book was entitled "Yet Another Crime of Fascism" and argued that Italian authorities had robbed Giulio Canella of his civil identity.[71] Published shortly before Brazil sent troops to the join the Allies in Italy, it was dedicated to Brazilian law students and urged them to embrace "liberty, justice, and humanity" and to restore "truth, trampled by a violent and corrupt fascism."[72] The drama of the Canella family was not a private tragedy but a political crime.

After the war, the family continued to lobby unsuccessfully to overturn the ruling that had declared the Sconosciuto Mario Bruneri, on the grounds it was the illegitimate verdict of a fascist tribunal. A quarter century later, Catholic authorities provided a measure of vindication when an ecclesiastical court declared that the mysterious amnesiac was in fact Giulio Canella. According to this pronouncement, Giulia had lived with her husband and not a lowly impostor, and the three children born after their reunion were the legitimate issue of an honorable marriage.[73] Yet the shadow of doubt never fully dissipated. In 1977, Giulia Canella died, still seeking absolution for her family.

6

JEWISH FATHERS, ARYAN GENEALOGIES

Who was my father? I was utterly indifferent about whom my mother had selected as my progenitor. . . . I only wanted to know whether he was an "Aryan" or a Jew.

—Hanns Schwarz

IN 1933, THE NAZI REGIME came to power and in succeeding years would promulgate some 2,000 ordinances pertaining to race. Virtually every dimension of an individual's public and private life—professional, educational, commercial, legal, civic, familial, sexual—would come to depend on his or her official designation as Aryan or Jew. This fact left Hanns Schwarz in limbo. A resident of Berlin, Schwarz was an accomplished professional psychiatrist, a husband, and a father of two young daughters. But his racial identity remained indeterminate, at least to the authorities, and became the subject of a bureaucratic and scientific investigation that would last a decade. The outcome of this investigation would determine not only his own fate but that of his wife and children.

The uncertainty surrounding his racial status arose from a question of paternity. Schwarz was born in 1898 to a young, unmarried mother, a seamstress not quite twenty years old, who had not listed a father on his birth certificate. While his mother's racial status as an "Aryan" was undisputed, his father's racial identity was unknown. If his progenitor was Aryan, then

Hanns himself qualified, in the fantastical racial taxonomy of the Nazis, as a full Aryan. His relationship with his wife, Eva, the daughter of a rabbi, would then qualify as a "privileged mixed marriage" and their two children as *mischlinge,* children of mixed descent.

But if this absent father was Jewish, then Schwarz was himself a *mischling,* half-Jewish, half-"Aryan." His two daughters, with three of four Jewish grandparents, became Jews, and his wife's Jewishness was no longer partially mitigated by her mixed marriage. The identity of Hanns Schwarz's absent father—a man he had never met, whose identity he did not know—thus determined whether they were considered citizens, whether Hanns could practice psychiatry, and whether his daughters could go to school. It would determine where Hanns and his family could sit on a bus, which benches they could use in the park, and whether they could ride bicycles. Soon, it would determine whether they lived or died.

In the summer of 1938, Hanns received a genealogical certificate from the Reich Kinship Office, the agency charged with determining racial status. The certificate arrived two days after a new law ordered Jews with "non-Jewish" names to adopt "Israel" or "Sara" as identifying monikers. The Kinship Office had been investigating Hanns's paternity and had discovered a record in which a man named Nathan Schwarz recognized Hanns, then two years old, as his son. Hanns had turned forty-two just weeks earlier, and for the first time in his life, he had a biological father: that father, however, was a Jew. Of the certificate of paternity he mused, "Death sentences are always short. The official note comprises all of seventeen lines."[1] He hired a lawyer and appealed the finding.

Having begun in the archives, the question of Hanns Schwarz's paternity now moved to his body. Hanns's mother came forward and claimed that the Kinship Office's information was incorrect: the father of her son was not the Jew Nathan Schwarz but an Aryan man named Robert Koch, who was no longer alive. The genealogists of the Reich Kinship Office remitted the case to the Polyclinic for Hereditary and Racial Care for further study. Over the next few years, the Reich's heredity experts would examine Hanns, his mother, and his putative father Nathan Schwarz in an attempt to determine whether the Jewish man was his progenitor or not.

By the early 1930s, paternity tests had become a thoroughly routine activity for the growing numbers of institutes dedicated to anthropology, genetic science, and racial hygiene across Germany and Austria. With the rise of National Socialism, paternity science was reoriented toward a radically different purpose. Methods originally designed to find fathers for the fatherless became tools for distinguishing Jews and Aryans. Paternity testing became, in short, an instrument of racial governance. Race and paternity were inextricably linked because the Nazis' genealogically based system of racial classification required having two known biological progenitors. The experience of Hanns Schwarz, a man whose parentage was pursued with singular zeal by multiple state authorities over the course of a decade, was thus hardly exceptional. Under National Socialism, scientific assessments of ancestry proliferated, acquiring both immense political significance and dramatic life-or-death stakes.

Because it was the father who was considered inherently uncertain, the authorities of the Third Reich were especially preoccupied with paternity—its social definition, its legal regulation, its scientific investigation. Spurred by their racial fanaticism, they advanced a radical version of modern paternity, one in which the father was strictly biological, absolutely knowable, and always necessary to know. They made the unprecedented claim that the science of heredity had solved the millennial problem of paternal uncertainty and was able to unequivocally identify the father through the body. In keeping with this claim, they undertook a sweeping revision of paternity law so that the father's identity could be freely investigated in all circumstances, opening up to contention identities and relationships that were once considered stable and irrefutable. Yet in so doing they inadvertently opened a peculiar avenue to escape their murderous contrivance: Jews could strategically challenge their own paternity in order to change their racial classification. Rather than making parentage and race certain, then, the Nazis' radical reworking of the science and law of paternity had the opposite effect. It bred uncertainty and instability that victims like Hanns Schwarz would seek desperately to exploit.

THE STRANGE AND TERRIBLE life of racial ancestry testing under National Socialism grew directly out of the search for illegitimate fathers initiated the decade before. In the mid-1920s, just as the serologist Fritz Schiff was introducing blood group testing into the courts of Berlin, in Vienna an ambitious racial anthropologist and a crusading judge were fixated on similar issues, albeit using different methods. Anton Rolleder, judge of the Vienna district court, had become fed up with the conventional legal proofs offered in paternity disputes. Echoing the lament of many of his peers at this time, he asserted that perjury, especially female perjury, was rampant in paternity proceedings. Yet the new techniques of blood group analysis then making their way into German and Austrian courts were of no value in positively identifying fathers. The judge wondered if it would be possible to assess parentage through morphological traits rather than serological ones. Several scientists he consulted were skeptical, but in September 1925, Rolleder approached Otto Reche, racial anthropologist and director of the Vienna Institute of Anthropology and Ethnography, who agreed to study the problem.[2]

Reche was an enthusiastic student of heredity and somatic analysis long before he embarked on a professional career in anthropology. As a high school student in the 1890s, he began studying the traits of his own family, thanks to an uncle who was an avid genealogist. "Even then I was inclined towards scientific thought," he would observe many decades later, musing that perhaps his curiosity itself "might have been determined by heredity, as many members of my family have been active in science-oriented professions." He tried to learn as much as he could about his ancestors, about their "life, work, worldview, and mindsets," and carefully studied old photographs in search of facial features and family resemblances.[3] For Reche, heredity encompassed both physiognomic features and "intellectual-mental" ones, and the first school of heredity was one's own family.

As a professional anthropologist, Reche would move from the faces of ancestors to the bodies of "natives" and criminals. In 1908 he accompanied an ethnological expedition to the South Seas, where he developed his first professional experience measuring humans and brought back the skulls and skeletons of some 800 people for the Hamburg Museum of Ethnology (this collection may have served as the basis for the thesis on Melanesian jaws that Josef Mengele authored several decades later). Later Reche lectured on criminal anthropology and Bertillon's anthropometric methods for the Vienna police and also helped establish the Viennese Eugenics Society.[4]

Reche set to work on the task Judge Rolleder had put before him. Over the course of almost a year, he worked out a genetic-anthropological method for establishing paternity between a putative parent and child based on an analysis of nineteen hereditary traits, including the usual suspects (ear, nose, head and face shape; skin, hair, and eye color) as well as fingerprints.[5] In the summer of 1926, Reche launched his new technique and soon after began furnishing expert reports for the courts on a routine basis. Typically, examination began with a blood test and if it was inconclusive proceeded to the more laborious and expensive somatic analysis. And expensive it was: the cost of Reche's new paternity examination ranged from 700 to 1,000 Austrian schillings, an amount that struck even some sympathetic judges as high, given that the child support payments at stake in the proceedings ranged from 20 to 40 schillings.[6] Still, as one advocate of the method argued, "if the method is conducted regularly, the costs will normalize, and in any case, in such important matters as paternity cases, the question of costs cannot be a serious consideration."[7]

Reche's method did not enjoy universal acceptance among his colleagues, some of whom believed there was insufficient knowledge about hereditary traits to determine paternity in the body. Even the lay public apparently regarded it with skepticism: at one point his paternity test became the subject of a humorous variety show. But Judge Rolleder considered the new method "ground-breaking." He characterized a series of opinions Reche submitted to the court as "each subsequent one more beautiful and detailed than the former, the last one comprising 32! pages with superb photographs."[8] Reche himself proselytized on behalf of his new "science of resemblances," mailing his publications to other practitioners to encourage its use. The international press also reported favorably on his new method.[9] Within a short time, the higher courts in Germany were accepting Reche's paternity analyses, and in 1931, the Austrian Supreme Court ruled that failure to accept somatic examinations in a paternity investigation amounted to a procedural defect. Reche later claimed that his method was also taken up across Scandinavia, Poland, and Hungary, but this is likely an exaggeration given that similar somatic methods were being developed independently in many places around this time.[10] Reche's self-aggrandizing conviction that his anthropological method could "prove paternity with all certainty in a large majority of cases" echoed the wild confidence of eccentrics like San Francisco doctor Albert Abrams or, later, Brazilian odontologist Luiz Silva. But in Reche's case it presaged what would become standard policy under National Socialism.[11]

Reche's interests in genealogy and heredity, race and paternity, were political as well as intellectual. In the years after World War I, he expressed an increasingly *völkisch*, or racial nationalist, worldview and by the 1920s was a committed anti-Semite. Shortly after meeting Rolleder, he cofounded the German Society for Blood Group Research, a collective devoted to promoting the study of racial differences in blood types. The group reflected Reche's interest in racial serology but also his political aspirations for its study: one of the group's objectives was the "elimination of the majority of the Jewry from this field of research."[12] In keeping with this position, the organization excluded prominent Jewish serologists, most notably Fritz Schiff. In 1933 Reche joined the Nazi Party and in succeeding years would become an important racial ideologue for National Socialism.[13]

HANNS SCHWARZ WAS BORN at the turn of the twentieth century, too early for the scientific paternity tests to which illegitimate children would be increasingly subjected two decades later. Besides, although he was born to a single mother and his birth certificate lacked a father, as a child his paternity was never a source of dispute or preoccupation. His mother, Adelheid, soon gave up the baby, and he was reared by her older sister, Maria. He grew up calling Adelheid "auntie." Maria married a freelance journalist, Max Schwarz, who was of orthodox Jewish descent but was thoroughly secular. The marriage was "bohemian," in Hanns's later assessment, plagued by financial troubles and constant instability. Still, whatever the couple's faults, they were Hanns's parents, and in 1913 they legally adopted him. In his words, "I was born again when I was 15 years old." Hanns's personal history provided a counterperspective to the growing fetishization of genetic ancestry in scientific and political thought. "Because of [being adopted] I learned about a fact that became influential for my professional life as well: nurture is more important than blood connections."[14]

Hanns finished high school and decided to study medicine, less out of any deep interest in science than from a humanistic desire to serve others. He eventually settled on psychiatry because "the doctor for psychological diseases comes closest to the core of a human being. . . . I did not want to treat an organ but a whole person."[15] As a young doctor, he met his future wife, Eva Maybaum, the daughter of a well-known Hungarian-born rabbi and scholar. Eva's father was dead, her sister was a member of the Communist Party, and Eva herself had apparently abandoned Judaism, so that when the couple married in 1928, Hanns observed, "I was, almost without knowing it, married to a Jewish woman."[16]

THE YEAR HANNS AND Eva got married, Otto Reche was settling into a prestigious new academic position as director of Germany's oldest ethnological institute at the University of Leipzig. Though he had left Vienna, he continued to supply expert paternity opinions to the Viennese courts because the reports turned out to be very lucrative. His residence in Leipzig did not allow him to perform the physical examinations of the Viennese subjects in person, so Reche contracted out the exam to a local hygienist and then used the data to author his expert opinions for the court.[17] Clearly the job of forensic expert paid well enough to justify such an arrangement. It also meant that Reche had never actually met the parents and children whose bodies were the subjects of his reports.

In Reche's absence, the Vienna Institute continued to receive a steady stream of requests for bodily assessments of paternity. But Reche's successor as the Institute's director, racial anthropologist Josef Weninger, was skeptical of the method and initially declined to perform them. His objections did not last, however. Whether because he reevaluated the science or was won over by the money, Weninger soon set aside his reservations and the Vienna Institute again began conducting somatic paternity tests. Judge Rolleder was thrilled. In 1931 the Supreme Court's decision requiring anthropological examinations in paternity suits created even greater demand for them. Soon Weninger was conducting up to six of the laborious examinations a week, and other colleagues began to join him. The resulting income was invested in the Institute's teaching and research activities and was used to purchase photography equipment.[18]

The following year, paternity became even more firmly ensconced in the Institute's portfolio when Weninger formed a Working Group on Genetic Biology, a cluster of researchers within the Vienna Institute working on problems of heredity. The group performed paternity assessments for the court—some 200 by 1934—but its other purpose, perhaps in response to Weninger's lingering doubts about the merits of the technique, was to make it more scientific.[19] The group sought to do so by collecting data on the incidence, distribution, and transmission of certain inherited traits, as well as patterns of morphological variability across ages.[20]

Its most ambitious effort in this regard was the Marienfeld Project, a massive anthropological undertaking carried out by the members of Weninger's Working Group in 1933–1934. Marienfeld was a German-speaking village in Romania and therefore, in the prevailing logic of racial anthropology of the time, a convenient opportunity to study the racial character of a "German enclave" that had remained "pure." The researchers set about recording the physical features of some 250 families and more than 1,000 individuals. The scientists installed a study center in the village. Eight workstations processed distinct parts of the subjects' bodies: one focused on the ear, another on the hand and foot, still others on the face and eyes, the head and body, hair, fingerprints, eye color, noses. At each station, technicians compiled photographs and drawings of head and facial features (the villagers resisted having their bodies examined). The value of these laborious activities was not merely theoretical. "Far from serving science alone," Weninger declared, "our work also pursues a practical purpose of the collective good of the people," namely, enhancing the methods of somatic paternity assessment. Funding agencies also assessed the project in these terms, and Judge Rolleder, whose request to Reche inspired the first paternity tests, enthusiastically followed the Working Group's progress.[21]

Paternity science was not just of scientific or practical interest; as Reche and then Weninger discovered, it was lucrative as well. The Marienfeld Project's equipment had been purchased in part with funds the Vienna Institute generated by providing expert reports to the courts.[22] The Institute also considered selling data on patterns of genetic incidence and transmission, of the kind collected at Marienfeld, to other paternity examiners in Germany and abroad.[23] Soon, however, political developments would provide paternity testers with other ways to capitalize on their expertise.

THE AVID PURSUIT of paternity science reflected the heightened scientific, social, and political interest in heredity and racial anthropology in the 1930s. But paternity testing did not merely reflect these agendas; increasingly, it helped justify them. The distribution of ear traits or the hereditary transmission of nose shape might seem esoteric to the nonexpert, but not if that knowledge improved practical methods of parentage assessment. The significance of the massive and expensive Marienfeld study was framed in terms of its application to paternity, as was a large-scale study of blood group distribution that Reche proposed.[24] In the years before race science was formally harnessed to the state, paternity testing contributed to the "general consensus . . . that anthropology could . . . serve the larger interests of society."[25] Knowledge of heredity had immediate social relevance when it was put to use in the millennial search for the father. The problem of paternity, and the conviction that science could solve it, thus helped to set hereditarian and racial research agendas and to justify their value to funders, the government, and the public.

The Vienna Institute's provision of expert paternity opinions helped to make it the second largest anthropological institute in German-speaking Europe.[26] In the late 1920s, other scientific institutes also embraced paternity testing, including Reche's institute at the University of Leipzig, the University of Munich, and the Kaiser Wilhelm Institute of Anthropology in Berlin.[27] And then 1933 arrived, and these scientific procedures were suddenly invested with appalling new political significance.

ALLUDING TO THE TITLE of his memoirs, *Every Life Is a Novel,* Hanns Schwarz drily observed that 1933 "marked the beginning of a new chapter."[28] He was married with two small daughters and practicing psychiatry when, in January of that year, the Nazi Party assumed control of the German state. Two months later the Law for the Restoration of the Professional Civil Service barred Jews and communists from civil service positions, university teaching, the bar, and the practice of medicine. It was the opening salvo in a series of laws that would purge Jews from science and medicine as well as many other professional fields. For Schwarz, the impact of the laws was not immediately clear because his racial identity was still uncertain. Soon, however, he received a letter asking for verification of his paternal ancestry.

The categories "Jew" and "Aryan" (and related terms used by the National Socialist state, like "German-blooded" or "racially alien") were understood to be bioracial designations. They were based not primarily on self-identification, social or religious practice, or community membership but on the conceit that there existed some essential biological difference between these groups. The task of distinguishing between "Aryans" and "non-Aryans" was complicated by long histories of secularization, conversion, and intermarriage. Jews who were no longer observant members of a religious community, had married non-Jews, or had converted to Protestantism or Catholicism sometimes generations earlier were not easily identified. And then there were people like Schwarz, whose biological origins were obscured by adoption or illegitimacy.

One answer to the conundrum lay in genetic science: if Jews and Aryans were racial groups, then surely some bodily marker, whether blood type, fingerprints, or a particular somatic trait, could differentiate between them. The identification of racial markers was of course a long-standing aspiration of transatlantic racial scientists, but National Socialism's experts were no more successful in locating them than their forebears had been. In the absence of a physical trait revealing racial identity, the Nazi state adopted a definition of race based on ancestry. As elaborated in the Nuremberg Laws of 1935, racial status was determined by the number of Jewish grandparents. An individual with three or four Jewish grandparents was Jewish. One with two Jewish grandparents was a *mischling* in the first degree or simply a Jew (depending on certain additional considerations, such as whether the person's spouse was Jewish). A person with one Jewish grandparent was a *mischling* in the second degree. The fact that Hanns Schwarz did not know two of his four grandparents explains the uncertainty of his racial status.[29]

Racial citizenship required a bureaucratic architecture and new administrative proceedings to assign a race to every individual. Both the civil courts and a new state agency, the Reich Kinship Office (RKO), were tasked with this purpose. Racial descent was established in the first instance by making a declaration about one's lineage and then through corroborating documentation, such as birth, baptismal, and marriage certificates. Such sources revealed the "Jewishness" or "Aryanness" of parents and grandparents through notations regarding their religious affiliation, membership in the Jewish community, Jewish-sounding names, and the like. A large majority of ancestry cases were decided based on documentary evidence alone. But in cases where documentation was missing, ambiguous, or disputed, the investigation shifted from the archive to the laboratory. The RKO requested a racial

determination from a scientific institute based on an examination of the individual's body.

SHORTLY AFTER PASSAGE of the Law for the Restoration of the Professional
Civil Service, Hanns Schwarz received the first official request for proof of
his paternal ancestry—a letter from a state insurance agency compiling a
registry of medical practitioners. For reasons that are unclear, the query
stalled, three years passed, and Hanns's paternity remained unresolved. In
1936, the RKO opened a formal investigation into his racial ancestry. Its
conclusion two years later contradicted his mother's claim. Archival research
had turned up a document in which a Jewish merchant named Nathan
Schwarz had recognized Hanns as his son when the boy was a toddler. The
news came as a shock to Hanns: Nathan Schwarz was none other than the
brother of his adoptive father Max. If true, not only was Hanns's adoptive
mother his biological aunt (sister of his birth mother), but his adoptive father
was his biological uncle (brother of his newly discovered progenitor). Hanns
claimed never to have heard of Nathan Schwarz or to have received any economic support from him.[30]

His birth mother and Nathan Schwarz were summoned for interviews,
but both denied his paternity of Hanns. Adelheid claimed instead that
Robert Koch, a deceased lawyer and "Aryan," had fathered her child four
decades before and that she was already pregnant when she first met Nathan
Schwarz. For his part, Schwarz claimed that his brother, Max, had persuaded
him to recognize the boy as his son. While the social evidence appeared to
contradict the documentary record, the racial state had no interest in either
social or bureaucratic paternity. It wanted to know Hanns's biological progenitor, who in the face of the contradictory evidence remained uncertain.
The RKO expanded its investigation to include the deceased Robert
Koch, requesting documents and photographs in order to examine this
other possible father.[31]

Hanns's case reveals how two seemingly distinct objectives—revealing
race and establishing paternity—were in practice inseparable. Racial assessment, like paternity assessment, sought the essential, immutable truth of
ancestry on the body. The techniques to reveal these distinct truths were

likewise indistinguishable. Sidestepping the lack of a conclusive marker of biological race, scientists repurposed techniques that up to then had been used to find the father—the analysis of blood groups, fingerprints, eye and hair color, face shape—and directed them toward a new object, Jewishness. The result was neither strictly a paternity test nor a racial assessment but an amalgamation of the two: an analysis of racial paternity.

Paternity science was now immediately relevant to state racial practice. When the Vienna Institute's Josef Weninger applied for funds to continue the Marienfeld study in 1936, an enthusiastic grant reviewer expressed the value of the project in the following way: "The more of these finding[s] that are furnished, the greater our certainty in preparing paternity opinions not only for civil cases but upon requests of the Reich Sippenamt [the Kinship Office]; the latter serving as a basis for decisions on Aryan or non-Aryan descent of extra-marital children, children of adultery, foundlings, and so on."[32] The Marienfeld Project was originally framed as a boon to paternity investigation in the context of family disputes; now it was touted for its value to the RKO's racial determinations.

The two assessments also drew on parallel legal framing. The German law of paternity revolved around the concept of "obvious impossibility," in which the man who had had an intimate relationship with the mother was automatically the father of her child unless such a relationship was obviously impossible. This conceptual vocabulary was imported into racial appraisals, where scientists expressed their assessments of Jewishness and Aryanness in terms of "great or extraordinary probability" or else "a possibility or obvious impossibility."[33] The fact that certain methods and procedures had already gained forensic acceptance for the purposes of paternity determination facilitated their application to racial investigation.[34]

In the 1980s, German geneticist Benno Müller-Hill interviewed scientists who had once worked for the Nazi racial project. As the elderly and usually unrepentant ideologues sipped coffee, nibbled cakes, and shared their recollections, scenarios involving uncertain or spurious fathers recurred with extraordinary frequency in their stories. Paternity was central to state practices of racial determination and therefore to the everyday practice of science in the Third Reich. This was true of ancestry assessments like the one involving Hanns Schwarz, but it could be true in any proceeding where race was at stake. For example, one common defense strategy by Jews accused under the Nuremberg Laws of "race dishonor"—sexual relations with an

Aryan—was to challenge the identity of their father in order to reclassify themselves racially.[35]

Nazis did not invent the association between race and paternity, which is a recurring feature of modern paternity. But they did perfect it. They fetishized paternal uncertainty, which was thought to create opportunities for racial indeterminacy and therefore racial pollution. Because true paternity could be hidden or falsified through illegitimacy, adultery, or adoption, these scenarios presented a specifically racial threat. In the early 1920s, Otto Reche had pioneered his scientific test to find fathers for the fatherless. During that era, a new welfare law brought illegitimate and foster children under closer state control in an effort to improve their well-being and reduce infant mortality. Ten years later, Reche initiated a police registry of illegitimate children in Leipzig in order to track potentially racially foreign individuals living among Germans.[36] Illegitimacy and absent paternity were now important matters not because of their relationship to child welfare but because of their association with racial infiltration.

After 1933, racial paternity assessment boomed. The practice was a seamless outgrowth of earlier kinship science: racial paternity inhabited the same laboratories and was performed by the same experts who had pioneered parentage testing in the 1920s. The Kaiser Wilhelm Institute for Anthropology, Human Heredity, and Eugenics (KWIA) was a prestigious scientific center founded in 1927 that began performing paternity tests for the courts a year later. Its preeminent paternity expert was director Eugen Fischer, one of the most influential race scientists of his generation (as a medical student, Hanns Schwarz had studied anatomy with Fischer). After 1933, the KWIA's ancestry assessments ballooned.[37] Wolfgang Abel, a racial anthropologist internationally known for his research on the genetics of nose shape, joined Fischer as the KWIA's chief paternity assessor. They were assisted by medical technicians who performed blood group tests and by colleagues such as Engelhard Bühler, who studied the genetics of face wrinkles, and the Norwegian Thordar Quelprud, an expert in ear shape. Abel estimated that over the course of the Third Reich, the KWIA completed some 800 ancestry reports for the RKO.[38]

The expansion of racial paternity at the Institute closely tracked political and legal developments. Following the Nuremberg Laws in 1935, demand for the examinations rose dramatically. Legal reforms in 1938 precipitated another spike. By 1939, the Reich Interior and Justice Ministries had tasked eleven institutes of racial biology and an unknown number of individual practitioners with performing racial paternity assessments.[39] By 1942, there were some two dozen, in addition to individual experts.[40]

As racial paternity became part of the workaday repertoires of scientific institutes across the Reich, hundreds of doctors, anthropologists, and lab technicians participated in them. Among them were some of the best known race scientists and racial ideologues of the Third Reich, including Otto Reche and Eugen Fischer as well as Fischer's protégé, Otmar Freiherr von Verschuer. At the University of Frankfurt, von Verschuer had an assistant who worked on racial paternity cases several years before he was dispatched to Auschwitz as camp doctor. That assistant was Josef Mengele.

Further reflecting its institutionalization, racial paternity became part of basic medical education. The three-day course on race science that the KWIA offered to jurists in 1934 included topics like sterilization, degeneration, and race hygiene as well as blood group analysis and anthropological paternity assessment.[41] Abel described an exercise he developed in the introductory lecture course he taught to 800 first-year medical students. He presented four photographs taken from actual ancestry cases and asked the students to identify the correct father and the individual's racial category. "It was cause for great amusement when the wrong combinations were picked out," he mused. The purpose of the exercise was "to educate them to be careful in judging a man by his appearance."[42] The lesson was presumably that racial ancestry assessments were difficult and complex, suited only to a trained expert like Abel. An alternative lesson—that perhaps such assessments were absurd to begin with—seems not to have occurred to him.

As racial paternity boomed, the techniques associated with it became increasingly complex. Where Reche's first somatic examination in the 1920s relied on a series of measurements and evaluations of nineteen somatic traits, later examinations expanded to a hundred traits or more.[43] Researchers continued to search for a single marker of racial paternity—perhaps the nose, Abel's area of expertise, or the ear, studied by his colleague Quelprud. But as long as that holy grail eluded them, they fell back on the kitchen sink method, the analysis of as many traits as possible. As the apotheosis of the "more is better" approach, the German examinations may well qualify as the most complex paternity tests ever performed. But practitioners followed no single, standardized method. Despite the incorporation of racial paternity into medical curricula and attempts by the

RKO to impose standards of practice, scientists tended to develop their own, disparate methods.[44] Little wonder that these assessments, with their fuzzy criteria and impossibly high stakes, would become frequent subjects of contention.

Beyond scientific institutes and medical school classrooms, racial paternity was also the subject of state propaganda. A short film produced by the Reich explained the scientific procedures to a lay audience. It narrates the story of Frau Weber, an anxious mother who has become convinced that her son was accidentally switched at birth and consults a kindly doctor for assistance. He conducts a full hereditary work-up on Frau Weber, her husband, and "little Georg," drawing blood, studying hair and eye color, measuring ears, nostrils, and heads, recording fingerprints. The doctor is gentle and authoritative, the nurses efficient and capable. As he works, he explains each procedure, asking the apprehensive mother, and by extension the film's audience, "Are you beginning to trust us now?"

Even as the film normalizes techniques of hereditary assessment, portraying them as scientifically sound and socially useful, it evacuates those methods of all racial and political content. Scientific expertise resolves a private, familial drama (one that, the viewer suspects, may have its origins in the overactive imagination of Frau Weber). And predictably, the story ends on a happy note: the scientist concludes that "little Georg" is the couple's biological son after all. The benign story could not contrast more dramatically with the life and death procedures then being conducted in genetic institutes throughout the Reich. The film was made in June 1944 as the deportations and mass murders were in full swing.[45]

THE RACIAL PROJECT of the Third Reich required not just normalizing paternity science but radically reforming paternity law. In 1938, the Reich Law for the Alteration and Amendment of Regulations Pertaining to Family Law dramatically expanded the grounds for challenging paternity. Until then genetic testing had been primarily used in paternity proceedings like that of Hanns Schwarz, involving illegitimate people and unmarried parents. The 1938 law opened up a new terrain of contention: legitimate paternity. The purpose of the reform was to clarify "true" racial parentage in cases where the legal fiction of marital legitimacy might have otherwise obscured it.[46]

Traditionally German law extended careful protections to the paternity of the child born in wedlock. The child's legitimacy could only be refuted by the husband and only within one year of its birth.[47] The 1938 reform stripped away the time limit so that husbands could refute their paternity whenever they learned of a child's supposedly adulterous origin. More radically, it empowered wives, children, and the state itself, as represented by a public prosecutor, to mount such challenges. The paternity of a legitimate child could now be contested whenever it was deemed in the interests of the "volk" or of the child to do so.

"Interest" in this context referred to racial status. As the law's coauthor explained, if the child born to an Aryan mother married to a Jewish man had been secretly fathered by an Aryan paramour, that child would have an interest in establishing her biological father in order to establish her more favorable status as a "pure" Aryan. Conversely, it was in the interest of the state to expose the mixed status of a child "born to . . . a married Aryan couple" whose true father was not the Aryan husband but a Jewish paramour.[48] The law located racial indeterminacy in uncertain paternity and attempted to resolve that indeterminacy by making paternity more certain. It mandated "genealogical and racial examinations" as the method for doing so.

The presumption of marital legitimacy was traditionally robust, and sometimes irrefutable, in Western law. The Nazi state repudiated that presumption, wresting biological paternity from the fictions of marital decorum. It sacrificed marriage and social propriety on the altar of racial purity and thus charted a radical departure from both German and global legal tradition. Another departure was the law's introduction of compulsion in paternity proceedings. Wherever physical evidence was accepted in paternity disputes, vigorous debates arose over whether courts could force minors, putative progenitors, or other family members to submit to blood tests or bodily examinations, particularly in civil proceedings. In Germany as elsewhere, the traditional consensus was that they could not.[49] The 1938 law changed that, obligating "parties and witnesses" to submit to tests of body and blood.

Yet a law that threw open the gates to paternity contention in order to clarify true racial identity could cut both ways. It permitted the state to bring suits, challenging the legitimacy of children in order to expose "true" Jews hidden behind the shroud of marital paternity. But it also opened up an opportunity for Jews, "full" and "partial," to change their racial status or that of their children by contesting the identity of the father. This is precisely what they did. After 1938, paternity suits flooded civil courts, from Munich to Vienna to Hamburg.[50] Women came forward to repudiate the paternity of Jewish husbands; children filed suits to challenge the paternity of Jewish fathers. The Reich's victims took advantage of the logic of racial paternity to save themselves and their family members.

Perhaps the most ironic story in this regard is that of Friedrich Keiter. Keiter was a racial anthropologist and avid anti-Semite associated with the Institutes for Racial Biology in Hamburg and Würzburg. He also happened to have a Jewish grandfather. In 1938, Keiter's grandmother declared that her husband, Keiter's grandfather, was not the father of her children, whose true biological progenitor was Aryan. Keiter underwent an anthropological assessment at the KWIA whose conclusion supported the grandmother's claim. Thanks to the examination, Keiter was able to claim his status as a pure Aryan and upstanding Nazi. He likely went on to perform on others the very racial paternity examination he himself had undergone.[51]

Filing a paternity suit required hiring a lawyer and paying for an expensive anthropological assessment. But for those who could afford them, the petitions were remarkably successful. In a majority of cases in Hamburg, for example, petitioners succeeded in "upgrading" their racial status.[52] Disputing paternity became common enough that officials became wise to the subterfuge, warning that "only in an extremely small percentage of cases" were the challenges bona fide.[53]

As Jews' status deteriorated, racial paternity petitions increased. They spiked with the Nuremberg Laws in 1935 and again after 1938, because of the marital paternity reform but also because of the pogroms. A month after Kristallnacht, the Institute of Racial Biology at the University of Hamburg complained about the surge of requests for racial paternity examinations. They rose yet again in 1941–1942, as the deportations got under way.[54] Clearly Jews strategically used paternity proceedings to stave off first discrimination and then death. The fact that the genetic institutes could not keep up with the surging demand worked to petitioners' advantage. Family members, to the Gestapo's chagrin, often could not be deported as long as an inquiry was in process. Incredibly, the courts sometimes requested that fathers who had already been deported be temporarily released from the camps to offer statements or to undergo paternity examinations.[55]

Contesting paternity was ultimately a resistance strategy based on the family rather than the community. Individuals sought to racially reclassify themselves and certain family members but left intact the essential logic of uncertain paternity and racial classification. The procedure was fraught for moral reasons as well. Exhaustive physical inspections could be humiliating (recall the Marienfeld villagers' resistance to full-body examination), perhaps all the more so for middle-class people in the context of mortifying legal proceedings. After all, challenging paternity required that a wife "admit" to adultery, that a husband self-identify as a cuckold, that a child disavow a father or grandfather and impugn the morality of a mother or grandmother.

Conversely, the matrilineality of Judaism may have helped some people to accept this bitter remedy. According to Jewish law, Jewish identity is passed from mother to child. Thus, when Jews and *mischlinge* impugned the paternity of Jewish fathers and grandfathers, they did not divest themselves of their Jewishness. Perhaps this fact provided some small comfort in what must otherwise have been an ignominious experience.

It is difficult to know how Jews experienced or understood this strategy because it appears to have been expunged from collective memories of the era.[56] A possible clue comes from Polish serologist Ludwik Hirszfeld, whose research on professors in Heidelberg and soldiers in Salonika first established the heredity of blood groups and helped usher in serological paternity tests. In 1943, Hirszfeld, a Catholic of Jewish ancestry, was hiding in the Polish countryside under an assumed name, having narrowly escaped the horrors of the Warsaw ghetto. He began to compulsively dictate his memoirs to his wife.[57] He dwelled on happier times, including the immense fulfillment of his blood group work a quarter century earlier. "In one respect, though, this research has been unpleasant to me," he observed, returning to the present. "The Germans use it for racial purposes. If a child from a mixed marriage proves that the husband of its mother is not its father, it receives all citizen rights, probably as a reward for not hesitating to slander his mother to personal advantage." Hirszfeld condemned not just the Germans with their "racial purposes" but the Jewish victims who "slandered their mothers." Yet the "personal advantage" he disparaged was a desperate and degrading attempt at survival.[58]

OTTO RECHE WAS INITIALLY pleased with the remarkable expansion of the method that he had pioneered. For the scientists and institutions providing racial paternity assessments, the dramatic increase and new importance of these procedures was a potential windfall. Their role in state-sponsored racial "cleansing" conferred on them great political importance. "In these times of rebuilding our völkisch state," the KWIA's Eugen Fischer noted in 1935, such tasks are given "our highest priority." The political importance of racial paternity gave scientists access to public funds and additional personnel, and they generated fees that padded institutional budgets.[59]

But the relentless growth of testing also had a downside. As Fischer soon lamented, the laborious expert reports took time away from basic research. Public funding failed to materialize, at least in the quantities anticipated, and Reche groused that only certain institutes and individuals benefited.[60] Scientists debated who should bear the cost of the assessments, particularly when they were requested not by state agencies but by Jews seeking to improve their status. The examinations were expensive: according to an estimate by the Interior Ministry in 1936, they averaged 90 marks—about a month's salary for an average worker.[61] Fischer asserted that "it is unfair that our fellow countrymen, who happen to have the means, should enjoy the benefit of being able to obtain an expert opinion certifying them to be Aryan, while our poor fellow countrymen are denied this opportunity." But others, like Theodor Mollison, chair of anthropology at the University of Munich, believed Jewish petitioners were exploiting the serious business of racial paternity assessment. "It is not advisable to provide such time-consuming investigation free for those who claim Aryan origins when they know they are not entitled to do so."[62] In what may well have been a self-serving claim in an interview decades later, anthropologist Wolfgang Abel asserted that the KWIA did not benefit financially from the reports. While other institutes attempted to profit from them, charging 700 marks or more, "Fischer and I refused any payment" from modest petitioners, charging only for the costs of paper and photographs, some 20 marks, or sometimes nothing at all.[63] Only belatedly, in 1942, did the Ministry of the Interior attempt to resolve the issue by setting prices for the analyses: 175 marks for an ancestry examination involving three people; 220 marks for four people. Each additional person cost 30 marks.[64]

Ultimately, some scientists came to regard the racial paternity procedures as more burden than boon to their institutes. Abel claimed that neither he nor Fischer wanted to perform the "Jewish expert reports" and that Fischer repeatedly offered to train officials within the RKO to do them instead.[65]

In 1939, Fischer complained that the KWIA received four times as many evaluations as it could perform, which also suggested that many more people sought racial reclassifications than obtained them.[66] With the outbreak of war, laboratory personnel were called to military service, staff shortages worsened, and backlogs grew. In March 1940, Reche, proud father of the somatic paternity test, disavowed his spawn, declaring he would only perform new assessments "in exceptional and very urgent cases."[67]

Scientists' inability to meet the demand for expert opinions was deeply ironic because in part it stemmed from the purging of Jewish practitioners from science and medicine. In fact, with the obvious exception of Otto Reche, most of the pioneers of early paternity science were of Jewish descent. They included serologist Fritz Schiff, who first introduced blood group testing into courts for paternity establishment; forensic expert Georg Strassmann, Schiff's collaborator; the mathematician Felix Bernstein, who correctly identified the inheritance pattern of blood groups; Ludwik Hirszfeld, codiscoverer of the hereditary transmission of blood groups; Karl Landsteiner, who won the Nobel Prize for his discovery of blood types and advocated for their use in the courts; Heinrich Poll, who published on the use of fingerprints in paternity cases; and Leone Lattes, the Italian serologist who advocated for the expansion of paternity testing. The fact that these individuals and their families did not necessarily practice Judaism or identify as Jews—Landsteiner and Hirszfeld were Catholic converts; Strassmann had been baptized by his mother—was of course irrelevant in the face of a definition of Jewishness based on blood and ancestry. All of these experts (save Landsteiner, who had moved to the United States decades earlier) would have been obvious experts to perform the ancestry tests so in demand after 1933. Instead all of them were removed from their professional positions.

Fritz Schiff's experience was especially poignant. Just as racial paternity exploded, his own career was plunged into crisis. Even before the Law for the Restoration of the Professional Civil Service began purging Jews from the medical profession, Schiff was deprived of his credentials as a forensic court expert. It was a severe blow to a professional life dedicated not only to serological research but to its practical application in court. The loss of the credentials was also financially calamitous. In a 1933 letter to Karl Landsteiner in New York, a mutual friend, the Danish doctor Hermann Nielsen, described Schiff's struggle to support his extended household of seven without his forensic credential: "As the vast majority of his income derives herefrom, he is now almost without money for his living." Unable to maintain his apartment, Schiff sought to move his family to some cheap rooms. Nielsen had taken Schiff's eldest son, eleven-year-old Hans Wolfgang, to Denmark, apparently to relieve the strapped household of a member. "Our friend, who sees his whole existence threatened, cannot write directly to you. Letters are opened and censured by the Nazis."[68]

Landsteiner and others mobilized on Schiff's behalf. Possible job opportunities emerged in Mexico, Switzerland, and Canada but fell through one by one. Landsteiner wrote back to Nielsen: "I cannot judge how unf[av]orable the conditions are for Dr. Schiff in Berlin and consequently it is difficult for me to advise him," but he counseled Schiff to "think the matter over carefully" before abandoning Berlin for a position at the Institute of Hygiene in Mexico City. "The drawback of the appointment in question lies possibly in the political uncertainty which obtains in countries like Mexico."[69] In a searing irony, Nazi Germany was still considered preferable to a "country like Mexico." The following year, 1934, a possible position that had emerged at Columbia University evaporated when the Rockefeller Foundation, which was approached for financial support, declined to contribute. Ironically, Schiff's role as the paterfamilias of a large household, which made his economic situation in Berlin all the more dire, proved a liability in his attempts to emigrate. "He has a large family, which would make it very difficult for him financially," wrote one benefactor.[70]

Schiff remained in Berlin. In February 1935, he wrote Landsteiner that his research had stagnated due to lack of lab personnel. Six months later, thanks to the Nuremberg Laws, he lost his university appointment. Thereafter his teaching certification was revoked entirely. His laboratory limped along, hobbled by lack of personnel and funds, and his scientific productivity plummeted.[71]

Finally, thanks to his contacts abroad and the wide international regard for his paternity work, Schiff landed a plum position as Chief of Bacteriology at Beth Israel Hospital in New York. In August 1936, Schiff, his wife Hilda, and their three sons boarded a ship. A photograph shows Fritz and Hilda on the deck of the SS Champlain, windblown and smiling, en route across the Atlantic.[72]

The timing of their departure was ironic. The very week Schiff left Europe, the Ministry of the Interior issued a decree recognizing the evidentiary value of blood testing in courts across the Reich. The decree gave formal state recognition to a method that had become standard practice in courts over the preceding decade. This was Schiff's blood test, the one he had first proposed to the Berlin forensics society in 1924, the one he had introduced into forensic practice, the one he had fought for in both courts of law and the court of public opinion. It was the test that had revolutionized the science of paternity in Germany and globally. The Nazis had driven Fritz Schiff out at the very moment that they were putting his life's work to use for their own nefarious ends.

IN THE YEARS FRITZ SCHIFF was seeking a way out, Hanns Schwarz was trying to fly under the radar. He had lost his job at a sanatorium but managed to keep his private practice. The question of his ancestry remained unresolved, a fact he skillfully exploited to his advantage. When he was drafted in 1940 for military service, he responded that because he was Jewish he could not serve. He then promptly wrote to the medical association that because of his military deployment, the state need not supervise his medical practice.[73] Meanwhile, his case continued to creep through the Reich's kinship bureaucracy. The backlog in ancestry assessments that maddened Otto Reche was, for the victims of these assessments, a lifesaving boon.

Finally, in June 1942, the RKO summoned Schwarz for a full racial paternity examination. The procedure, for which he paid 200 marks, was conducted at the Polyclinic for Hereditary and Racial Care, one of Berlin's busiest forensic labs, but the final report was authored by the RKO's own director, Kurt Mayer.[74] A rabid Nazi, Mayer was neither a doctor or scientist; he had a doctorate in history.[75]

His report, issued in the summer of 1942 in the shadow of the deportations, concluded Schwarz's paternity was indeterminate and his racial status ambiguous but also declared Schwarz was "racially mixed" and "probably Jewish."[76] Schwarz's lawyer distanced himself from his doomed client, but Schwarz quickly hired new counsel, appealed the finding, and was granted a second exam. More months slid by. A second exam was conducted, this time by Fred Dubitscher, the Polyclinic's director, who was active on various state health agencies and an enthusiastic proponent of eugenic sterilization.

He was also, it so happened, a psychiatrist, and the two men—examiner and subject—knew each other from professional circles.[77]

What must it have been like for Hanns Schwarz to stand unclothed before this colleague, whose examination of his body would determine whether he, his wife, and his daughters lived or died? As Dubitscher took the laborious measurements and dictated them to a secretary in the room with them, he remained aloof and offered no acknowledgment that he knew his subject.

But as soon as the exam was over and the secretary left the room, Dubitscher relaxed, expressed sympathy for his colleague, and offered to lend him 1,000 marks. Schwarz, whose professional opportunities were by then severely constrained, took him up on the offer.[78] Dubitscher, whose career otherwise reflects a steadfast commitment to the project of racial cleansing, may have purposefully saved the life of his colleague. His report retained large portions of Mayer's unfavorable assessment but came to a diametrically opposed conclusion. Because of Spanish ancestry on the paternal side, it concluded, Schwarz might look foreign but he was not necessarily Jewish. Dubitscher pronounced him of predominantly "German blood"—the fateful phrase that would save Schwarz and his family.[79]

It seems likely Dubitscher wrote a purposefully exculpatory report on his colleague's behalf. But how could two experts' examinations of the same man reasonably support two diametrically opposed conclusions? In fact, such contention was common, among scientists and between scientists and Reich officials.[80] One reason was that given the extraordinarily high stakes, corruption was endemic in racial paternity proceedings. Money changed hands; photographs were switched in files. Wolfgang Abel of the KWIA acknowledged the problem in order to set himself above it. He proudly remembered that he and Fischer gained reputations as "incorruptible." In one instance, they even ran afoul of authorities when they ejected the head of the German paper industry from the KWIA after the man tried to bribe them for a favorable ancestry report. "Leniency," declared Abel, "was not a scientific concept."[81] If for some Jewish petitioners salvation came through bribery, for Hanns Schwarz it came through professional comity. Other examiners may have crafted reports out of quiet sympathy for their subjects. One paternity tester at the Vienna Institute was denounced by a lawyer of the Supreme Court for consistently finding in favor of examinees.[82]

But the principle reason why two reports could reach opposing findings was not corruption, fraud, or sympathy. It was because of the fundamental lack of consensus surrounding racial paternity in the first place. The assessments were consistently vague in their observations and generic in their conclusions. One seasoned paternity tester, Hans Koopmann of the Hamburg health authority, concluded his reports with observations like, "according to intuitive overall impression, it is considerably more likely that the legal father is not the progenitor."[83] There were no obvious standards by which to judge a good assessment from a bad one. Wolfgang Abel reminisced that he had been recognized as "the expert in the field of human facial and cranial shapes" but then cryptically added, "no one could check my decisions, and so it was not usual for other institutes to write expert reports on the same cases. We made ours, they made theirs."[84] He tried to tout his status as a peerless expert but inadvertently revealed the lack of expert consensus on racial paternity.

Conflicts among experts were common. In a 1940 case, Fritz Lenz, a leading racial scientist at the KWIA and longtime collaborator of Fischer, refused to provide an expert assessment for the Cologne district court on the basis of photographs alone. The Ministry of Justice questioned Lenz's position, querying five other racial anthropologists, including Fischer and Reche, all of whom contradicted him, asserting that photographs alone were indeed a valid basis for paternity assessment.[85] It is striking that years after they had begun, no consensus existed among leading paternity scientists on this basic methodological question.

In spite of the lack of scientific consensus, somatic methods achieved an authority in Nazi Germany that was historically unprecedented. Brazilian odontologist Luiz Silva and oscillophore inventor Albert Abrams were no less wildly confident in their methods than Otto Reche and his colleagues. But Silva and Abrams were mavericks; in Germany, Reche was a torchbearer. Even as German experts argued about basic methods, they made far-reaching claims for paternity science. They were afforded positions of prestige and authority, and their expertise was accorded unprecedented power and significance in the context of a state project. What distinguishes paternity science under National Socialism, then, was not the science but the politics. Only here did the state recognize somatic paternity assessment and, for the first time, institutionalize scientific methods for the purpose of mass ancestry determinations. The paternity test became a routine procedure and an instrument of public policy.

In 1939, the Supreme Court (Reichsgericht) declared that scientific methods had developed to the point that they could positively identify the illegitimate father. Up to that moment, the global scientific and legal consensus held that blood group tests could definitively exclude some impossible fathers; that in rare instances hereditary pathologies might strongly indicate kinship between a man and a child; but that science had not yet unraveled the fundamental mystery of the father. Departing radically from that consensus, the German Supreme Court's ruling held that paternal identity could indeed be positively revealed.[86] Though many racial scientists no doubt agreed with this conclusion, the verdict expressed as much a political contention as a scientific one.

THE POLITICAL LIFE of paternity under National Socialism appears historically anomalous. Nowhere, not even in the present, when DNA testing could be seen to justify it, has a state so resolutely embraced a definition of paternity that is so emphatically biological. Nowhere has it so cleanly cleaved biological paternity from its legal, economic, or social foundations. Nowhere has a society opened the doors so widely to scientific modes of fixing paternity or invited so decisive a challenge to notions of familial honor, sexual morality, and marital order. Nowhere has a critical mass of scientists believed so uncritically in the power of their expertise to resolve the problem of the father. Nowhere did paternity testing become a matter of life and death.

Yet the unfathomable depravities committed in the name of race science should not obscure what was emblematic rather than exceptional about racial paternity in the Third Reich. Like state authorities for centuries, the Nazis regulated paternity so as to protect public and private interests. What differed was how they defined those interests. They privileged racial truth and racial purity over the protection of marriage, children, honor, propriety, morality, or the public purse. While the Nazi state was especially heavy-handed in imposing its vision of paternity, like states everywhere it defined paternity in the service of political, moral, economic, or, as in this case, racial goals.

The Nazis likewise shared the age-old presumption that paternity was intrinsically uncertain and therefore perpetually vulnerable to concealment or suppression. Their quest for new modes of scientific and legal scrutiny to find the father was of course a matter of broad transatlantic interest. The almost complete absence of maternity in the Nazis' kinship revolution is likewise emblematic rather than exceptional. Nazi definitions of race depended on how many Jewish parents or grandparents one had; it did not matter whether the Jew was a mother or father, a grandmother or a grandfather. Yet aside from the relatively rare scenario in which petitioners alleged they were foundlings or had been adopted, in which case potentially neither biological progenitor was certain, challenges to racial status almost always involved a contested father (or grandfather). The Nazis worried about stealthy Jewish progenitors hidden in Aryan bloodlines; Jews themselves redrew family trees to introduce serendipitous Aryan begetters. Either way, maternal identity was almost never at issue because of the assumption that it was always certain. The Nazis radically remade the science and law of paternity, yet they left unquestioned abiding assumptions about maternal certainty and paternal uncertainty.

The reason that uncertain paternity mattered, of course, was that it created racial indeterminacy: the true father must be known to order to determine the true race of his offspring. While this was a very different motive for finding the father than in, say, an inheritance dispute or hospital baby swap, in a key respect it too was more emblematic than exceptional. After all, modern paternity was everywhere imbued with racial meanings and agendas. Paternity and race came together in the crystallography of Reichert, the oscillophore of Albert Abrams, the sero-alchemy of Manoiloff and Poliakowa, and the somatic test of Lehmann Nitsche. In all these instances, racial knowledge helped to reveal paternity. For the Nazis, paternity helped to reveal race. By identifying an individual's "true father," paternity assessment exposed his or her "true race." In terms of both scientific technique and legal logic, the determination of parentage and race coalesced into a single practice. Racial paternity under National Socialism was the apotheosis of the transatlantic race–paternity nexus.

This blind pursuit of the chimera of racial paternity had a paradoxical outcome. So convinced were lawmakers, jurists, and scientists of the truth and power of their scientific methods that they recklessly threw open to contention kin relations that were typically considered settled. Their hubris created openings for the persecuted: Jews responded by filling the courts with thousands of petitions challenging the identity of their fathers. The fetishization of biological truth paradoxically destabilized it. Uncertainty and instability exploded at the very moment that the timeless quest for the father was supposedly resolved. In this regard, paternity in the Third Reich anticipated, in a curious way, the DNA future.

FOLLOWING THE FAVORABLE RKO decision, Hanns Schwarz quickly moved out of Berlin and settled with his family in a provincial town outside Berlin. His daughters were able to return to school, and the family attempted to integrate in the community after spending a decade under a looming shadow of danger. Fearful that neighbors might be aware of their story, they kept a low profile: "I simply Aryanized my entire family, just without the RKO."[87]

By then, most other Jewish doctors were long gone. Of those associated with paternity science, Felix Bernstein, the mathematician, left for New York. Georg Strassmann, Schiff's forensic collaborator, emigrated to Massachusetts, fleeing not only the usual persecutions but also the terrifying prospect that state authorities would seize his adoptive son, whose birth parents were "Aryans" (Nazi law prohibited the adoption of Aryan children by Jewish parents). Fingerprint expert Heinrich Poll, unable to find a position in the United States, went to Sweden but died of a heart attack a week after emigrating. As the racial laws descended on Italy, Schiff's friend and collaborator Leone Lattes left for Argentina, where he continued to write about and perform paternity tests. Josef Weninger, who had inherited the directorship of the Austrian Anthropological Institute from Otto Reche, found his career blocked because his wife and fellow scientist, Margarete Weninger, was Jewish. He attempted, unsuccessfully, to obtain a position in Britain on the basis of his reputation as a paternity expert.[88]

As for Fritz Schiff, he arrived in New York in 1936 with his wife and three sons. After years of uncertainty and hardship, the family was able to leave Germany as middle-class immigrants. They brought with them their furniture, objects that Schiff had collected during his travels in the Levant, and a collection of Shakespeare plays. Following some neighbors who had lived in the same building in Berlin, they moved to the suburb of New

Rochelle, to a two-story house. Schiff's aged mother had not wanted to leave Berlin, and his sister remained behind to care for her. His new job pushed him into other research, but he was able to resume his forensic work too, performing paternity tests in New York courts, where they had only recently been adopted. The happy resolution took a tragic turn when, less than four years after arriving, the fifty-one-year-old Schiff died suddenly from complications in surgery. Leone Lattes penned an obituary for Schiff from his own exile in Argentina: his death was especially tragic given that Schiff had so recently overcome such "painful trials" and "achieved the possibility . . . of dedicating himself once again to the problems that were the reason and purpose of his life."[89] The man who had pioneered a method to match fathers and children now left behind three sons; once again Schiff's colleagues rallied to support the family.[90] Sometime in 1944, Schiff's sister and mother, who had remained in Berlin, perished in concentration camps.

We know little about how the experts who had pioneered ancestry testing felt about the fact that their scientific work had been hijacked for such barbarous ends. Polish scientist Ludwik Hirszfeld was one of few who commented. Hidden in the Polish countryside, he observed simply, "what can be done? This is one of the few cases where scientific discoveries are used for unworthy purposes."[91] Hirszfeld and his wife would survive the war but lost their only daughter in the Warsaw ghetto.

And what about their onetime colleagues, the geneticists, racial anthropologists, and other scientists and doctors who dedicated themselves to the "unworthy purposes" of the Third Reich? After the war, most of them quietly returned to their professional lives. A few were deprived of their academic appointments and opportunities for research, but with the notorious exception of Josef Mengele, who disappeared to Argentina, every individual discussed here was able to resume a professional career in some form. All of this is well known. Less appreciated is the central role that paternity testing played in race scientists' transition to postwar life.

With the fall of the Third Reich's racial state, the political enterprise of racial paternity collapsed, but the war created new demands for parentage testing. The dislocations of armed conflict and the separation of children from parents created scenarios in which kinship was in doubt, as did the absence of soldier-husbands from home.[92] The techniques that during the

Third Reich had been harnessed to the project of identifying Jews now returned to their earlier function of identifying kin and reuniting families.

Denazified race scientists did not just navigate this transition; they capitalized on it. In their work consulting for courts, in universities, and in private institutes, paternity testing provided a convenient refuge for genetic agendas and a socially and politically palatable cover for continued research on heredity. Providing expert paternity reports to civil courts also proved an excellent way for ex-Nazis to earn a livelihood. A long list of racial ideologues and anti-Semites, sterilizers and genocidal collaborators, now reinvented themselves as paternity testers. In Vienna, Dora-Maria Kahlich, Karl Tuppa, and Josef Wastl, suspended from the Institute of Anthropology because of their scientific collaboration with the Third Reich, found a new professional calling that brought them full circle back to activities the Institute first embraced in the late 1920s. Into the 1960s, they conducted research on paternity testing and authored forensic reports for Vienna's courts.[93] The notorious Otmar Freiherr von Verschuer, Eugen Fischer's successor at the KWIA and Josef Mengele's advisor, was banned from the KWIA but later opened his own private institute, where he pursued genetic research, funded in part by his forensic paternity reports.[94] Friedrich Keiter, the race scientist whose grandmother had Aryanized him by claiming her children were fathered by an Aryan lover and not her Jewish husband, wrote in the 1960s with apparently no trace of irony about the great social benefits of paternity testing: "to free spouses from doubts about infidelity may restore the natural family relations to the great benefit of children."[95] The list goes on: Peter Kramp, Wolfgang Lehmann, Fritz Lenz, Theodor Mollison, Heinrich Schade, Johann Schauble, and probably dozens of others who had dedicated their scientific skills to racial paternity under the Nazis found postwar vocations conducting research on paternity assessment and providing forensic opinions to the court. In the early 1950s, the German Anthropological Society provided courts with a list of qualified paternity specialists, and special academic meetings were devoted to the topic. Some experts touted their advanced scientific methods to the public and to colleagues in the United States, where the science of somatic paternity assessment was "almost entirely unknown." In short, paternity science in Germany in the 1950s and 1960s was a well-defined field—one populated by former Nazis and collaborators.[96]

The history of paternity under the Nazis remains largely unacknowl-
edged. It is not surprising that ex-Nazis who took up paternity research
after the war conveniently expunged the earlier political uses of their techniques; the amnesia of the Reich's scientific handmaidens is well docu-
mented. More noteworthy is the fact that paternity experts elsewhere in
the world also erased this history. The international scientific literature
praised the advanced state of paternity science in German-speaking Europe
but rarely acknowledged the heinous life of racial paternity under the Third
Reich. Perhaps scientists who sought the expansion of paternity science in
their own countries were reluctant to taint the field.[97]

And finally there is Otto Reche. Arrested after the war, Reche was interned, subject to the denazification process, and exonerated. He was one
of very few race scientists permanently barred from academic positions, although he continued to collect professional accolades.[98] Indeed, the "Nestor"
of German anthropology, as the press called him, referring to a wise and
venerable king, found ways to return to the somatic craft he had cultivated
since adolescence.[99] Over four decades, Reche's professional career had gone
full circle. Having pioneered the first anthropological method of paternity
testing in German-speaking Europe in the 1920s, he had participated in
the promotion of his method for racial ends and become an active spokesman
for racial paternity in the Third Reich.[100] In the 1950s and 1960s, he brought
his methods back to their original application, finding errant fathers and
perjuring mothers for German and Austrian courts. His hubris was untem-
pered by the events he had witnessed and helped to propel: "our 'data' do
not constitute average 'evidence,' the value of which is open to dispute," he
declared of his paternity method. Rather, "our data are . . . guarantees that
were written by Nature itself."[101] Reche believed he had solved the millennial
quest for the father.

Hanns Schwarz had a different assessment. Decades after his brush with
death in the examining room of the Polyclinic for Hereditary and Racial
Care, he observed: "This examination, if one can call it that, of a doctor by
a doctor who is not supposed to diagnose a disease or provide therapeutic
aid but has the gruesome task of providing an expert opinion on racial belonging that is connected to either life or death—such an examination is
inhuman, impossible, and has feet of clay."[102]

7

TO THE WHITE HUSBAND A BLACK BABY

> Where the bull and cow are both milk-white,
> They never do beget a coal-black calf.

—Shakespeare, *Titus Andronicus*, Act V, Scene 1

IN OCTOBER 1945, five months after the Allies declared victory in Europe, a housewife from a small town outside Pisa named Quinta Orsini gave birth to her second child. She named the baby boy Antonio di Remo after her husband, Remo Cipolli. In retrospect, it was a hopeful name, perhaps a kind of charm. When her husband went to the hospital to visit his son, he was immediately taken aback by the baby's appearance. Though both husband and wife were white, "even an inexpert eye" could see the child had "more black than white." Orsini and her mother tried to reassure the skeptical father that babies were sometimes born that way—Orsini's nickname was "Bruna," brown, because of her dark complexion—but Cipolli was unconvinced. He called in a doctor, who pronounced the infant to be "negroid" and confirmed what Remo now suspected, that the child must have been sired by a black father. Baby Antonio was recorded in the civil registry as the legitimate son of Remo Cipolli and given his surname, but Cipolli filed suit against his wife for adultery, conjugal separation, and repudiation of paternity. The infant was sent to a local orphanage.[1]

The previous year, Pisa and the surrounding region had witnessed the final bitter stages of the war, as Allied troops struggled to push the Germans northward out of the Italian peninsula. Instrumental to this effort were the African American troops of the Ninety-Second Infantry Division, the so-called Buffalo Soldiers of the segregated U.S. Army. After the Allies liberated Pisa in September 1944, the region became a vast supply depot for the U.S. Army, and the troops settled in as occupiers.

In the investigation initiated by Cipolli's suit, two stories emerged. One was the version Orsini told the court. One day a few months after the liberation, she was in the house of her sister, who did laundry and ironing for the GIs, when two African American soldiers arrived to pick up their clothes. Realizing she was alone, they overpowered and raped her as she fainted from fright. Afterward she kept the traumatic secret even from her husband and family. The second version of the story began circulating in the neighborhood soon after Orsini gave birth. The home of her sister had a notorious reputation as a place where Italian women consorted with black GIs, and Cipolli had prohibited his wife from going there. But the neighbors had seen Orsini fraternizing with soldiers, and she was even reputed to have a boyfriend who had gone around showing off a photo she had given him. A witness testified to hearing the couple arguing about Orsini's pregnancy. Another said she had confided she wanted an abortion.

Over the next few years, as Remo Cipolli's suit against his wife and her child wended its way through the courts, in Rome an assembly was busy writing a new constitution for the post-Fascist republic. As the delegates hammered out the document, they spent considerable time contemplating the importance of the family to the new nation. On this basic point, the fractious delegates, which included Communists and Socialists as well as the Catholic Christian Democrats, could agree. "The State is not just a group of buildings, ministers, bridges, roads, and judicial or legislative chambers," one observed. "The State is before all else an aggregate of families living on the same soil, connected by ties of affection, trust, culture, and domestic traditions."[2] As such, the lawmakers imagined themselves as architects of a "new Italian family," one that would "provide a concrete and effective moral foundation" for the new Italian nation.[3]

The challenges of this new Italian family were brought dramatically to the fore by couples like Quinta Orsini and Remo Cipolli. A recurring allegation, repeated in the Constitutional Assembly and across other ambits of national life, was that years of war and occupation had produced a rash of female adultery. In a scenario "that nowadays happens everywhere in Italy," according to a common refrain, veterans and prisoners of war returned home after years of separation and suffering only to find "the family increased, perhaps with children of another color."[4] If the children of the occupation were problematic, so-called *mulattini* like baby Antonio provoked an especially tortured public reckoning. The hundreds and perhaps thousands of biracial children born in these years as a result of Italian women's relationships with nonwhite soldiers were regarded as conspicuous symbols of their mothers' moral degradation and the nation's defeat in the war.

The precise circumstances of Antonio's conception might have been in dispute, but the basic fact of his adulterous paternity seemed irrefutable. After all, it was manifest in the dark skin and curly hair of the baby himself. Accordingly, the court found Orsini guilty of adultery, a crime that for women (but not men) was punishable by up to a year in prison.[5] She was saved from jail thanks to an amnesty for postwar crimes promulgated two weeks before her conviction. The fact that Orsini was pardoned by a political amnesty law seems fitting. After all, illicit sex with a nonwhite occupation soldier was symbolically, if not formally, a political transgression as much as a moral one.

If the adultery conviction was straightforward, Remo Cipolli's attempt to repudiate legal paternity of baby Antonio turned out to be more complicated. In August 1947, shortly after the Constitutional Assembly concluded its lively discussions about the Italian family, a civil court in Pisa reached what to many observers was an astonishing conclusion: because he was Quinta Orsini's husband, Remo Cipolli was the father of the child to which she had given birth. The court acknowledged that to the layman the decision "might appear absurd," but the "rigorous logic of the law" was intransigent.[6] His name would remain on the baby's birth certificate, and he was invested with all the rights and responsibilities that fatherhood entailed.

Cipolli had run headlong into the granitic weight of the so-called presumption of legitimacy. This was the principle derived from Roman law that *pater est quem nuptiae demonstrant*—the husband is, legally, the father of his wife's children. If paternity was inherently uncertain, monogamous marriage made it unequivocal. Marriage, in short, made the father. Here was a definition of paternity that was social and legal, defined by matrimony rather than by biological procreation. The principle was enshrined in the 1942

Italian civil code, which featured strong protections for marriage and the legitimate family that followed in the nineteenth-century Napoleonic vein and reflected the heavy hand of the Vatican (the code's overtly fascist elements, such as the prohibition of marriage between Jews and non-Jews, had been mostly expunged after the fall of the regime).[7] The marital presumption was consistent with the code's conservative orientation, but it was by no means unusual: it had been a feature of diverse legal traditions historically.

The story of how a white husband became the father of his white wife's black baby in postwar Italy suggests a new twist in the history of modern paternity. In a nascent republic imagined as an aggregate of families, where "the State and the Nation take the sweet name of Patria, land of the fathers," as one delegate to the Constitutional Assembly put it, the quest for the father was deeply politicized.[8] Italians debated who the father was and how and when he could be known in the long-winded soliloquies of the Constitutional Assembly, in newspapers and film, in arcane juridical discussions, citizen petitions, and legislative bills. While doctors and scientists also chimed in, these public exchanges reveal the limits of scientific modes of knowing filiation.

The notorious case of "the little Moor of Pisa" became part of this broader reckoning with paternity. But in contrast to many stories of modern paternity, this one did not revolve around an unknown father whom nature had concealed and science was called on to identify. To observers, and to Remo Cipolli himself, nature itself had already revealed the truth of baby Antonio's (non)paternity. The problem was the law's disregard for that truth.

In Italy as elsewhere across the transatlantic, modern paternity's impulse to define the father in biological terms and to make him known via science was tempered by the potent consequences of this knowledge. In cases of illegitimate paternity, it could be disruptive, as the cases of the Arcardinis in Buenos Aires and Charlie Chaplin just a few years earlier had shown. In cases involving married couples like Remo and Quinta, it was potentially catastrophic. The postwar circumstances that had produced the "little Moor of Pisa" were no doubt singular, but the case resonated much further, for nonpaternity as revealed by a baby's race raised the analogous specter of nonpaternity as revealed by a scientific test. The power of science to demonstrate that the husband was not the father had the potential to shatter a foundational legal fiction. The protection of marriage would prove a powerful incentive for keeping modern paternity and its technologies in check.

THE FIRST ALLEGATIONS of interracial sex between Italian women and foreign troops arose with the Allied occupation of Naples. The occupation, which began in October 1943, presented the demoralizing spectacle of a population brought to its knees by war and hunger. That some of the occupying troops were nonwhite merely added salt to the wounds. Among the Allies were French and British colonial forces from West and North Africa, South Asia, and the Middle East; Brazilians; and some 15,000 African American troops mobilized by the segregated U.S. Army. Steeped in Fascist anti-Semitism and colonial racism, many Italians regarded the prospect of liberation by racial inferiors with alarm and aversion. "I couldn't understand it," remembered a Neapolitan street vendor. "Could it really be that they came to liberate me? We always commanded them, enslaved them, and now they come here and act like the masters of our house?"9

His domestic metaphor—the nation as household—was telling. Italian women quite literally embodied the occupied nation invaded by dark and savage foreigners. Axis propaganda included virulently racist representations of lascivious black men seizing Italian women. "Defend her!" exhorted one poster. "She could be your mother, your wife, your sister, your daughter."10 Such rhetoric was hardly peculiar to Fascists; the Allies also repeated it, and it persists even in recent histories of the occupation.11 An instance of mass rape by colonial French troops from North Africa—the so-called *marocchinate*—is part of the living cultural memory of the war.12

But perhaps even more troubling than violent relations were consensual ones. Following the Allied invasion, local populations in places like Naples were reduced to hunger and misery. Through delicate and desperate exchanges, some women bartered sex for money, food, security, and affection. The moral degradation of Neapolitan women was a near constant refrain of the occupation period, with interracial relations singled out for special reproach.

Such tropes captured the ambiguous politics of occupation. In 1943, shortly after the invasion of Naples, Italy declared war on Germany, and the Allied troops that had arrived on Italian territory as invaders were abruptly transformed into liberators. Was Italy's capitulation coerced or consensual? Representations of the deeply ambivalent experiences of invasion and occupation dwelled not just on violation but also on betrayal. Italian women who willingly participated in their own despoliation personified Italy's own humiliating surrender to the Allies.

The Allied campaign continued further north into Tuscany. In the fall of 1944, the Germans took refuge behind the Gothic Line, their final major line of defense in the Italian peninsula. Allied troops, including the Buffalo Soldiers, drove them further north and then settled in the liberated territories as occupiers. The Tuscan coast became the command center of U.S. military operations, housing a vast supply depot and American troops, both black and white. A lively black market and allegations of widespread prostitution emerged, and a wooded area between Pisa and Livorno known as the Tombolo gained notoriety as a redoubt of vice and a hideout for deserting soldiers of all nationalities.[13] Tensions between occupying troops and locals were often directed at nonwhite soldiers. One night in the summer of 1947 in Livorno, tensions boiled over when roving bands of Italian youths began attacking black GIs and Italian women. The soldiers took refuge in their barracks. The women were stripped and displayed on a merry-go-round in the public square.[14]

MEANWHILE, THE CHILDREN of the Allied occupation, including biracial ones, had begun to appear. In Naples, a song called the "Tammurriata Nera" narrated the birth of a black child to a white mother. The *tammurriata* was a genre of southern folk music played to a special drum, the tammorra. The "black" tammurriata was composed by a songwriter who worked in a local hospital and was inspired by a story he had heard: "Sometimes I don't understand what's happening / What we see is not to be believed! Unbelievable! / A black baby is born." The song was released in 1946, just as the court in Pisa was debating Antonio Cipolli's paternity.[15]

Italy had historical experience with biracial people as a result of its colonial exploits in east Africa. First a liberal government and then the Fascist one had promulgated laws and policies regulating interracial sex and the citizenship status of Afro-Italian children.[16] But the children of empire were born of relationships between Italian men and African women, and for the most part they remained in distant colonial possessions. Occupation children, in contrast, were born of interracial sex of a variety Italians regarded as far more disturbing, between Italian women and nonwhite soldiers, and they appeared not in faraway Africa but right there in the bombed-out rubble of Italy's own communities.

As the upheaval and desperation of war and occupation subsided, the children of these unions became a conspicuous reminder of the traumas of wartime and the ambiguous gifts of liberation. No reliable census was ever taken of Italian occupation children generally or biracial ones in particular, and different sources gave wildly diverging figures, ranging from one hundred to 11,000.[17] Whatever the actual numbers, the popular, scientific, and political attention they inspired far exceeded their numbers. "These little innocents, the color of caffe-latte" were at once objects of paternalism and symbols of degradation.[18] A Constitutional Assembly delegate who was also a left-wing medical doctor lamented that "this Italo-Black color on the cheeks of these children represents the Patria's sense of abjection; and we all feel . . . an anguished sense of responsibility for them." But he went on to note that mixed-race people were constitutionally weak and "don't better the human type."[19] A diffuse sense of charity easily coexisted with racist beliefs and eugenic sensibilities.

While some biracial occupation children remained with their mothers, the social opprobrium heaped on these women meant that many were unwilling or unable to keep them. It was not just the children's racial status that was problematic but their natal status as well: their color immediately marked them as illegitimate. Italian society had long stigmatized extramarital sex and the children born of it. In the nineteenth century, a massive system of orphanages received illegitimate children who were systematically and often coercively removed from their mothers. At its height, some 39,000 children a year were entering these institutions.[20] Such practices formed the backdrop for policies toward occupation children, many of whom wound up in church-run orphanages.

Institutionalization was also a metaphor for the children's relationship to the nation. Children without a legitimate paternity could not easily claim membership in the patria. Sequestering them in orphanages was a physical expression of this fact: it placed them outside the family and thereby outside that "aggregate of families," the nation itself. Afro-Italians in particular might be objects of charity, but they were rarely regarded as citizens.[21]

As an infant, Antonio Cipolli was placed in a local orphanage, reared first by the nuns of Santa Chiara and then by priests in institutions where he was one of a handful of "mulattini." When charitable women came to the orphanage, they touched his hair and treated him, in his words, "like a poodle." On trips to the beach, he would scrub his face with sand in order to take away the darkness, asking himself, "Why am I like this?"[22] His mother, Maria Orsini, visited him a few times a year, but he knew nothing of his origins or the dispute surrounding his paternity, and he could not understand why he was not allowed to live with her. When the two walked down the street during their periodic visits, people would hurl slurs at her, so she would retreat with her son into the Tombolo. It was in this forest reputed as a den of vice and lawlessness that Maria Orsini and her son Antonio could find relative peace and safety.

The fate of biracial children inspired considerable public debate. One well-known race scientist conducted a lengthy study of so-called *metticci di guerra,* or "mixed children of war" in public orphanages and concluded they would be best off in special, segregated institutions: racial segregation rationalized as benevolence and legitimated by "science."[23] Another proposal was to remove the children from Italy entirely, via international adoption. Inspired by dire reports about biracial GI babies in the African American press, both Black and Italian American civic groups organized various schemes to send aid or place the children in adoption with African American families. An Italian priest, founder of a special asylum for so-called "children of the sun," advocated sending them to new families in a society imagined as multiracially tolerant: Brazil.[24] At one point, Antonio Cipolli lived for several months with an African American military family stationed at a base near Pisa and was asked if he wanted to go live with them in the United States. He was not an American citizen and did not know his biological father was an American GI. Yet for the adults around him, his race meant that he belonged somewhere besides Italy. Antonio was not so sure. His true family, he felt, were the children in the orphanage. In the end, he chose not to go with the American family, yet the question remained: where did he belong?[25]

In the summer of 1950, Italian moviegoers were invited to explore that question in a film entitled *Il Mulatto.* Opening with the superfluous message that the story was "inspired by real events," the movie recounts the story

Antonio Cipolli (left) at the beach with unknown companions, circa 1950.
Courtesy of Dunja Cipolli.

of Matteo Bellfiore, a busker who returns home after a long jail term for stealing food during the war. His wife has died in childbirth during his imprisonment, and he is eager to meet for the first time the young son she left behind. But when he arrives at the orphanage, a nun presents him with Angelo, a "negretto." The child was born, he learns, after his wife's rape by a drunken black soldier.[26]

Bellfiore is horrified. Initially he feels intense revulsion for the little boy, but gradually he overcomes these sentiments and develops first compassion and then love for him. His path to redemption is not easy. Bellfiore's coworkers play the "Tammurriata nera" to humiliate him; little Angelo is the subject of cruel taunting and violence at the hands of other children. Eventually Bellfiore is able to imagine himself as the boy's father, raising him together with his new fiancée. He—and by extension the Italian nation—have been redeemed.

An improbable plot twist allows Bellfiore, and Italy, to claim compassion for the mulattini without having to contend with their integration. The boy's "true" uncle suddenly arrives from the United States in search of his nephew, and Bellfiore's hard-won sentiments yield to deeper and more primordial affinities. As the Black American stranger belts out a Negro spiritual, Angelo is instinctively drawn to him. Bellfiore reluctantly accepts that, for Angelo's own good, he must let the boy go. The film ends as Angelo and his uncle depart for the United States. Blood has won out over affection, and race has trumped national identity.[27]

AFRO-ITALIAN CHILDREN POSED pressing questions to postwar Italian society not just about how they would be treated but about the relationships that produced them. The fictional Matteo Bellfiore becomes the accidental father of a "mulattino" because of his wife's rape. But what about men like Remo Cipolli, who had been handed this ignominious paternity by unfaithful wives? Many observers were convinced that war and occupation had produced an epidemic of adultery, of which mixed-race children were merely the most conspicuous evidence. "When a world war like the one that occurred happens, adultery, too, runs off its normal tracks," noted the minister of justice during deliberations of the Constitutional Assembly.[28] References to "survivors, prisoners, ex-combatants who upon returning to the Patria find the family honor destroyed" by an unchaste wife—to cite another delegate—recurred time and again in public discussions of the family, including the Assembly and the press.[29] In 1946, a veterans' group claiming to represent hundreds of former war prisoners from Bari wrote to the prime minister of their plight:

> After serving the Patria for many years, after the . . . dangers of the trenches, after the torturous Calvary of imprisonment . . . the signs of the enormous devastation and miseries left by the War were revealed to each of us in the moment when we crossed the threshold of the home itself, the no less sad signs of the profanation of the domestic sanctuary, the moral ruination of the family itself due to our irresponsible women, who during our long absence have displayed a conduct that is anything but that of honest wives and mothers.

The veterans' committee went on to single out these women's relations with nonwhite soldiers.[30]

Allegations of widespread adultery were part of the broader complex of "diminished manhood" that resulted from women's increased autonomy and independence in war, the occupation by foreign troops—first Germans, later Allies—and the ways that these experiences compromised men's roles as patriarchs, providers, protectors, and procreators. An additional political development may have contributed to the perceived crisis: in 1945, Italian women gained the right to vote.[31]

The aggrieved cuckold did not simply inspire moral indignation. He demanded a political response. Husbands and sympathetic statesmen clamored for a remedy that, in Catholic Italy, was deeply controversial: the legalization of divorce. The law allowed for conjugal separation, but it did not permit marital dissolution or remarriage. Fausto Gullo, minister of justice and Communist Party member, made the case for legalizing divorce. When adultery "becomes such an extraordinary and remarkable social phenomenon," it requires "a new law that is likewise extraordinary and remarkable." He and others argued that spouses should be allowed to end their marriages "at least in this case, when, for reasons of war, the husband has been away for years and [returns to] find the family upended and the number of its members increased."[32]

The indissolubility of marriage was just part of the problem; the other was the incontestability of paternity. Not only could husbands not divorce their duplicitous wives, but as Remo Cipolli discovered, they could not legally disavow their wives' children. Fausto Gullo told the Constitutional Assembly of the "multiple and continuous requests from all parts of Italy asking me [as minister of justice] to facilitate the procedure for repudiating paternity."[33] One report cited 400 requests for separation or repudiation of paternity in one small southern province.[34] The unyielding presumption of marital legitimacy created dilemmas for the legal ordering of families and also for child welfare. If a couple informally separated and the wife formed another relationship, her child was still considered her husband's. Meanwhile a child born of adultery and abandoned by its mother could not be given in adoption because it still had legal parents.[35]

The presumption of legitimacy permitted the husband to challenge paternity of his wife's children in just three specific circumstances. He had to show that he had not cohabitated with his wife for the period from 300 to 180 days prior to the birth; that he suffered from physical impotence; or that his wife had committed adultery and had then hidden the pregnancy and birth from him.[36] The conditions were so narrow that not even Remo Cipolli's petition—to the casual observer such an obvious and transparent case of nonpaternity—could be shoehorned into them. Cipolli and Orsini had lived together during the period specified by the code, and there was no allegation that Cipolli was physically unable to engender a child. While the somatic features of the child amply proved adultery in the court's view, Orsini had not attempted to hide her pregnancy as the clause required. In fact, a witness had overheard the couple engaged in a contentious discussion about it, and her husband had taken her to the hospital to give birth. Not even Orsini's own admission that Antonio had been sired by another man sufficed; a mother's declaration could not remove her husband's paternity. The facts of baby Antonio's conception and birth simply did not fit into the restrictive provisions of the law.

Shortly after the Pisa verdict that found Remo to be the father, and as the case headed to an appeals court, a young associate at the University of Pisa's law school named Antonio Carrozza suggested a novel approach to the case. Perhaps the key to a just resolution could be found not in the clause on adultery but in the one concerning impotence. Impotence was a much-discussed concept in canon and civil law as well as forensic medicine because it was grounds not only for the repudiation of paternity but also for the annulment of a marriage. The law contemplated two forms: "the incapacity of the individual due to anatomical or functional defects of the genital organs . . . to accomplish copulation (impotentia coeundi)" and the incapacity "to conceive offspring (impotentia generandi)."[37] But what if impotence referred not to the inability of the husband to engender any child but to his inability to engender a specific child? Carrozza suggested that Remo Cipolli was physiologically unable to father a dark-skinned baby with his white wife. What he exhibited, in other words, was "racial impotence," in which "the genital organs . . . are not capable of procreating a child of a different race."[38]

Racial impotence shifted the term's meaning from generation to genetics. Paternity, or rather nonpaternity, could be found not in the man's body—in his reproductive organs—but on the child's body—in somatic or other physical markers. In investing the age-old forensic concept of impotence with new meanings related to race and heredity, Carrozza attempted to bring the civil code's archaic treatment of the father into accord with the precepts of modern paternity. For as he lamented, in assuming that paternity was unknowable, "not subject to the senses," the Italian code "is still the Code of

Napoleon!"[39] That is, it relied on outdated nineteenth-century assumptions about how nature concealed the father, ignoring scientific advances that increasingly revealed him. It was thus law, not biology, that stubbornly obscured paternity. Meanwhile, the idea of racial impotence squared perfectly with modern paternity's enduring racial fixations. It was not by chance that a young jurist seeking to remedy the tribulations of "a society poisoned by a serious multicolor invasion" would rethink the law of paternity.[40]

Carrozza's creative argument was embraced by a number of other jurists, not least of whom was the public prosecutor of the Florentine appeals court.[41] But ultimately, it did not sway the judges themselves. In July 1949, three and a half years after Remo Cipolli had first filed suit to disavow his paternity, the appeals court handed down its final verdict. Baby Antonio was now a little boy, almost four years old, being raised by the nuns of Santa Chiara. The judges upheld the decision of the lower court: Remo was Antonio's legal father. His name would remain on the boy's birth certificate, and the two would share, for the purposes of the law at least, all the mutual rights and responsibilities inherent in the father–child relationship.

THE VERDICT EXPLODED into public. "To the white father is given a black son," announced the national headlines.[42] A local newspaper baptized young Antonio the "child moor of Pisa" and lamented that "this poor husband betrayed by his wife was also mocked by the law."[43] Antonio Carrozza, inventor of racial impotence, discussed the case for a popular Italian weekly.[44] The chief judge of the Pisan court felt compelled to explain the seemingly bizarre verdict in a long letter to the local newspaper.[45]

Perhaps anticipating this reaction, the Florentine judges explained that the legal issue at stake was "very simple." Despite Cipolli's manifest nonpaternity of the black child born to his wife, racial difference "is not in itself among the circumstances in which civil law gives the husband the right to challenge his paternity." Neither racial impotence nor any other creative sleight of hand could change the fact that "the law as presently conceived" placed strict limits on challenges to the marital presumption in the interest of "familial integrity."[46]

The ensuing legal commentary was divided over whether the judges could have found some way around the law's constraints, but everyone, even the magistrates themselves, shared a deep dissatisfaction with the outcome. "To the man on the street, accustomed to reasoning based only on good sense, the judges' conclusion may seem totally aberrant," noted one legal commentator, worrying that the case would undermine respect for the law.[47] A few years earlier the "Tammurriata nera" had given voice to the incredulity with which many Italians regarded the birth of a black child to a white mother: "I don't understand what's happening / What we see is not to be believed!" Now a new spectacle, this one produced not by nature but by a human tribunal, inspired similar incredulity: a white man had been declared the black baby's father.

Earlier the Pisan prosecutor had declared that the outrageous outcome would perhaps have a silver lining. It might attract the public's attention to Italy's archaic law of paternity, "a pressing problem that captivates the sensibility of anyone still committed to the idea of honor, honesty, and the family."[48] Public attention indeed surfaced, not only in the pages of the newspapers but also on movie screens: *Il Mulatto,* released a year after the Florentine verdict, was a direct response to the case. Despite its title, the film is not actually about the young boy Angelo (who is largely mute during the film). It is about the father, Matteo Bellfiore, and the affective paternity that, in the absence of biological paternity, he grudgingly comes to embrace.

The movie is also a commentary on the law of paternity. Even as it commends Matteo Bellfiore for accepting a paternal role toward Angelo, it bitterly denounces the law that has foisted a false paternity on him in the first place. When Bellfiore first glimpses the child at the orphanage and turns uncomprehendingly to the nun ("but he's black!"), she replies, "Listen, before God, that little mulatto is nothing to you. But before the law, he is your son . . . because he was born to Maria Dosella, wife of Matteo Bellfiore." The realization that his wife has given birth to another man's—a black man's—child occurs simultaneously with the revelation that the law holds him to be that child's father.

Bellfiore and his friend consult a lawyer. On the wall of his office the scales of justice are conspicuously off kilter. The friend asks wonderingly, how can color not be proof of the child's nonpaternity? And gesturing to the legal tome on the lawyer's desk he inquires comically, are you sure this is the latest edition? The officious lawyer then reads from a decision rendered by the "Court of Appeals of Florence" concerning a man "whose wife

gave birth to a child . . . clearly of the negroid race." The decision concludes: "difference of color is not among the circumstances that give the
husband the right to reject the child as his own." The sentence is a perfect
paraphrase of the wording of the Florentine appeals court in the Cipolli–
Orsini case. *Il Mulatto* is clearly intended as a comment on this specific
legal cause célèbre.[49]

The essential dilemma of the story is thus the calamity of law. To drive
the point home, the film opens with a message on the screen exhorting "the
JUSTICE OF MEN, to consider and resolve one of the most delicate
HUMAN PROBLEMS created by the war." This "human problem" is not
the plight of children like Angelo but the iniquitous law that holds white
men like the fictitious Matteo Bellfiore, and the real Remo Cipolli, to be
their fathers. The film, like the Orsini–Cipolli case, suggests that the postwar
preoccupation with the black children born on Italian soil to white mothers
was also about white men and their vexed paternity.

The film's righteous call to the justice of men did not fall on deaf ears.
Shortly after the Cipolli verdict, a Socialist legislator presented a bill to reform the offending article of the civil code. To the three cases in which a
husband could repudiate paternity, he proposed a fourth: when "it is shown
from manifest and undeniable somatic traits—scientifically ascertained—
that the child conceived in marriage belongs to a race different from that of
the presumed parent."[50]

Some jurists commended the proposal. Others agreed with its spirit but
found it "schematic and simplistic."[51] Ultimately the bill went nowhere, but
not for lack of reformist sentiment. In the wake of the Cipolli verdict, the
clamor for reforming paternity law was remarkably universal, among jurists
and also the general public. The chief publication on family law declared
the "radical reform of the extremely archaic norms governing filiation" a juridical and "social-political" necessity.[52] The journal was a perennial critic
of the overweening influence of the Catholic Church and the conservatism
of Italian family law. But even Catholics agreed that the law's defense of marriage and the legitimate family went too far. The Union of Catholic Jurists
called paternity reform a "keenly felt necessity" and endorsed the Socialist
proposal.[53] Meanwhile, it turned out that the popular movie version of this
call to reform—the film *Il Mulatto*—had been released under the auspices
of the International Catholic Cinematographic Union.[54] But while everyone

agreed that paternity law needed to change, the question remained: how far should that reform go?

IF THAT QUESTION HAD concerned only the Remo Cipollis of the republic, white men whose wives bore nonwhite children, its symbolic stakes would have been great but its practical stakes relatively limited. After all, relatively few brown babies were born to white couples in Italy; fewer still would be born after the occupation ended. But from the time the story first emerged, scientific and legal observers recognized that its implications went far beyond the specific scenario of the "mulattini."[55] For if somatic markers of race were allowed to unseat marital paternity, what about other kinds of physical proof? What about, for example, blood group tests? Would husbands be allowed to challenge paternity with the help of science?

Faced with such questions, the reformist ardor cooled. "If an investigation of this type were licit," observed the Florentine judges, "everyone knows what sorts of disputes it would lead to."[56] Allowing Remo Cipolli to disavow paternity of Antonio could throw open the gates not just to other white men with brown babies but potentially to all husbands who doubted their wives' fidelity. All babies, after all, were born with inherited blood types that could reveal the nonpaternity of their fathers. A simple blood test could trigger the collapse of the sacrosanct presumption of marital legitimacy. Were this to happen, a reform meant to enhance justice and truth in the family could unwittingly annihilate its very foundations.

Such considerations were clearly on the judges' minds as they considered the Cipolli case. Blood group testing was discussed repeatedly in relation to the case, including by the Pisan and Florentine courts, even though the test was never used. It was never used, of course, because it was considered superfluous. The child's racial appearance was nature's own paternity test.

Precisely because it involved a nonpaternity visible and obvious even to the nonexpert, the case exposed more powerfully than a scientific blood test ever could the apparent illogic and injustice of marital paternity law. And yet even though no serological test was conducted in the case, it was steeped in references to blood. In their commentaries on the case, jurists invoked not just "racial impotence" but also concepts like "hematological impotence" and "serological races." Such concepts captured how serology tests and race could expose biological nonpaternity.[57]

Even those who believed that the courts should have allowed Remo Cipolli to disavow his paternity contemplated such scenarios with trepidation. It was one thing to let a white husband repudiate his black baby, quite another to allow all husbands to test, and perhaps disavow, their wives' children. By rejecting Cipolli's petition, the judges sought to avert that troubling prospect.

ITALIAN SCIENTISTS, JURISTS, and the general public tended to regard the drama of Remo Cipolli, Maria Orsini, and baby Antonio as a sui generis case, a tragedy resulting from a unique and unprecedented chain of events "stemming from the unfortunate contingencies of war."[58]

But the birth of a dark-skinned baby to a white wife and her white husband, Natus Ethiopus, recapitulated a trope that reached back to antiquity.[59] Ancient Greek and Hebrew texts narrated the story (and less frequently its reverse, Natus Albus, the dark-skinned mother who births a light-skinned child). Renaissance authors repeated it, too. One oft-cited anecdote held that Hippocrates saved a white princess from an adultery accusation after her baby emerged "black as a Moor."[60] Early modern narrators told the story to illustrate the phenomenon of maternal impression, in which a pregnant woman's imagination could mark the body of her infant. The princess whom Hippocrates saved had conceived the child in a bed over which hung a portrait of a dark Moor. In Shakespeare's *Titus Andronicus,* in contrast, the wife Tamora gives birth to a tawny infant as a result of adultery with her dark lover.

In the modern era, Natus Ethiopus was often attributed to racial atavism. According to this belief, blackness could never be expunged from a white bloodline and could make an inconvenient reappearance even generations later. Atavism was a favorite motif of nineteenth- and early-twentieth-century American fiction.[61] In 1913, eugenicist Charles Davenport called scenarios of racial reversion "a matter of great social moment to hundreds of our citizens." But while it provided fodder for novels and newspapers, he was skeptical that it was a true biological phenomenon.[62]

For ancient and modern observers alike, Natus Ethiopus provided fertile grounds for pondering the nature of gestation and generation. Twentieth-century scientists, however, tended to reject maternal impression and racial atavism as myths. What caused a white woman to bear a dark infant was a black father: a child's racial appearance was a transparent manifestation of paternal identity. "We don't need lab science, just the vulgar experience of physical reality in comparing the color of skin, to call black black and white white," quipped the jurist Carrozza of baby Antonio's ancestry.[63]

Modern paternity's avatars constantly invoked Natus Ethiopus as a story about paternal revelation. According to German scientists, mixed-race children represented "the simplest case" of paternity determination. An English jurist, noting the common law's ambivalent regard for physical resemblance as evidence of parentage, argued that nevertheless, "by universal consent, ancient and modern, resemblance may be taken into . . . serious account, whenever the controversy lies between two alleged parents who belong to different races of mankind." Courts in South Africa, Scotland, and some U.S. states that rejected evidence of filial resemblance nevertheless permitted it if the alleged parents were of different races.[64]

It depended, however, which parent was of which race. Tellingly, some jurisdictions, like Nebraska, admitted evidence of racial resemblance only in cases involving white mothers and putative fathers who were black.[65] Race was nature's paternity test—but only in the case of Natus Ethiopus. Paternity testers virtually never referenced the case of Natus Albus—the light-skinned child born to a darker woman. If resemblance as a tool of paternity investigation followed a conspicuously racial logic, it was because modern paternity and its techniques were preoccupied with surreptitious black fathers, not white ones.

As for baby Antonio, Italian observers asked not what explained his complexion or who his biological progenitor must be. Instead they asked how the self-evident fact that he was born of his mother's "colored adultery" could be reconciled with the white father assigned him by law.[66]

IF THE SPECTACLE of the black child born to a white mother was not peculiar to this time or place, neither was the presumption of marital paternity.

While the Italian civil code entertained a strict version of this principle, the principle itself was extraordinarily widespread. "A time-honored legal institution" enjoying "the sanction of the centuries," as a Philippine jurist characterized it in the 1930s, the presumption runs wide and deep, traversing a variety of legal traditions ancient and modern, religious and secular, Western and "non-Western."[67] Present in canon, Jewish, and Islamic legal traditions, the presumption of marital paternity exists in Anglo-American law and the civil law of continental Europe, as well as Latin American and Middle Eastern legal systems. It is "as close to a cultural universal in law as we get."[68]

In some contexts, the principle has been virtually irrefutable. Under Roman law, neither proof of adultery nor the declaration of the mother regarding the paternity of her child was sufficient to contest the presumption that her husband was the father. According to seventeenth-century English common law, if a husband was located within the "Four Seas" of the English empire at the time of his wife's conception, he was legally the father of her child.[69] Shar'ia law held that "the child is affiliated to the [marriage] bed," a presumption so strong that Islamic jurisprudence held that pregnancy could last four or five years, ensuring the legitimacy of children born long after a marriage ended in death or divorce.[70] However nonsensical it might sometimes appear, the legal presumption was "so natural and decorous, so useful and just, and above all so necessary," according to a well-known nineteenth-century Spanish jurist, "that outside of it society itself would not be possible."[71] Or in the words of two Ohio Supreme Court justices a century later, "even the apparent justice of any particular case [does] not justify a departure from the rule so necessary and salutary to the best interests of society."[72]

To be sure, jurists periodically denounced the presumption's more egregious fictions. In a 1930 case, the American jurist Benjamin Cardozo criticized a court's finding that the children of a woman in a long-term relationship with a man were nevertheless her husband's. This was a modern version of the "Four Seas" rule, and Cardozo had little patience for such "follies and vagaries." But he did not intend to uproot the presumption altogether; he merely wanted to "prune" it of its "extravagances."[73]

AS IN THE CIPOLLI–ORSINI case, juridical discussions of the presumption of legitimacy consistently described it using racial referents. Jurists invoked

Natus Ethiopus to demonstrate just how far and deep the presumption extended. Anglo-American law had a dictum that captured this idea. In an 1849 inheritance dispute involving an Irish couple, the English jurist Lord Campbell declared, "So strong is the legal presumption of legitimacy, that, in the case of a white woman having a mulatto child, although the husband is also white and the supposed paramour black, the child is to be presumed legitimate, if there is any opportunity for intercourse."[74] Significantly, the case at hand did not actually involve a Natus Ethiopus scenario; this was simply the most extreme hypothetical Lord Campbell could muster to illustrate the presumption's power. "Lord Campbell's dictum" was invoked frequently not just in relation to paternity but in discussions of the laws of evidence, the treatment of legal presumptions, and the function of legal fiction. Certainly not all Anglo-American jurisdictions accepted it, but some did, and not just metaphorically.[75] In a case in New Zealand in the 1930s, a court declared a "Mongoloid half-caste child" to be legitimate though its mother and her husband were both white and she had a lover who was Chinese.[76] A child's racial appearance could signal biological impossibility without signaling legal impossibility. Race was thus the bright line that distinguished biological and legal notions of the father; in this sense, race helped to define what paternity was.

The other kind of impossibility, as observers of the Cipolli–Orsini case noted, was serological. While Italians worried that blood tests would pose a radical challenge to the marital presumption, in practice that challenge proved fitful. In the 1930s and 1940s, as blood group testing made inroads in transatlantic forensic practice and beyond, courts remained reluctant to embrace it in the case of married couples. A study of blood testing in European law found that the technique was applied much more commonly in cases of illegitimate paternity than to contest legitimacy.[77] In Egypt, courts also hesitated to permit blood testing to challenge the paternity of a child born in marriage.[78] In the United States, there was significant variation between higher and lower courts and across states, but into the 1960s judges were in general tenaciously committed to the marital presumption even in the face of serological evidence rebutting it.[79] Sometimes they treated biological testing as just one piece of evidence to be considered among others; in other cases, they rejected it entirely, as contrary to what they considered a conclusive presumption.[80] One California judge declared that to remove the "protective shield which the law has heretofore placed around children born to married couples" was to "subject their status . . . to the vagaries of test tubes and chemistry."[81] The 1967 case concerned a marriage that had lasted four days.

If steadfast allegiance to the presumption of marital legitimacy was the transatlantic rule, there was one dramatic exception. In Nazi Germany, this presumption was sacrificed to the fanatical pursuit of racial truth. Especially after 1938, the state threw open marital paternity to contestation by husbands, wives, children, and state officials. According to Nazi lawmakers, it was in the interest of both the child and the "volk" that legal fictions be stripped away to reveal bioracial truth. The reform making this possible was of course part of a wider revolution in law and science that rendered paternity radically biological.

The contrast between the radical racial paternity of the Nazis and the unyielding postwar Italian commitment to paternity by marriage is instructive. In the years prior to Antonio's birth, Fascist Italy had elaborated its own racial laws, of course, including prohibitions on mixed marriages. But where Nazi jurists launched a full-throttle assault on traditional norms of family law in the service of its racial project, the Fascist civil code Italy promulgated in 1942 largely deferred to the Vatican and its impulse to protect the legitimate family.[82] It was this code that in post-Fascist Italy made Remo Cipolli the father of baby Antonio, privileging marital paternity even when it was in manifest contradiction to cultural notions of biological truth.

Yet despite the radically different ways that Nazi law and Italian post-Fascist law defined paternity, in both contexts paternity was central to the construction and reconstruction of national order. In postwar Italy, as in National Socialist Germany, the question of the father—who he was, how he could be known, and the role of law and science in these determinations—emerged with striking urgency in very different projects to construct new national communities.

THE STORY OF THE "young moor of Pisa" as told in legal sources, film, and the press featured three protagonists: the child Antonio; his mother, Quinta;

and her husband, Remo. Conspicuously missing from the drama was a fourth figure: Antonio's biological father, the American soldier. Nowhere in the voluminous discussions of the case was his identity ever revealed or even commented on.

His absence was not accidental. Rather than being simply unknown, his identity was actively erased, in the first instance, by Italian law. From the law's point of view, Antonio could have only one father, and the fact that his mother was married dictated who the father was. Had his mother been unmarried, the law would have created a different set of obstacles to knowing his father. Following in the Napoleonic vein, Italian civil law narrowly restricted the investigation of illegitimate paternity, making it difficult or impossible for a child born outside marriage to establish the identity of its father. Such children had their birth registrations and other identification papers marked with the stigmatizing designation "N.N."— father unknown. The law's restrictive treatment of illegitimate paternity, like its ironclad definition of paternity in marriage, attracted vigorous debate among jurists and the public as well as perennial proposals for reform in the years after the war.[83]

If Italian law erased Antonio's biological father, so too did the U.S. military. The statutes governing the American occupation made it difficult or impossible for local women to pursue paternity suits against soldiers.[84] As for GIs who wanted to recognize their children or marry their children's mothers, bureaucratic hurdles impeded them at every turn. Soldiers and their girlfriends wishing to marry had to submit to a paternalistic and arbitrary review by military authorities who pronounced on their compatibility and moral character. Not surprisingly, the segregated U.S. military regarded petitions for marriage by biracial couples with special consternation, when not open hostility. Some African American GIs found their requests to marry their white European fiancées automatically denied if they hailed from states with antimiscegenation laws (of which there were twenty-nine).[85] By impeding marriage, the U.S. military rendered children illegitimate. By erasing illegitimate paternity, Italian civil law then denied them a legal father. If Antonio's progenitor was unknown and unknowable, it was because he had been strategically veiled, not by nature but by law and politics.[86] Indeed, from the perspective of both civil law and public opinion, the man's identity was beside the point. What mattered was that he was a black man and that he was not Remo Cipolli.

Modern paternity posited a keen impulse to know the father and the steadfast promise that he could be known. But not all fathers were worth knowing.

ANOTHER POWERFUL INSTITUTION that shaped ideas about paternity in postwar Italy was the Catholic Church. Various Catholic-identified groups in civil society weighed in on the Orsini–Cipolli affair and the issues it raised. The Union of Italian Catholic Jurists lamented the sad affair and debated the laws that had produced it. The International Union of Catholic Film-makers, producers of *Il Mulatto,* staged a dramatic call for their reform. But even as lay Catholics chimed into the public conversation, the church itself remained silent. No Catholic official appears to have spoken publicly about the Orsini–Cipolli case or about the questions it raised concerning marital paternity. In fact, the church was conspicuously silent about the broad chal-lenge to marital paternity that new scientific methods posed. Beginning in the late 1920s, scientific techniques to establish paternity expanded across the transatlantic, including the Catholic countries of Latin America and southern Europe. During these very decades, the Catholic Church spoke out in vociferous opposition to a variety of scientific interventions into the realm of sexuality, reproduction, and the family, taking vocal positions against eugenics, sterilization, artificial insemination, and birth control.[87] But it appears never to have taken an official stand on scientific methods of paternity determination. Perhaps this was because, in contrast to those other technologies, paternity testing appeared not to artificially manipulate the natural course of reproduction but merely to reveal the natural facts ema-nating from it.

Yet as Italians' trepidation about blood testing in marriage reflects, natural facts themselves could threaten marriage and family. For this reason, testing posed difficult questions for the Catholic conscience. The dilemma was felt most acutely by Catholic doctors who in the course of their professional duties were asked to perform the procedures. In France and Italy, in Argen-tina, Brazil, and elsewhere in Latin America, doctors who identified as Cath-olic, and even some who did not, wrestled with the moral repercussions of paternity testing. Usually testing happened when courts ordered it in the context of legal proceedings, and it fell to a judge to assess the potential implications. As long as testing occurred under the aegis of judicial authorities, its uses could be supervised and its most explosive consequences held in check. But what happened when a doctor was asked to perform a blood test not officially by a judge but privately by, say, an estranged husband?

Early on, physicians confronted the question of private testing. Ludwik Hirszfeld, the Polish serologist who was a Catholic convert, wrote in the late 1930s that when he received such requests he sent the couple to a priest, "who is often better at bringing about a reconciliation and a reconstitution." "The serologist should not abuse the weapons he possesses," he advised his colleagues, "but should become a defender of family happiness and never a snooper into adultery nor a provoker of divorces."[88] Hirszfeld's position became the professional consensus among doctors in Catholic countries. Paternity testing was a potent technology with potentially dire consequences for marriage, children, and the family. As such, doctors had an ethical duty to wield it wisely.

In some contexts, private requests for testing were rare. At São Paulo's Instituto Oscar Freire, almost all paternity analyses were ordered by local courts. The pattern is noteworthy because some other common procedures, such as virginity examinations, were often requested by private parties, typically a young woman's family. The fact that they did not ask for paternity tests may reflect the cost of the laborious procedures. Or perhaps few people felt the need to obtain the information a scientific paternity analysis provided or shared the vision of kinship it promoted.

In 1950s Italy, in contrast, requests for private testing were common. Perhaps this demand reflected the afterlife of the postwar crisis of adultery. Or perhaps continued restrictions on the use of paternity science in legal proceedings encouraged individuals to bypass courts and approach labs directly. By this time, the public was familiar with the uses of hereditary blood groups to determine (non)paternity, thanks to the pedagogical role of the press.[89] In this context, doctors like Vincenzo Mario Palmieri, director of the Institute of Legal Medicine at the University of Naples and a devout Catholic, noticed a troubling trend. Men "assailed by doubt over the fidelity of their wives, or desirous of knowing if the children procreated by their own lovers or concubines were effectively theirs" were turning up with increasing frequency in their labs asking for a paternity test.[90] Forensic specialists from Bari to Bologna shared this impression.[91]

An analysis of blood group inheritance was not just another laboratory procedure, to be casually performed as if it were "a syphilis or glycemia test," Palmieri warned his colleagues.[92] A procedure that could exclude biological parentage raised serious social, legal, and ethical questions, above all when it involved a married couple and their children. Knowledge of the father was thus dangerous—and irresistible: "the very phrase 'proof of blood' has, undoubtedly, a promise that stimulates the desire for confirmation."[93] An Italian ecclesiastical lawyer observed that given the growing frequency of such requests, many laboratories "have had to put a brake on the unhealthy curiosity of the public."[94] Following this logic, French scientists roundly rejected a policy, introduced at the Brazilian Instituto Oscar Freire in the late 1930s, of including ABO blood types on university students' identity cards. They worried that vulgarizing blood group knowledge could inadvertently expose family secrets and corrode "social peace."[95]

The most obvious potential victims of unchecked testing were children. The Neapolitan Palmieri posed the scenario of a doubting husband and a wife who consented to testing "more or less voluntarily." What about the couple's children? "Who will be able to consent in their name?"[96] A few drops of blood could render them illegitimate, adulterous, and legally fatherless. Because of this possibility, in Palmieri's opinion, doctors should decline to perform the tests in any private case involving minors. Flamínio Favero, head of the Instituto Oscar Freire, was so moved by Palmieri's thoughts on the ethics of private testing, which he read in an Argentine journal, that he felt compelled to address them in the column he wrote for a São Paulo newspaper, even though his own institute rarely received such requests. Favero declared himself deeply committed to making medical knowledge "accessible to all." But after "reading and rereading" his colleague's article, he was surprised to find himself in agreement with it. Under Brazilian law, he mused, doctors who unwittingly exposed a legitimate child's nonpaternity might even be liable for defamation.[97]

Others raised objections to private testing based on its potential consequences for wives. A well-known Italian civil jurist suggested that the children of married parents could never legally petition for blood testing because doing so "would mean allowing the child to cast suspicion on the mother, thereby attacking family honor, of which the husband is the sole judge." Another observer warned that the results of the test could expose women to "extremely grave dangers" in the form of a "bloody vendetta" on the part of jealous husbands.[98]

Oversight of this potentially dangerous technology fell in the first instance to law, which in Italy, of course, placed tight strictures on the husband's ability to challenge his paternity. But private requests bypassed judicial checks, and doctors must therefore be wary of them. Because husbands might lie about their motives or resort to subterfuge—Neapolitan doctor Palmieri imagined a scenario in which a husband requested tests for himself, his wife, and his child at three different labs in order to conceal his intentions—he advised that all private requests for paternity testing should be considered only exceptionally, for a petitioner personally known to the doctor "and whose motives are very clear." Favero concluded that "the danger of a proof of exclusion is so great, its personal, familial, and social consequences so serious," that doctors should decline such requests altogether.[99]

Implicit in all of these commentaries was the conviction that paternity testing, above all in the case of married couples, had potentially catastrophic implications, not only for men, women, and children but for society generally. Knowledge of paternity should not be bandied about willy-nilly, subject to capricious impulses or the public's "unhealthy curiosity."

While these issues were most pointedly discussed among Catholic doctors, they raise a question that has undergirded parentage testing from its beginnings: who has the right to know? In the 1950s, doctors in southern Europe and Latin America concurred that the right to know was not unconditional. The need to protect female honor, children's rights, and above all the integrity of the family meant that no individual enjoyed an inalienable right to knowledge of biological (non)paternity. Scientific testing posed a contest between "personal interest" and "public interest."[100] It was up to public authorities, specifically to law and the courts, and only secondarily (and reluctantly) to doctors themselves, to negotiate the tension between these interests.[101]

Paternity testers saw themselves not just as avatars of scientific truth but as guardians of moral order: "The new information obtained through science should enhance social order rather than undermine it."[102] Biogenetic truth was just one value to which they owed professional obeisance; morality, propriety, and societal order were of equal, perhaps superior, significance. The doctor who recklessly exposed information might be guilty of an ethical or criminal transgression even if the interested parties had requested it—and even, remarkably, if the information revealed was true.[103] In assessing the social value of truth, context was everything. In the case of immoral relationships and illegitimate paternity, truth-telling technologies could discipline the reprobate and punish the guilty. In the case of married couples and legitimate children, it might have the opposite effect, and morality might best be served by discretion, fiction, and suppression. Even German scientists, who tended to defend the much more expansive use of blood testing, argued that it had a "prophylactic" effect by discouraging people from committing indiscretions that science could now reveal.[104] In all these scenarios, testing the link between child and parent always reverberated beyond the family, posing moral implications for society at large.

THE ANXIETIES OF POSTWAR Italy provoked vigorous public debate about family law in general and paternity law in particular, debates that came to a head in the Orsini–Cipolli affair. Jurists, scientists, the press, and cinematographers all called for reform, but in the end reform came slowly. The stigmatizing designation "N.N." was removed from Italian identification papers in 1955. Female adultery was decriminalized only in 1968; divorce was introduced in 1970; illegitimate children were declared equal to legitimate ones in 1975.[105] As for the famous article that had provoked such outrage by protecting the presumption of marital paternity, it was only in 1975—thirty years after Antonio Cipolli's birth—that blood and other genetic tests were permitted in disputes over marital legitimacy. Even then, genetic evidence could serve as corroboration of prior social facts but could not on its own delegitimate a child: the husband had to present evidence of his wife's adultery before scientific evidence would be permitted. It was only in the 2000s that Italian courts began to permit DNA evidence in the absence of prior proof of female adultery.

If the cause célèbre did not catalyze legal reform, it nevertheless had a long legal and cultural afterlife. The "celebrated case of the mulatto of Pisa" is still cited in contemporary Italian juridical texts and discussed in relation to legal definitions of impotence, DNA testing in judicial proceedings, and the increasing trend in Italian family law to consider biological truths alongside legal ones.[106] Versions of the story are even revisited in contemporary novels.[107]

There is, however, one person who until recently was entirely unaware of the famous case: Antonio Cipolli himself. For a few years, I trolled the Internet looking for traces of the baby born and abandoned in 1945, uncertain whether he had even survived to adulthood. Then, in 2017, I found a Facebook profile for an Antonio Cipolli. My tentative message—are you by chance Antonio Cipolli, born October 9, 1945, son of Maria Quinta Orsini of Pisa?—at first went unanswered. Then, at four one morning, the cell phone began buzzing: "Sì, sono io," read the message. "Yes, it's me." After a lifetime wondering about the mystery of his origins, Antonio was eager to talk.

Today the baby whose paternity caused such a scandal is a retiree who lives in a town not far from Pisa. I spoke with Antonio and his daughter, Dunja, the family's historian, and they told me parts of the story that my historical sources—the court cases, the newspaper accounts, the legislators' responses—had not revealed. Remo and Quinta were never reconciled, and Quinta remained socially ostracized by the community. Antonio grew up in the orphanage, where he could not understand why his mother had abandoned him. In her occasional visits, she promised him, "Tony, when you grow up, I will explain everything." The law of paternity was designed to protect the integrity of the legitimate family, the honor of its members, and the rights of its children. In this case, it failed on all counts. In 1962, at age seventeen, Antonio left the orphanage to return "home." As he jumped off the bus, he ran into an acquaintance who told him his mother had just died in a car accident. He arrived just in time to attend her funeral. Whatever information about his origins she hoped to share with him went with her to her grave.

Antonio's relationship with his other family members, including his older half-brother Marco, son of Quinta and Remo, remained strained. In the 1940s, the Italian public and Remo Cipolli had balked at his paternity of a dark-skinned baby. As a young man, Antonio Cipolli came to reject his paternity as well. He regretted the last name he bore, the name of his nonfather, wishing instead that he could be Orsini, his mother's surname. And he never claimed the inheritance due to him as the legal son of Remo Cipolli. Italian law had given him a patronym and a patrimony, but he rejected both. If his relationship with the family that had cast him out as a black baby remained strained, as a young black man, his relationship to the nation itself was likewise complicated. Over six feet tall and dark-skinned,

Antonio found it impossible to live in a racially homogeneous and deeply racist Italy. Joining an economic exodus of his compatriots, in the 1960s he emigrated to Switzerland, where he worked, married a German woman, and had two children. Occasionally he would return with his family to visit Pisa. His daughter Dunja remembers that her cousins were aloof, and she always felt conspicuous with her biracial family. After three decades abroad, Antonio returned to Italy to retire.

At the time of his birth, most observers adamantly believed that the truth of Antonio Cipolli's biological paternity should be publicly exposed and legally acknowledged. It is therefore deeply ironic that Antonio himself was never privy to this knowledge. For more than seventy years, he confronted a suffocating silence concerning his origins. No one—not the nuns and priests who reared and educated him, the foster families with whom he lived during summers, or anyone in his family or the local community—ever revealed a word to him about his paternity. Antonio knew nothing of the cause célèbre that surrounded his birth and his family. He did not know that he had been baptized "the little Moor of Pisa" on the front pages of national newspapers. He did not know that the events surrounding his birth were the subject of a popular movie. The mystery of his father's identity was always just a private question, and it was merely one piece of a larger, deeply painful experience of family abandonment and estrangement.

The history of paternity testing is also an antihistory: a history of when law and social norms conceal the identity of the biological father and circumscribe, discourage, or even prohibit the use of scientific methods to reveal it. It is a history of truth, however defined, left strategically unspoken. The dispute over Antonio Cipolli's paternity reveals the tensions between cultural, legal, and biological definitions and in particular the power of marriage to define the father. Scientific truths but also social ones ran aground on the shoals of the law's marital presumption. Antonio Cipolli and his children bear the surname of a man whom law declared his father, but they do not consider him to be that any more than did observers at the time of his birth.

8

CITIZEN FATHERS AND PAPER SONS

> A blood test has failed to back up the claim of a young Chinese that
> he is the son of an American and entitled to United States citizenship.
> [A judge's] decision to bar citizenship on the basis of this test may
> affect citizenship petitions by an estimated 20,000 other Orientals.
>
> —*Chicago Daily Tribune,* June 27, 1954

THE THREE SIBLINGS descended at a New York City airport on a warm June evening in 1952. They must have been both thrilled and nervous. It had taken six days to travel from Hong Kong, a journey that included stops for refueling and plane changes in Vancouver and Montreal. They spoke no English and had never been to the United States.

Twenty-one-year-old Lee Kum Hoy, thirteen-year-old Lee Kum Cherk, and their sister, Lee Moon Wah, just three days shy of her twelfth birthday, were coming to be reunited with their parents. They had not seen their mother since the day she had boarded a bus in their village in Toishan District, in the Pearl River delta, more than three years before. As for their father, a Chinatown grocer long resident in the United States who made occasional trips back to China, he was a stranger to the two youngest children. Their father had become an American citizen a quarter century earlier, and it was through him that the three siblings, as children of a naturalized American father, arrived in New York claiming U.S. citizenship.

But the reunion would have to wait. Upon landing, immigration authorities reviewed the passengers' papers and quickly pulled the Lees aside. As was common in these years, the siblings were traveling not with passports or visas but with special affidavits issued by the American consulate in Hong Kong.[1] The affidavits allowed them to travel to the United States, but having arrived on U.S. soil, they would now have to prove to immigration authorities their identity as children of a citizen father.

Instructed to gather their luggage, the Lees, together with another thirteen-year-old boy traveling from Hong Kong on an affidavit, were taken to Ellis Island. Here they would be held until their immigration status was sorted out. Its heyday as a storied gateway to America long past, Ellis Island had aged into a decrepit Cold War immigration detention center. After dwindling for decades, its population suddenly ballooned to more than 1,000 people in the early 1950s, as the United States began aggressively detaining new arrivals suspected of ties to totalitarian or Communist organizations.

In the shadow of the 1949 revolution, Chinese immigrants were one group regarded with suspicion. Among the detained at Ellis Island were a few hundred Chinese, many of them, like the Lees, young people seeking to prove their status as the China-born children of Chinese American fathers. The siblings' detention reflected not only the Cold War climate but also more than half a century of contestation over Chinese citizenship in America. In the nineteenth century, Chinese laborers first entered the United States en masse to work in gold rush California and then on the transcontinental railroad. But within a few decades, a series of racist exclusion laws barred them from entry. There was a crucial exception: like other American citizens, Chinese who were citizens, whether through birth or naturalization, could pass their citizenship on to children born abroad. Over the course of succeeding decades, dense transnational kin networks proliferated, as two generations of Chinese-born people, predominantly young and male, claimed American citizenship as the sons, and then grandsons, of those nineteenth-century immigrants. Between 1920 and 1940, some 71,000 Chinese entered as "derivative citizens," the legal term for people like the Lee siblings, born outside of the United States but claiming American citizenship by way of a parent.[2]

By the time the threesome arrived in New York City in 1952, the Chinese exclusion laws had been abolished. But an absurd quota admitting no more than 105 Chinese annually did the same exclusionary work, and derivative claims were one of the few ways of gaining lawful entry. Meanwhile, civil war and the Chinese revolution of 1949 caused a dramatic spike in derivative claims. Tens of thousands of people clamored to establish their status as children of American citizens.

Consular and immigration authorities were overwhelmed by the petitions, especially since they believed that many, perhaps most, of them were fraudulent. Over the decades of exclusion, Chinese immigrants developed practices of so-called paper immigration, in which people gained entry to the United States by falsifying their identities as derivative "sons" of Chinese American "fathers." U.S. authorities had long been at a loss as to how to expose fraudulent claims. In the shadow of the Cold War, the stakes had grown, as officials became convinced that Chinese Communists could use paper kinship to infiltrate the country.

Two weeks after their arrival, the Lee siblings were taken from Ellis Island to the U.S. Marine Hospital on Staten Island. There, a Chinese interpreter instructed each of them to hold out an arm while a technician drew several vials of blood from a vein in the forearm. The oldest brother Kum Hoy later recalled that they had no idea what the procedure was for.[3] Days later, their parents, Lee Ha and Wong Tew Hee, also traveled to the facility to give the mysterious blood samples. The tests, as the Lees would soon learn, were a method that the Immigration Service had recently adopted to assess whether they were the true children of Lee Ha or paper imposters. Hereditary blood groups had been used in paternity proceedings for more than two decades. Now modern paternity entered immigration and citizenship proceedings, as the government drew on scientific methods to establish not belonging in the family but belonging in the nation.

At a hearing in August, an Immigration and Naturalization Service (INS) inspector interrogated each of the family members and asked the children and their father if they had given blood samples. But it was only when it was their mother's turn that the inspector drew out a pink card with the results. One wonders if he reserved the moment of revelation until Wong Tew Hee's examination because of the assumption that only she, the mother, could truly know the father of her children. One wonders too if this subtle dig on her fidelity as a wife would have translated across the cultural and linguistic divide that separated her from the inspector in this hearing room on Ellis Island.

The inspector now announced that, according to the blood test, "You and Lee Ha could not possibly be the parents" of Lee Kum Hoy and Lee Moon Wah. Whatever Wong Tew Hee was thinking at that moment, through the translator she responded simply, "the applicants are my children."[4]

THE RESULTS OF THE BLOOD test would dramatically upend the lives of the five people whose status as a family was suddenly thrust into doubt. The Lee siblings would spend twenty-eight months in detention on Ellis Island and then five more years in litigation as their claim of kinship and citizenship was examined in successive administrative hearings and then in federal court. For the government, blood testing was a powerful new weapon in the half-century battle against paper immigration. It provided a cheap and easy way to assess cases quickly. It appeared to yield evidence that was accurate and incontrovertible, objective and therefore unobjectionable.

Yet the truths that genetic tests of parentage told were never neutral. State officials made strategic decisions about what questions testing would answer, how the results would be interpreted, and who would be subjected to testing in the first place. In using these methods, the government insisted on biological kinship for Chinese derivative citizens, whereas its understanding of kinship in other circumstances and for other racial-national groups was more flexible. Blood tests were introduced into immigration proceedings in an era when a century of anti-Chinese racism was being reworked and reformulated. Increasingly (though by no means entirely) excised from law, racial and racist attitudes were alive and well in everyday administrative practice. Scientific tests could not, as immigration officials believed, be conducted apart from that context, not least because the perceived need for them arose out of the very history of Chinese exclusion. Paternity testing was a seemingly neutral bioadministrative technique, but it epitomized the modernization of racist immigration practices, rather than their elimination. Modern paternity laid bare the tensions between social and biological paternities and also revealed how those paternities were stratified, applied by state authorities to different groups for different purposes.

From its beginnings in the 1950s, kinship testing of immigrants was dogged by two legal questions: Were Chinese people the only ones subjected to the tests? And if so, was this constitutional? The questions were timely ones.

The very month the Lee siblings arrived in New York, the Supreme Court agreed to hear *Brown v. Board of Education*. The debate over the new blood tests would unfold against the backdrop of these momentous developments in civil rights.

When immigration and consular officials began drawing the blood of immigrants, they applied a now familiar scientific technique to an entirely new set of social and political questions. By this time, blood group tests had become universally accepted among international scientific experts as a reliable method for excluding paternity, and U.S. courts, initially slow to accept them, were increasingly receptive. A technique for establishing the borders of family was now put to work safeguarding the physical and metaphysical borders of the nation.[5] In fact, these two applications were very much linked because relations of filiation structure membership in both the family and the nation. The child enjoys the benefits of familial membership—a name, parental support, inheritance—by virtue of his or her status as son or daughter; belonging in the nation may likewise derive from one's status as son or daughter of a citizen.[6]

The principle of descent was especially central to the architecture of Chinese American citizenship. In a long-standing pattern of transnational family formation, Chinese American men residing in the United States returned to China to marry but were prohibited from bringing their wives to the United States, so their children were born in China. Claims to American citizenship among successive generations of Chinese Americans thus rested not on birth in the national territory *(jus soli)* but on descent from citizen fathers *(jus sanguinis)*. The history of exclusion explains why kinship testing was first applied to the Chinese: for American officials, the scientific method was necessary to verify their transnational families. Meanwhile, even as the testing of the Lees grew out of this history, it also anticipated the future. Genetic tests of parentage would come to play a key role in modern immigration practices globally.

THE LEE CHILDREN'S FATHER, Lee Ha, was born in China around 1903 and in 1923 married their mother, Wong Hee Tew. Soon the first of the couple's five children, Lee Sang Hai, was born. Three years later, Lee Ha left his wife and son and immigrated to the United States, where he suc-

From Lee Ha's immigration file: an application to return to China in 1929, at age twenty-three.

Lee Ha, 111/52, Box 359, Chinese Exclusion Case Files, National Archives and Records Administration, New York.

cessfully claimed derivative citizenship through his father, Lee Poy, who according to INS records was a native-born U.S. citizen. Over time, several of Lee Ha's brothers and a sister would also enter as derivative citizens. He took up residence in New York's Chinatown to work as a laborer, but the exclusion laws did not allow his wife to join him.[7]

Over the years, Lee Ha returned several times to China to visit his family, and each time he reentered the United States, he reported to immigration authorities the birth of additional children back in China. The twins Kum Hoy and Kum Ock were born in the fall of 1930. Kum Cherk and Moon Wah followed in 1939 and 1940. When Lee Ha returned to the United States after his second trip, he brought one of the twins, nine-year-old Kum Ock. After almost a quarter century of transcontinental family separation and with the exclusion laws abolished, in 1949 Lee's wife joined him in Chinatown, leaving the other twin, Kum Hoy, and the two youngest children in China. Finally, that summer night in 1952, the three remaining siblings arrived at the airport in New York. It was the first time the family had ever been together on a single continent. By this time, one of the twins, twenty-one-year old Kum Hoy, had his own family: like his father a quarter century earlier, he left a wife and infant son in China.

This, at any rate, was the Lee family on paper. But the transnational family patterns produced by exclusion inspired creative adaptations and sometimes bred lucrative opportunities for fraud. On each trip to China to visit wives, men resident in the United States could claim the birth of a child who did not actually exist, thereby creating a paper identity, or "slot," that could later be sold to a would-be migrant. Because men garnered higher wages in the United States, male immigration slots could be sold for a higher price and therefore sons predominated in these paper families. The ruse could then be perpetuated across generations, as one paper son conferred his citizenship to his own paper son. Paper slots were not just sold on the market; they were used by nephews, cousins, and other relatives. As webs of paper kinship became increasingly complex over time, fathers with paper identities had to use them to immigrate even true sons. Over the first half of the twentieth century, tens of thousands of people entered via paper immigration. By 1950, an estimated one-quarter of the Chinese population in the United States were paper children, people whose legal status relied on fictitious relations with a citizen parent.[8]

Government officials were flummoxed in their attempts to expose paper families. The identity of people born in rural villages was difficult to verify because they typically had no birth or marriage certificates. The Lees presented a certificate of identity that Wong Hee Tew had solicited from a village elder attesting to the identities of her five children, but U.S. officials were hard-pressed to judge the veracity of such documents. Instead, they relied on the testimony of migrants and their witnesses, together with the INS's own records. Over time, the process of assessing identity claims became increasingly complex and time-consuming. Chinese migrants and their kin were subjected to detailed interrogations with hundreds of questions about their family histories and homes. In the weeks after their arrival, the Lee siblings and their parents were each interrogated about details of their house in Lung Woy Village, including the color of the shutters, the structure of the windows, the layout of the rooms, and how many skylights it had.[9] Their answers were then compared with each other and with the interrogation records of Kum Ock, the twin who had immigrated more than a decade earlier.

The inspector concluded that the Lee family's testimony was "reasonably harmonious" and "reasonably consistent" with INS records. Yet this was hardly conclusive, for as officials constantly lamented, applicants were frequently coached to produce their answers; consistent responses, no less than inconsistent ones, could therefore indicate fraud. Even the most thorough interrogation therefore yielded information of questionable value.

BEGINNING IN THE EARLY twentieth century, U.S. authorities frustrated by the limitations of oral testimony and inadequate documentation had begun to experiment with another archive of identity: the physical body. They applied the full battery of late nineteenth-century technologies of bio-identification to Chinese immigrants, including fingerprinting, photography, and anthropometry. They also attempted to assess physical kinship. As early as the 1910s, immigration inspectors looked for resemblance among putative kin. One preprinted form used in the 1920s in Chinese derivative citizenship cases asked: "Is there resemblance between alleged father and applicant? Is there resemblance between applicant and prior-landed brother? Is there resemblance between prior-landed brother and father?"[10] In one inspection, three immigration officials examined an applicant and his alleged father, finding "the general contour of the face and head is quite similar" as well as "an exact duplication of the right ears of the two" and "a great deal of similarity in the left ears."[11] These assessments echoed the use of resemblance in paternity suits, except here it was an immigration official rather than a judge or juryman who scanned the body in search of telltale likeness.

Immigration authorities were consequently primed for modern paternity, with its promise that parentage could be made certain through new scientific techniques. They followed developments in kinship science then emerging across the transatlantic world. In 1925, a San Francisco-based inspector wrote the INS central office in Washington in response to an article he had read in the newspaper, "Fingerprints as Proof of Paternity," concerning the research of a Norwegian expert. He wondered if the immigration service might fingerprint "all Orientals admitted as citizens of the United States" to assess their kin ties. Foreign governments could hardly object to such a method, he reasoned, since the individuals in question were alleging American citizenship—a tacit acknowledgment that the method might be found objectionable. The central office replied that the proposal was "inadvisable" for reasons that anticipated the controversy surrounding blood tests a quarter century later: it "would be looked upon as discrimination against these people because of their race and undoubtedly would cause a great deal of criticism."[12]

A decade later, the San Francisco inspector tried again. Forwarding a thick file documenting Chinese immigration fraud that he had compiled to justify his request, he again directed his superiors to a news article, this one about a scientific method recently introduced into paternity proceedings in New York City courts: blood group testing. Using the technique to assess derivative citizenship claims "would greatly facilitate examination to the best interests of all concerned." Again he proposed that immigration officials borrow from paternity proceedings, and again his knowledge of the method came, as such knowledge so often did, from the press.[13] Again, however, his superiors were unconvinced. "You are advised that this office is not prepared to give consideration to this method of testing paternity at this time," the INS commissioner replied curtly.[14] Other correspondence reveals why: the central office worried that in their zealousness to exclude Chinese, San Francisco inspectors were running roughshod over U.S. citizenship law and due process.[15]

BUT BY THE 1950s, such reserve was gone. Paradoxically, methods and procedures that the authorities had rejected as overzealous and racist in the 1920s and 1930s, when Chinese exclusion was the law of the land, were embraced several decades later, after the exclusion laws were abrogated and when formal racial discrimination was on the defensive. What had changed? First, modern paternity had become established convention. In the 1920s, the idea that science could and should find kinship on the body was novel and, for many observers, improbable: after all, the INS inspector was writing from San Francisco, home of Albert Abrams's oscillophore. By the 1950s, not only did the technique of blood grouping enjoy universal scientific consensus, but the broader conceit that science could find parentage on the body was both more familiar and more acceptable.

If the science had changed, so too had the perceived need for it. By the early 1950s, the guardians of America's borders believed that Chinese paper citizenship had morphed from a perennial challenge to a full-blown crisis. The Chinese Revolution had spawned a large emigration from mainland China and sparked a surge of petitions by derivative citizens and the wives of Chinese Americans. Consular officials in Hong Kong and INS officials in U.S. cities found themselves in an administrative emergency. The traditional methods for vetting applicants—laborious investigations, time-consuming interrogations—simply could not meet the deluge.

The nature of the problem had also changed. Officials believed that paper immigration had evolved over the decades into a formal criminal operation stretching from Hong Kong to California to New York. They claimed that some 124 citizenship brokerage firms operated in Hong Kong, matching would-be immigrants with available slots, coaching them for interrogations, selling pre-prepared family histories to be memorized by "fathers" and "sons," and greasing the palms of dishonest officials.[16] American citizenship, they charged, was bought and sold on a $3-million-a-year black market.[17] "An alien Chinese can purchase American citizenship for $3,000," alleged an oft-quoted report by the American consul at Hong Kong. "Terms: $500 down, balance after arrival in the U.S."[18]

Officials charged that the citizenship racket was tied to narcotics trafficking, prostitution, and labor exploitation. Worst of all was the threat that it posed to national security. Consular and immigration officials sounded increasingly shrill warnings that Chinese Communists could insinuate themselves into paper genealogies to gain entry into the United States as sons and citizens. Or they might blackmail the many paper families already resident in the United States to further their subversive designs. The long-standing anti-Chinese thrust of law and policy, with its impetus to bar an undesirable and "unassimilable" racial group, now acquired a national se-curity motivation. The "yellow peril" became a red one.

What is more, officials were convinced that the paper imposters were win-ning. Hamstrung by a system that always seemed to favor the applicant's claim to citizenship—as a judge concluded in a 1920 case, "better that many Chinese immigrants should be improperly admitted than one natural born citizen of the U.S. should be permanently excluded from his country"—they watched as apparently fraudulent petitioners waltzed through. Of 6,000 individuals who like the Lee siblings arrived in the United States on travel affidavits, only 2 percent were eventually turned away by immigration au-thorities. Meanwhile, because applicants whose citizenship claims had been denied by the consulate in Hong Kong had the right to appeal their case before a federal tribunal, the crisis also spilled into the courts. Some 1,100 such actions were filed, mostly in New York and San Francisco. The gov-ernment found itself "practically powerless to produce evidence to offset the self-serving testimony of the plaintiff" and rarely succeeded in doing so. "All this," groused one exasperated official, "merely to establish the identity of some obscure Chinese claimant."[19]

Nowhere was the sense of crisis more acute than in the American con-sulate in Hong Kong. By 1950, American consulates in mainland China were shuttered, leaving Hong Kong to absorb all their diplomatic and con-sular duties, including a massive backlog of citizenship applications. That year, 117,000 people applied for U.S. passports as derivative citizens, a 67 percent increase from the previous decade. On top of the pending citi-zenship cases, the consulate began accumulating a backlog of an additional 150 cases per month due to staff shortages. Officials complained of the "in-cessant demands of Chinese applicants," "inundating the office from all parts of China"; three staff members handled 150 calls a day from people inquiring about their applications. As the calls and inquiries reached "stam-pede proportions," the management of the consulate's building complained about the "large numbers of Chinese persons" spilling out of the waiting room and into the corridor.[20]

In the spring of 1950, two State Department inspectors arrived in Hong Kong to assess the crisis. Observing the deluge of derivative citizenship ap-plications, they made a proposal: why not use blood group tests to vet them?[21] An idea that in the 1930s had been summarily dismissed now seemed a godsend. If officials harbored any reservations about the proposal, they did not leave any trace in the consular records. Within a month they requested permission to implement the new procedure. It is probably not by chance that the blood testing of Chinese American citizens and would-be citizens began not at the behest of immigration inspectors in San Francisco, nor of authorities in Washington, DC, but under the auspices of consular officials halfway around the world. Far removed from the scrutiny of the U.S. public and the courts, the new method could be imposed on people who were not U.S. citizens but desired to be. It was through the American consulate in Hong Kong that modern paternity first entered immigration proceedings.

Shortly after the State Department inspection, consular officials recruited two local doctors to perform the tests. One, the Australian Lindsay Ride, was vice-chancellor of the University of Hong Kong and had conducted research on the racial inheritance of blood groups in the 1930s.[22] The other, Eric Vio, was an Italian doctor who moonlighted as honorary vice-consul of Italy. It now became standard procedure to require blood tests of all "persons of the Chinese race claiming United States citizenship and applying for United States passports at Hong Kong" as well as their parents, even if they resided on the Chinese mainland. By mid-1953, Vio had tested more than 3,000 individuals. At one point, consular officials even entertained the possibility of establishing an in-house laboratory (the proposal was scrapped as expensive and impractical).[23]

The technique of hereditary blood typing was uniquely suited to the task at hand. By the 1950s, the discovery of additional blood characteristics gave the tests greater powers of exclusion, but the logic of the method first developed in the 1920s remained the same: testing could exclude impossible fathers but never positively identify actual ones. It was thus a test of nonpaternity. But what in some contexts was a limitation in citizenship proceedings was an asset: if the principal goal of government officials was to identify fraudulent claims, this was precisely what the test could do. Both U.S. immigration policies vis-à-vis the Chinese and the technique of hereditary blood typing rested on the logic of exclusion.

THE INITIAL RESULTS of the new method were, from the point of view of U.S. officials, spectacular. Testing revealed that 40 percent of applicants could not be the children of their alleged parents. Given that available serological tests could identify only about 50 percent of false paternities, this meant that as much as 80 percent of claims to derivative citizenship among applicants in Hong Kong were fraudulent. Based on certain patterns of derivative claims—dramatically skewed ratios of male to female children, very high birth rates, suspiciously low child mortality rates—government authorities had long deduced that paper kinship was rampant. Now they had an actual statistic to capture its extent. The 80 percent figure was cited triumphantly by officials and other observers.[24] The test not only demonstrated collective patterns of fraud but allowed them to identify which specific families were bound by paper, not blood.

Blood group testing not only yielded dramatic results but did so in an efficient manner, thereby promising to reduce "the burden on the government."[25] Where interrogations and other investigative techniques were crude, laborious, and costly, blood testing was conclusive, quick, and cheap—especially since applicants paid for it themselves. It allowed strapped officials to make rapid determinations on large numbers of cases and gave judges a way to partially relieve jammed court dockets. A simple blood test, in short, appeared poised to vanquish a half century of paper duplicity. "The striking results of the blood testing program . . . in uncovering the significantly high percentage of fraudulent citizenship claims," gushed the American consul, "must be a source of satisfaction to all of us in this work."[26]

The impact of the policy on the thousands of people applying for derivative citizenship was less salutary. Those who refused to take the test were automatically denied a passport. Those whose blood type was found incompatible with people they claimed as kin not only found their applications denied but also faced criminal charges for what consular officials now called "blood fraud." When a blood test determined that applicant Loui Toy could not be the "blood son" of his alleged mother, Leong Yuet, local police searched their residence for incriminating evidence. Leong Yuet claimed her son was adopted but was convicted of conspiring to obtain a passport by fraud and given a sentence of three months in prison or a fine of HK $1000. Charges against Loui Toy were dropped after he declared he did not know he was adopted.[27] The consulate eagerly pursued such cases, convinced they would serve as a deterrent to others.[28]

Back in Washington, State Department authorities were so impressed by the results that they soon recommended expanding blood testing across the agency and even pressed Hong Kong to test for additional serological characteristics to enhance the power of exclusion.[29] Meanwhile, the State Department began coordinating testing with the INS. Collaboration between the two agencies was indispensable for effective testing, since members of the claimant's family might live in Hong Kong or mainland China (under the consulate's purview) and others in the United States (under INS jurisdiction). Soon the Hong Kong consulate was working with regional immigration offices in New York and then San Francisco, Boston, and other cities. By late 1952, a little more than two years after it had first implemented the method, the consulate was forwarding one hundred requests for blood tests a month to the New York office, a task that required two full-time investigators (probably because officials physically accompanied applicants to be tested in order to verify their identity).[30] Officials at the INS and the State Department pronounced themselves thrilled with the "splendid cooperation" they had achieved.[31]

Meanwhile, having learned about the serological method from the State Department, the INS took up the "invaluable aid" for its own work stateside.[32] The three Lee children were among its first subjects. By the end of the year, the policy was already being written into the Operations Instructions, a manual that guided local officials on agency policies.[33] Shortly thereafter, the INS's highest administrative body, the Board of Immigration Appeals (BIA), declared blood test results conclusive, meaning that other evidence favorable to an applicant—photographs, testimony, even documents like birth certificates—were rendered moot by a negative blood finding.[34] Testing also expanded from applicants for passports to those requesting visas.[35]

The new procedure required complex coordination across three federal agencies, the State Department, the INS, and the United States Public Health Service (USPHS), whose doctors performed the tests.[36] But the logistics were most trying for families themselves. If in the Lees' case all five family members were located in a single city, in other cases the applicant or applicants were in Hong Kong, the mother in mainland China, and the father and sometimes other siblings in different parts of the United States. Testing mothers required that they travel from rural Chinese villages to Hong Kong, an arduous journey not least because of growing travel restrictions. Even within the United States, the logistics could be daunting and expensive. Not all USPHS facilities were equipped to perform the tests, occasioning long trips for applicants and their kin. Initially all applicants at the Miami District Office were required to travel to the USPHS hospital in New Orleans, physically accompanied by an INS official to verify their identity.[37] Recognizing the burden of the policy, the INS eventually allowed people to be tested in local blood banks or by private doctors closer to home.

Having begun in the Hong Kong consulate, taken root in the State Department, and then expanded to the INS, blood group testing also entered the federal courts. Judges in California and Washington saw the method as a quick way to clear dockets overwhelmed by citizenship applicants who sought redress in the courts after the consulate denied their petitions.[38] In just a few years, multiple federal authorities charged with immigration and citizenship matters embraced the technique. As for the INS, internal correspondence shows no qualms, legal or ethical, about the use of the method. Correspondence does, however, reflect careful study of the legal and scientific bases of testing, suggesting that officials anticipated judicial challenges to the new method—challenges that were not long in coming.[39]

A MONTH AFTER THE Lee family's negative test results, they were summoned for another hearing. This time they brought a lawyer who was asked if he wanted to present new test results challenging the first ones. Inexplicably he instead chose to enter into evidence two snapshots of the siblings taken at the Hong Kong airstrip the day of their departure. While photographs were routinely used as evidence to establish family relationships, the recency of these made them useless for establishing a long-term relationship. It was as if a visual image representing the children as a family might somehow conjure their kinship into being despite the negative blood group evidence.

Since the lawyer presented no new blood evidence, the official now pronounced on the meaning of the INS's own tests. Not only did they show that the two boys could not be biological children of Lee Ha and Wong Tew Hee, but they called into question the claim of their alleged sister. "Although the claimed parentage is possible in the case of you, Lee Moon Wah," the official told the twelve-year-old, "your credibility is impeached by the results of the blood tests."[40] Negative results for any member of a family challenged the truthfulness—and therefore the citizenship claims—of all of them.

This logic, which became standard practice in citizenship proceedings, demonstrates immigration officials' expansive understanding of the truths that blood testing revealed. From the point of view of scientific experts, the method's powers of exclusion were decisive but limited; even using all known serological factors, only about half of fraudulent cases could be uncovered because the test could not identify paper kin who happened to be compatible with one other. But the INS held that an incompatible result for one member vanquished the claims of the entire family, such that even biologically compatible applicants could be rejected if their siblings tested negative.[41] Read as a test of genetic possibility, blood had the power to exclude half of paper children, but when the government used it as a test of credibility to exclude a compatible child like Moon Wah, it had the power to exclude considerably more.

The only relationship legally material to a determination of derivative citizenship was that between the applicant and his or her putative parents, but following the logic of credibility officials began to ask for blood tests from a much wider circle of kin—siblings, aunts and uncles, grandparents. A negative finding between any of them impugned the claims of all of them. One State Department official advised a Los Angeles lawyer that because "credibility . . . is the prime factor" in the adjudication of citizenship cases, a client's father could "support the case and evidence his good faith" by submitting blood test results for himself and "all his alleged sons in the United States." The official also promised to "give further consideration . . . in the light of such blood-type evidence" to two other unsuccessful cases if the lawyer submitted blood results for all family members of the rejected applicants.[42]

Read through the scientific logic of blood group heredity, the official's advice makes no sense at all. Blood typing could never yield positive proof of paternity; the only conclusive result it could produce was that the alleged relationships did not exist, which is what the State Department had already concluded. Perhaps the official was trying to trick the lawyer into submitting evidence that could further damage his clients' cases. Perhaps he simply did not understand blood group science. But if what a blood test tested was not biological compatibility but certain social qualities—trustworthiness, veracity, "good faith"—then his advice made perfect sense. The applicant's willingness to submit to tests would be held as a badge of truthfulness.

In assessing credibility, genetic tests assessed precisely what, in the long-standing racist assumptions of immigration authorities, the Chinese lacked. Blood tests were supposed to be more objective, impartial, and accurate than interrogation, the traditional method of vetting Chinese citizenship claims, yet the two methods were more alike than different. Scientific tests no less than interrogation methods assessed the credibility of subjects allegedly inclined to deceit. Based on this logic and the negative tests of two of the three Lee children, the INS official now ordered all three siblings excluded from the United States and deported back to Hong Kong.

In succeeding months, the family spent additional time and money appealing this verdict up through the INS's administrative channels. In February 1953, the BIA affirmed in another case that blood group evidence was conclusive of nonpaternity and hence noncitizenship. The applicant in that case, a sixteen-year-old boy who had arrived to New York from China just eleven days before the Lees and was probably detained with them at Ellis Island, was almost certainly deported. By the time their own case reached the BIA three months later, the children had no obvious grounds for challenging the negative finding and soon lost their appeal.[43]

By this time, the siblings had languished in detention on Ellis Island for almost a year. With no further administrative remedies available to them, the INS began making arrangements to place them aboard the American President Lines and transport them to China. Out of time and nearly out of money, the family made a last-ditch effort to avert the deportation. They hired a young attorney, Benjamin Gim, whose office was just down from the family's apartment on Mott Street in New York's Chinatown. With deportation now imminent, Gim raced to file a writ of habeas corpus in federal court (in his—possibly apocryphal?—telling forty years later, the children had already boarded a plane and were pulled off at the last moment thanks to the filing).[44] With the writ of habeas corpus, the Lees' case now exited the INS and entered the federal courts, initiating a five-year saga that would take them all the way to the Supreme Court. For their young attorney, the case was the challenge of a lifetime. "I'd never had so much fun in my life," he remembered decades later, "and gotten so little money."[45]

Gim was four years out of Columbia University Law School and one of just three Asian American lawyers in private practice in the state of New

York. Having begun his career working for the government because no law firm would hire an Asian American, he had recently gone into private practice in Chinatown. As he later recalled, not even Chinese wanted to hire Chinese lawyers because they feared discrimination would disadvantage their cases. What is more, Gim had no experience in immigration law. For the Lees, hiring him was an act of desperation.

Their act of desperation turned out to be a stroke of luck. Gim was young, unmarried, and had no other clients, "so I just spent a lot of time in the library" and soon proved himself a brilliant litigator.[46] Born in rural Idaho and raised in Salt Lake City, Gim had lost both his parents as a child, and his sister had kept his four siblings together through the Depression. Perhaps the Lee children—long separated from their parents while in China and now completing a year in detention together—reminded him of his own parentless family.

GIM'S WRIT OF HABEAS corpus made two arguments. The first was that there remained serious doubts as to the accuracy of the blood tests and, second, that the method was being used in a discriminatory manner, applied exclusively to Chinese applicants. The first argument seemed like a nonstarter. The BIA had already affirmed the admissibility of blood groups in parentage determinations, and New York was an especially unfavorable place to challenge this principle. The state was the first to pass legislation admitting serological evidence into parentage proceedings (in 1935) and boasted almost two decades of case law amenable to testing. Federal courts hearing immigration cases originating in New York were subject to the state's statutory law on the matter. New York's blood-group-friendly legal landscape facilitated the INS's use of the test in the state with the second largest Chinese community in the country.[47]

This, of course, made Benjamin Gim's job much harder. Instead of questioning the legal validity of the method, he drew attention to certain irregularities of the Lees' tests. As it turned out, the agency had conducted two sets of tests on the family and obtained different results in each. The first set excluded all three children, while the second excluded two of the three. Somewhere there had been an error—and the young peoples' lives hinged on it. Throughout the hearings, INS officials had been remarkably dismissive of this fact, arguing it did not matter if errors had occurred if subsequent tests had corrected them. "It is not cause for derision that rechecks have revealed different results," noted the BIA defensively. "It is their purpose to reveal errors if there have been errors." Judge Dimock of the U.S. District Court, who reviewed Gim's writ, was less forgiving. "Is there any reason to believe that a third set of tests would not result in a still different conclusion?" he asked. He found the allegations serious enough as to constitute a violation of the Lees' right to due process and ordered the INS to reopen their case to explore the question. Another year of hearings ensued. The Lee children remained in detention at Ellis Island. From the yard, they could see sweeping views of the Manhattan skyline, visible through the high fence.[48]

The apparent mistakes in the Lees' case were by no means exceptional. Especially in its early years, the blood testing of immigrants was marked by frequent errors. USPHS technicians were not experienced blood testers and were especially ill equipped to perform certain tests, such as the highly delicate MN analysis.[49] Even the doctor who oversaw the Staten Island hospital where the Lee family had been tested admitted his facility was not equipped to follow certain medicolegal protocols and that "there is a possibility of error which may creep in."[50] And then there was the sloppiness of the INS officials themselves. It emerged that apart from the contradictory test results, the officials in the Lee case had incorrectly reported them by transposing the children's names. In the first hearing, they had announced that younger son Kum Cherk was the only child compatible with the alleged parents and that his two siblings were excluded; in a second hearing, they reported that daughter Moon Wah was compatible and the two boys were not. In fact, the second report was wrong and the first report right—assuming, of course, that the tests were accurate in the first place.

The INS's cavalier response to error in the Lee case characterized its attitude in general. In another case of alleged error, the BIA implied that retesting was a gracious courtesy on the part of the government because the burden was on the applicant to prove his or her identity and citizenship.[51] Despite its public dismissal of the problem, the INS must have recognized that allegations of incompetence would undermine the powerful new method because in June 1953, just as it was gearing up to deport the Lee children, it quietly discontinued testing at the Staten Island hospital and began sending applicants to private doctors. By then the facility had performed some 500 tests in citizenship cases.

TO ESTABLISH THE QUESTIONABLE competence of the USPHS, Gim deposed a well-known hematologist and seasoned paternity tester, Leon Sussman, and had him cross-examined by a lawyer named Sidney Schatkin, a leading legal expert on scientific paternity testing. Sussman and Schatkin would become familiar fixtures in the new landscape of citizenship testing, as would prominent blood group researchers Philip Levine and Alexander Wiener, both protégés of Karl Landsteiner (who had died in 1943). Blood testing seamlessly entered immigration and citizenship matters thanks to the collaboration of a medicolegal community with years of experience promoting and defending it in paternity proceedings. New York City was the ground zero for these developments because of its large Chinese community, favorable laws on blood group evidence, and its concentration of paternity experts. Beginning in the 1930s, New York City's Court of Special Sessions had begun accepting blood group evidence in what were called "bastardy proceedings." Sussman, Wiener, and Levine had testified in those cases, and for more than two decades, Sidney Schatkin had participated in thousands of them as a lawyer of the court. In the process, he became the nation's leading paternity lawyer and paternity testing's most outspoken public champion.[52] Because blood group testing involved scientific knowledge most lawyers lacked, Schatkin was brought into court to examine doctor witnesses, thus serving as a translator between these two fields of expert knowledge. In immigration cases, he and the other paternity experts would serve as translators of another kind: between "bastardy" and citizenship proceedings.

In the Lees' case, Sussman and Schatkin testified for the family in their appeal against the INS. Under Schatkin's questioning, Sussman testified to the "doubtful validity" of the family's blood tests due to possible "inaccuracies" at the USPHS lab.[53] The medicolegal community had spent two decades convincing judges and juries that blood group analysis was a reliable way to establish (non)paternity. The early patterns of error in INS testing were of concern to them, for unreliable tests could undermine the whole enterprise of paternity science. In the first instance, their loyalty lay neither with Chinese applicants nor with the INS but with their method itself.

Once the INS cleaned up its act, however, medicolegal experts were more likely to be found testifying for the government. Citizenship testing represented a whole new application of their science, one they enthusiastically embraced and defended. Sussman and Schatkin coauthored articles about Chinese blood testing together with an INS inspector named Dorris Yarbrough: the government entered the articles as evidence in their case against the Lee children.[54] The testing program brought representatives of science, law, and the immigration bureaucracy together to defend the integrity of the nation through a simple finger prick.

It also provided medical experts with opportunities for scientific research. Leon Sussman authored several studies of blood group distribution among people of Chinese ancestry, including the largest survey ever conducted, involving more than 800 people. As he openly acknowledged, he gained access to Chinese research subjects thanks to his collaboration with the immigration service.[55] Besides raising the unsavory prospect that people who submitted to INS testing had their blood used in scientific research, Sussman's work suggests that the INS's testing program contributed to racial serology, the specious study of racial difference in human blood types. Leading experts in the field, which endured into the postwar era, cited Sussman's studies of Chinese subjects.[56]

AS THE LEE FAMILY continued to pursue their case in New York, back in Hong Kong blood testing had run into a wall. The problem was not error but, paradoxically, the technique's remarkable success. Initially blood testing had been "highly productive" because of the "surprise element," but immigration brokers and applicants soon adapted to the new tests.[57] They bribed doctors and nurses for compatible results or switched the people being tested. By far the most effective adaptation was simply to test the blood of would-be migrants so that they could be matched with compatible paper kin. Pretesting became one more service offered by Hong Kong immigration brokers to those preparing their citizenship case. The high rates of exclusion that the consulate had initially achieved declined precipitously.

Officials were baffled by the speed of these adaptations. Back in New York, Inspector Dorris Yarbrough reported that on August 14, 1952—coincidentally, the very day the Lees were informed of their blood test results—a Chinatown travel agent suspected of involvement in the citizenship racket was speaking with Yarbrough about blood testing and mentioned that the Rh test was worthless for people of Chinese ancestry.[58] A few days later, Yarbrough came across a letter dated August 1 from the Hong Kong consulate forwarding a report by Dr. Eric Vio, one of the consulate's two official testers, making this very argument.[59] It had taken less than two weeks for this esoteric piece of scientific information to travel from a doctor in Hong Kong to a shady travel agent in New York City.

By mid-1953, the method appeared to have reached "its peak of usefulness" for the Hong Kong consulate. Where a year before rates of exclusion had run as high as 35 or 40 percent, they had since plummeted to just 8 or 9 percent.[60] The consulate's adoption of blood testing made it more difficult for brokers to match slots and buyers and probably drove up the costs of establishing citizenship, whether bona fide or fraudulent.[61] From the point of view of immigration and consular officials, this was a good thing. But what had seemed an administrative holy grail to expose paper kinship now became largely useless.

For officials, this turn of events was further proof of Chinese perfidy. In 1956, a British diplomat reported that he had recently lunched with the former American consul in Hong Kong over the following question: who was more devious, the Russians or the Chinese? The consul "ruefully" declared the Chinese and cited the blood pretesting operation as evidence.[62] In New York, Inspector Yarbrough expressed similar sentiments: "the Chinese doctors in New York City are now doing a big business in blood tests." He warned his colleagues that allowing Chinese applicants to choose their own doctors was "not advisable in a place having a large Chinese population as it invites collusion and fraud. In New York City, this procedure has resulted in one Chinese doctor making most of the tests required."[63] Yarbrough's fears played on long-standing tropes of the Chinese as cunning and deceitful and the Chinese community as insular and inclined to protect their own. An alternative explanation for why Chinese testing was concentrated in a few hands—that there were only two doctors practicing in Chinatown at this time—seemed not to have occurred to him.

Pretesting may have ruined blood as an administrative tool, but blood did not therefore disappear from citizenship proceedings. Instead, it moved from a government-supervised activity to a private one. The government's adoption of blood testing catalyzed the creation in Hong Kong of a private market in (pre)testing, in which local testers provided their services to citizenship applicants. With paper citizenship under assault by U.S. consular authorities and the local police, private labs offering testing services understandably kept a low profile, though they did advertise in local Chinese-language newspapers.[64] Hong Kong was thus home not only to the first government program of genetically testing immigrants but also, as a result, the world's first commercial market for paternity testing.

Pretesting raised the issue not only of how to defend scientific truth but of who should have access to that truth in the first place. Shortly after the INS began to test would-be citizens, Inspector Yarbrough wrote a memo to colleagues explaining the science, law, and practical uses of blood testing. He discussed the USPHS's troubling pattern of early errors but did so to point out what he believed was a greater evil: that those errors had regrettably led the service to accept blood tests performed by private doctors whom the INS could not monitor. In other words, the inspector was worried less about bad science—a problem revealed to be pervasive in the USPHS facilities—than about good science in the wrong hands. A Chinatown doctor could not be trusted, and a Chinatown travel agent had no business being privy to the latest serological research.

Yarbrough wanted the government to enjoy a monopoly over genetic knowledge. As long as this powerful technology remained a specialized technique in the hands of vetted experts acting under government auspices, it would be dedicated to legitimate purposes. But when it became accessible to travel agents, Chinatown doctors, or private labs in Hong Kong, it became disruptive and even dangerous. Who has the right to know? This was the question that pretesting raised, a question that has dogged kinship testing throughout its history.

THE LEES' CASE now shifted from the allegation that the INS's testing was riven by error to Benjamin Gim's second potent claim: that it was riven by prejudice. According to Gim, the Immigration Service was testing Chinese, and only Chinese, derivative citizenship cases, and this amounted to unconstitutional racial discrimination. The Lees' case had begun as an inquiry concerning the fate of the three putative children of a humble Chinatown grocer. Now it became a referendum on the constitutionality of the government's new blood testing practice—and of racial discrimination in immigration proceedings generally.

Advocates of citizenship testing asserted it was not only a cheap and efficient tool but also a neutral one, yielding information that was objective and conclusive. It was also supposed to be innocuous. When used in the context of paternity suits, blood testing raised vexing questions about children's rights, marriage, morality, and family integrity. But advocates argued that in citizenship matters, it was conveniently unencumbered by such quandaries.[65] Blood corpuscles simply revealed whether the applicant could or could not be the son or daughter of the citizen father.

This representation of testing as neutral conveniently sidestepped blood's most profound cultural association: its evocation of race. Given the potency and persistence of blood as a metaphor for racial identity, it was well-nigh impossible to talk about blood, particularly in relation to immigration, citizenship, and national belonging, without talking about race. By the time immigration authorities adopted blood testing, the architecture of Chinese exclusion had been dismantled, but immigration law was hardly race neutral. In 1952, Congress passed a new Immigration and Nationality Act that retained discriminatory national quotas and defined Chinese as a racial category (peoples of Chinese ancestry automatically fell under the Chinese quota regardless of where they were born or whether they were nationals of a third country). The new act was controversial; President Truman called it "un-American," and in a revealing turn of phrase, the African American press referred to its racial quotas as "blood tests."[66] That the same term referred to both racist quotas and hereditary analysis captures how race could not be eliminated from blood talk.

This was especially true, of course, given Gim's allegation that the INS was using blood testing itself in racially discriminatory ways. He now set out to prove this contention, deposing the medical director of the USPHS facility where the Lees had been tested. The witness testified that the INS had instructed him to test certain individuals of Chinese descent but not people of other races or nationalities. Of the 200 people who had given blood at his USPHS station, all were Chinese. Gim then deposed Dr. Leon

Sussman and asked about the 300 blood tests he had performed for the INS once testing was discontinued at USPHS facilities:

Q. Now, according to your own recollection what is the racial antecedent of all people tested?

A. They are of Chinese extraction.

Q. All Chinese extraction?

A. Yes.

Q. You don't recall a white person being sent by the Immigration service for the purpose of disproving paternity?

A. No, I don't recall.[67]

An INS lawyer disputed the allegation, pointing to four recent cases in which "non-Chinese" had been tested. On cross-examination, however, it turned out that in two cases the individuals were citizens of the British West Indies and hence "likely were negroes," and in the other two the witness could not identify the subjects' race, stating only that their names suggested they were not Chinese. Gim requested that the INS make available its case files so that he could examine racial patterns of testing, but the Service refused. Gim had queried local lawyers and learned of Chinese applicants who had been subjected to testing even though they had documentary evidence of their identity, including marriage and birth certificates generated by British authorities in Hong Kong. He asked to examine those records, but again the Service declined, citing privacy concerns. Finally, Gim asked the government to share the instructions the INS directed to its officials regarding blood testing. Again, the government refused.

Had Gim been allowed to consult the INS's records, he would have found abundant and explicit evidence of bias. The archives of the Service bristle with references to the blood testing of Chinese, both in the workaday communications of local officers and in the official instructions of central authorities. Three days after the Lee siblings arrived in the United States, for example, an official at the INS Central Office wrote to the Boston District director about the "new procedure" to be used "in connection with the issuance of U.S. passports to citizens of the Chinese race."[68] A telegram between two INS officials referred to the "Chinese blood typing program." An official in the Miami office noted that "all Chinese persons in this District" had been sent to New Orleans for testing. Inspector Yarbrough suggested his memo on blood testing "will be of interest to the other Immigration offices handling Chinese." Similar references appear throughout the files.[69]

They also appear in the Operations Instructions, official directives that sought to ensure uniformity of procedure across the agency. An instruction entitled "Action to be taken on visa petitions filed by Chinese petitioner when no documentary evidence of relationship of Chinese petitioner to beneficiaries submitted" refers to the "Chinese petitioner" no less than six times in four paragraphs. What is more, tucked in the archival files is an earlier draft of this same directive. This version refers to a generic "petitioner," but in a series of handwritten edits, someone had penciled in the word "Chinese" throughout the document.[70]

BENJAMIN GIM WAS NOT the first person to deduce that blood testing was being applied disproportionately, and perhaps exclusively, to people of Chinese descent. Because testing began in Hong Kong and its subjects were by definition not American citizens, the American public was not immediately aware of the new policy. But as testing arrived stateside and began to affect increasing numbers of Chinese Americans, it sparked growing outrage among immigration attorneys with Chinese American clients, the Chinese American press, and representatives of Chinese American civic organizations.[71] As early as the fall of 1952, a Los Angeles lawyer with a large Chinese American clientele complained to the State Department about the "untold indignities" that the Hong Kong consulate inflicted on derivative applicants and "the diabolical scheme forcing every person of the Chinese race who claims American citizenship by derivation to submit to blood typing." Such indignities were not inflicted on persons of other races, charged the lawyer, and in public hearings on the proposed new Immigration and Nationality Act, he called on Congress to prohibit these discriminatory practices.[72]

The Chinese World, the largest English-language Chinese daily in the country, emerged as the loudest voice of opposition to blood testing. In passionate front-page editorials, some of which were later published in book form in Chinese, its editor Dai Ming Lee criticized the anti-Chinese thrust of U.S. policy in general and the discriminatory practice of blood testing specifically. He sent a telegram to the newly appointed commissioner of the

INS, Joseph Swing, in May 1954. After congratulating him on his appointment, he called on Swing to end the testing program, "a practice that is extremely obnoxious to the Chinese people." Readers could follow the telegram exchange in *The Chinese World*.[73]

In the summer of 1954, such allegations were especially urgent. One week before Lee's telegram, the Supreme Court handed down the momentous *Brown v. Board of Education* decision. *The Chinese World* hailed it as "a sweeping victory for Negroes" that "elevates the spirit of democracy in America."[74] Lee called out the deep hypocrisy of racial discrimination in immigration policy at the very moment the Court had deemed racial segregation unconstitutional.

Immigration officials categorically denied Lee's charge. In a reply to his telegram, Swing insisted that blood tests were useful as proof of kinship, and their use was "not restricted to American citizens of the Chinese race."[75] A week later, the BIA issued a second decision in the case of the Lee children, finding the claim of racial discrimination unfounded. The Board argued that even if Chinese applicants were disproportionately or even exclusively subjected to blood tests, what the tests queried was not race but parentage. Its decision made implicit reference to *Brown v. Board,* even as it asserted that the case was entirely inapplicable to its testing program:

> When a case involves a question of whether or not members of a minority group in the United States shall vote or go to certain schools or be employed in certain jobs, and the allegation is that they are being prevented from voting, schooling or employment because of their race, then racial discrimination is an issue. This is not such a case. Here is only one question, that of *identity*—"Is this child the offspring of the claimed father?" There is no question of *race* involved.[76]

The history of modern paternity suggests how specious this argument was. In testing the blood of Chinese, the government used paternity science to establish the citizenship of a group defined by race, and long-standing discrimination against that group created the "need" for the scientific test in the first place. Immigration testing was thus a distinct iteration, but an iteration nonetheless, of the historical pattern in which modern paternity and its technologies are racialized.

The BIA argued that it was not racial bias but the peculiar characteristics of Chinese American citizenship that explained why the Chinese were subjected to blood testing. Chinese applicants lacked documentary evidence of their identity, and consular officials could not investigate their claims in a hostile communist country. Chinese American men lived in the United States separated from their wives and Chinese-born children, making it difficult for officials to discern the paternal link and easy for unscrupulous immigrants to fabricate it. "The cases which have so far arisen have involved Chinese," argued the INS, "because of their historic custom of leaving their children in China to be raised and educated before bringing them to the United States."[77] It was the "historic custom" of the Chinese, not the racial animus of the government, that made blood tests necessary.

This argument neglected the crucial point that this "historic custom"—the transnational character of Chinese families—was itself an adaptation to exclusionary policies. American law, not Chinese custom, made Chinese American families bi-continental and therefore inscrutable to the government. The INS's decision to test the blood of the Lee family could not be divorced from the preceding seven decades of exclusion. If the use of a new scientific technique to establish Chinese kinship and citizenship seemed sensible and necessary, it was because long-standing anti-Chinese policies made it so.

In the fall of 1954, the Lee siblings were released from Ellis Island on $1,000 bond each—an amount higher than in many criminal cases—and went to live with their parents on Mott Street in Chinatown. While most detainees at Ellis Island were held a week or two, the Lee children had spent twenty-eight months in what they called "jail," probably one of the longest detentions in the history of the island.[78] It was also one of the last. During its sixty years in operation, some 12 million immigrants had passed through the storied gateway, and Kum Hoy, Kum Cherk, and Moon Wah—now ages fourteen, fifteen, and twenty-five—were among the final two hundred to do so. A month after their release, the decaying facility was closed for good.[79]

IS THE APPLICANT the true child of this citizen? The conundrum at stake in derivative citizenship cases was a new version of an old question. According to the government, knowledge of the father was more difficult for the Chinese and required another method of proof, a different evidentiary paradigm, than for other groups. Chinese petitioners lacked not only documentary proof of parentage but certain kinds of social evidence as well. A government witness explained how paternity was normally established in citizenship cases. The typical American father who sought to pass his nationality to his foreign-born child was living overseas to conduct "social business or similar purposes." He was a member of the expatriate community and "would be well known both to the consulate and many other American citizens residing abroad and in business and in social circles." Consular officials had firsthand knowledge of the kin ties between this man and his child, and even if they did not know him personally, they could easily "ascertain from neighbors, friends, local government officials and other sources the true identity of the child." In such circumstances, blood tests were unnecessary because the relationship was considered obvious.[80]

The government claimed this was the scenario in "European" cases. In the typical Chinese case, in contrast, consular authorities did not know the father or his family. Legally he was an American citizen, but in culture, language, and race, he was Chinese, as was his wife. He presumably also differed from the consular community in terms of his humble, rural background. His child was born in a "remote interior" village that was physically but also culturally distant. The Chinese American father was thus unknown, and unknowable, to American authorities.[81]

Whether or not the "typical" case really existed in Europe or anywhere else, it was a convenient foil for describing different kinds of paternity. In normal conditions, expatriate paternity was certain thanks to its milieu; knowledge of the father was local and social. But because the Chinese American father was not a part of this community, his identity could be known only through other methods. Such an idea echoed Dr. Lehmann Nitsche in Buenos Aires several decades earlier. In a city transformed by immigration and urban growth, the community's knowledge of the father no longer sufficed, and the filial link must instead be identified through scientific methods. What Lehmann Nitsche had articulated, and what U.S. immigration authorities echoed as they clamored to examine the blood of Chinese families, was the idea of modern paternity. The case of Chinese American citizenship further shows that even as paternity was modernized, it remained stratified: some fathers were more knowable than others. While some could be identified through social knowledge, others required a scientific determination.

IN THE FALL OF 1955, a journalist from *The Chinese World* reported "rejoicing" at 115 Mott Street. The grocer Lee Ha, his "madonna-faced" wife Wong Tew Hee, and the three young people learned that Judge Dimock had ruled that in testing the family's blood, the INS had practiced unconstitutional racial discrimination.[82] All the evidence except the blood tests corroborated the paternity of the Lee children. Since a white applicant in a similar position would never have been subjected to such a test, the INS was guilty of applying different standards of proof for different groups. Dimock's ruling concerned racial discrimination, but it was also a ruling about paternity. It held that paternity must be race neutral, not only determined by the same kinds of evidence but defined in the same manner for all applicants. Thanks to the ruling, the Lee children were declared U.S. citizens. But its impact went far beyond them: some 20,000 Chinese applicants with outstanding citizenship applications were potentially affected.[83]

The celebration on Mott Street was short-lived, however. The government appealed, and a year later the Second Circuit reversed Dimock's finding. While the evidence showed that only Chinese were subjected to blood tests, the court found that the authorities were motivated not by "racial prejudice but by a proper police motive for their aid in the solution of difficult cases."[84] The distinct nature of Chinese paternity really did justify special methods of scrutiny. Gim appealed, and in January 1957 the Supreme Court agreed to review the case. An inquest originally focused on a new method for proving paternity had become a referendum on the constitutionality of racial discrimination in immigration law, one with the potential to fundamentally remake federal law and practice.[85]

AS THE LEES' CASE PROCEEDED, the government continued to test the blood of Chinese people. Because of pretesting, blood had lost its capacity to expose paper kinship in new citizenship applications. But it soon developed another potentially more extreme and far-reaching application: reassessing citizenship in cases where it had already been established. For if blood could no longer stymie new paper sons, it could still expose old ones. In fact, it could potentially be used to retroactively uncover paper kinship among people who had established their identities as derivative citizens months, years, or even decades earlier. The prospect was a radical one, and even the INS's general counsel cautioned, "the courts are not friendly to the invention of new ways to derogate already established rights."[86] Despite his warning, government officials began to do exactly that. In the fall of 1955, reports surfaced that State Department authorities in San Francisco were requiring blood tests from U.S. citizens of Chinese descent when they requested a passport. In one instance, a man who had acquired American citizenship twenty years earlier was told he and his family members had to provide blood tests before he would be issued a passport.[87] Dai Ming Lee, editor of *The Chinese World,* exhorted his readers to resist the new requirement: "We strongly urge all Americans of Chinese ancestry who are requested to take blood tests to refuse such violation of their rights."[88] The newspaper's condemnation of blood testing had become an open call to resistance.

These developments coincided with a new campaign on the part of U.S. officials to quash Chinese paper citizenship once and for all. In early 1956, the Department of Justice impaneled grand juries in New York and San Francisco to investigate citizenship fraud. The juries subpoenaed suspected paper immigrants, alleged brokers, immigration attorneys, and the records of Chinese American civic organizations.[89] The investigations threw Chinese American communities into a panic. The confessions of some individuals implicated others, and the subpoenas snowballed. Authorities began receiving anonymous denunciations of paper citizens. In Larchmont, New York, an inebriated laundress identifying herself as Mrs. Elizabeth Warner telephoned an INS office to denounce a local laundry she claimed was funneling hundreds of Chinese Communists into the area.[90] The air of a racially motivated witch hunt was unmistakable.

The grand jury investigators made liberal use of blood tests. Queens resident Ng Mong On received a letter from the Immigration Service summoning him for a test to affirm his kinship with the son he had brought to the United States six years before, but he already knew what it would show. A short time earlier, apparently tipped off to an impending subpoena, the two men had consulted a doctor on Mott Street. When a test showed they were incompatible, the agent who had helped arrange his paper son's immigration referred them to a lawyer, who advised them to refuse the government's testing request. The two men ultimately decided to confess.[91] The government could not force citizens to undergo testing, however, so in other cases its efforts were stymied. In response INS inspector Yarbrough, who was actively involved in the grand jury crackdown, subpoenaed the medical records of doctors in an attempt to show incriminating patterns of pretesting. Among those targeted was Dr. Arthur Liu, a beloved Chinatown doctor who had been mentioned by confessants and who may well have been the doctor Yarbrough had earlier accused of collusion.[92] Also subpoenaed was Dr. Leon Sussman, the serologist who had served as an expert witness in INS cases and had written an article in support of the program with Yarbrough. The inspector thus found himself in the strange position of subpoenaing the records of his coauthor.

The passport requirements and the grand jury investigations heralded a new front in the government's blood testing operation. Now it encompassed not only people like the Lee children—prospective derivative citizens hailing from China—but long-established citizens resident for years or even decades in the United States. The impact of blood testing on citizenship paralleled its impact on marriage. Membership in both nation and family was founded on social and legal filiation, presumptions that scientific testing challenged. Just as blood could reveal the false legitimacy of the child born in marriage, so too could it could strip the citizen of belonging in the nation. In so doing, it threatened to subvert long-established statuses, whether of kinship or citizenship. Testing could catastrophically destabilize both.

The Chinese World, immigration lawyers, and civic organizations denounced the growing assault on Chinese American citizenship. "It would be very interesting to know just what State Department bureaucrat is responsible for this attempted violence to the constitutional rights of American citizens," thundered editor Dai Ming Lee.[93] When the San Francisco-based Chinese-American Citizens Alliance requested clarification from the State Department about the policy, the agency responded that testing was required only when passport applicants lacked documentary proof of citizenship.[94] This was the same argument the government had used to justify testing in Hong Kong, yet as in that case, local lawyers could cite multiple cases in which blood tests had been demanded of passport applicants who presented government-issued certificates of citizenship. As the controversy grew, blood testing in Hong Kong itself was denounced. Members of New York's Chinese American chapter of the American Legion, an association of more than 900 veterans, protested the discriminatory use of blood testing as "highly unethical" and "unfair" and declared that it hindered "their efforts to bring members of their immediate families to the United States to live under the American ideals of democracy."[95]

The public reaction reflected the potentially cataclysmic impact of testing on Chinese American citizens. In a community in which an estimated one-quarter of people had paper identities, a method used to retroactively strip individuals of long established citizenship threatened quite literally everyone, particularly when the government expansively interpreted it as delegiti-mizing an entire family when one member was shown to be incompatible. While no critic ever spelled out the issue in these terms, taken to its logical extreme it raised the specter of the wholesale revocation of citizenship of an entire ethnic group. It is not hard to see in this scenario the echo of the Jews' experience in fascist Europe a decade earlier.

IN APRIL 1956, with the grand jury investigation in full swing, Chinatown resident Eng Gim Chong offered a sworn affidavit admitting that Eng Nee Shong was "not my blood son." Years earlier in China, his three-year-old son had died, he declared, and his wife had replaced him with a child from a nearby village: the adopted child had "continuously and uninterruptedly" resided with the family in China and then Hong Kong, and Eng "consid-ered him as a blood son and . . . always supported and maintained him to the present day." Eng's other children also considered their eldest sibling "their full blood brother."[96]

Steeped in references to blood, Eng's affidavit echoed the terms of dis-cussion of Chinese citizenship in the early 1950s. INS and State Depart-ment officials routinely referred to "blood brothers" and "blood mothers" to distinguish them from paper ones. They dubbed the practice of identity forgery "blood fraud." Such invocations reflected the logic of *jus sanguinis,* which understood parentage to be a legal relationship based on a "natural," biological tie. Blood talk was also reinforced by the use of actual serolog-ical tests. But these tests did not simply expose "blood fraud"; in promising to reveal the truth of kinship, the tests defined what counted as kinship in the first place. As Eng's experience suggests, a method that privileged "blood" could displace nonbiological forms of kinship practiced among Chinese immigrants.

Caught up in the grand jury dragnet, it is possible that Eng invented the adoption story to solicit lenience and to protect the citizenship claims of his younger, "blood" children. But his story was credible in the first place because in the southern Chinese communities from which most migrants hailed, adoption was a familiar cultural practice. A means of continuing one's lineage, it was fueled by high rates of infant mortality and the preference for sons, as parents were particularly likely to replace lost male children or to add them when they had only daughters.[97] Remarriage also produced nonbiological kin, as when widows brought children into their new relationships, and stepfathers presented those children as their "own" offspring. So too did polygynous arrangements, in which a man had multiple partners. In such cases, only the children of the first wife were recognized as legitimate and therefore eligible under U.S. law for derivative citizenship, but the man might attempt to secure paper citizenship for the children of his other partners as well. All of these scenarios defied the logic of blood kinship in U.S. immigration practice, a logic reinforced by blood testing.

U.S. government officials were not unaware of such practices. In 1955, the rabidly anti-Communist and anti-Chinese U.S. consul general in Hong Kong, Everett Drumright, wrote an incendiary and often fantastical eighty-nine-page exposé of the paper citizenship crisis. The report circulated widely and helped set the stage for the grand jury investigations (it was also submitted as evidence by the government in the Lee case).[98] In a section dedicated to "cultural aspects of the problem," Drumright acknowledged the variety of Chinese family forms, describing the adoption of boys as a "large scale" practice, for example. But he characterized such practices as "problems" rather than as explanations for some noncompatible blood tests. Through adoption, he claimed, "a common cultural occurrence becomes a perfect alibi"—a contention that may well have been true in some cases but failed to recognize that it served as a perfect alibi precisely because it was a common cultural occurrence. What blood tests revealed in some instances was not fraud but nonbiological kin relations. Meanwhile, Drumright called for an expanded battery of serological properties that would make the test more powerful.[99]

Even more striking than the divergences between Chinese and U.S. government notions of kinship were inconsistencies in the government's own definition. While blood defined derivative citizenship for the Chinese, this was not true for all groups. In the early 1950s, some American GIs married single mothers abroad and then introduced their stepchildren as derivative citizens, a practice that some officials rejected as perverting "true" (biological) paternity and therefore citizenship but that some other officials accepted. State Department officials in Europe sometimes issued U.S. passports "even though they are fully aware of the fact that the U.S. citizen father is not the true blood father of the child." In one case, an INS official, insisting on a biological understanding of paternity, advised a serviceman who had recently married a French woman that he could adopt her illegitimate child and file for a visa. The consul in France, however, embraced a more flexible social paternity. He overrode the official and immediately issued the stepchild an American passport as the child of a citizen father.[100] Not coincidentally, such scenarios, which occurred in France, Italy, and Germany, tended to involve American fathers, foreign wives, and children who were all white. Some officials were willing to accept nonbiological kin relations for white families even as they insisted on biological ones for Chinese Americans. The very distinct treatment of African American soldiers' families only reaffirms the racial salience of the state's kinship recognition. Who or what the government considered a family relation depended on the racial-national identities of the people in question.[101]

The citizenship of GI babies born of nonmarital relationships between American soldiers and foreign women is further instructive. Despite their biological descent from citizen fathers, illegitimate children were not automatically entitled to U.S. citizenship. In order for an American father to automatically confer his citizenship, he had to be married to the mother. What these two scenarios reveal—the first involving nonbiological stepchildren of married men, the second biological children of unmarried ones—is that marriage, not biological paternity, was the operative criterion of derivative citizenship, just as it was in the determination of legitimacy. If marriage made the father, it also made the citizen. For Chinese, however, the criterion was more complex: marriage was necessary but insufficient, and it was biological descent—blood—that was required, indeed fetishized, to make a citizen. Not only did the government use blood testing exclusively with the Chinese; it insisted on a narrow, biogenetic definition of paternity only in relation to this group.

Authorities' insistence on biological kinship in Chinese citizenship also contradicted new developments in U.S. refugee and immigration law. At the very moment the government was preoccupied with rooting out Chinese paper sons, Congress was in the process of creating a different kind of paper child: the international adoptee. Beginning in 1948, the Displaced Persons Act made the large-scale adoption of European refugee children possible for the first time. Several years later, the Refugee Relief Act provided an additional 4,000 visas for foreign orphans, about half of whom were from Asia; various temporary extensions facilitated the continued entry of foreign adoptees. In 1961, international adoption was permanently institutionalized when foreign-born adoptees ceased to be "eligible orphans" (aliens subject to quota restrictions) and were henceforth treated as family members (subject to the more privileged provisions of family reunification).[102] Over the course of the 1950s, international adoptees were thus transformed from unrelated foreigners into members of American families, a metamorphosis that reflected an administrative recognition of nonbiological kinship. Yet at that very moment, nonbiological kinship in Chinese American families was demonized as fraudulent and persecuted as criminal.

AT THE END OF 1957, as the Supreme Court heard oral arguments in the Lee case, in Arkansas troops were escorting the Little Rock Nine past angry white mobs outside of Central High School. Lee Kum Hoy, Kum Cherk, and Moon Wah had arrived in the United States five and a half years earlier, the same month the Court had agreed to hear *Brown v. Board*. Now the final chapter of their case played out against *Brown*'s dramatic reverberations. As civil rights struggles reached a fever pitch, the Court's decision in the Lees' case was potentially historic. If it struck down the discriminatory blood tests, it would go against decades of racist precedent in immigration law and recent congressional legislation, perhaps extending the *Brown* decision, whose reach the Court sought to cautiously manage, to a breaking point.[103] If it upheld the tests, it would undermine the promise of *Brown v. Board* and hand a gift to segregationists.[104]

Instead, the Court did neither. The government had shown to the Court's satisfaction that since the mid-1950s, it had expanded its testing beyond Chinese subjects. Its evidence for this assertion was a "Digest of Blood Tests of White Persons," a list of around sixty white aliens who beginning in the mid-1950s had been subjected to testing by the New York office.[105] The fact that the government began to keep the list only in response to the Lees' suit strongly suggested an attempt at retroactive remedy, but the list also suggested it had in fact modified its discriminatory practice. In light of this fact, the Court declared the question of racial discrimination moot, thereby avoiding a potentially provocative ruling about a racist practice that was no longer in effect.[106] At the same time, its three-sentence opinion left a door open to the Lees and a way for the government to end the litigation once and for all. The justices determined that because of the contradictory results of the first blood tests, the siblings were entitled to a new hearing based on a new set of tests. At this point, the government decided to fold. The Lees were not retested, and their U.S. citizenship was affirmed.[107]

Within a few years, the conditions that had spurred the blood testing of Chinese would disappear. The harrowing grand jury investigations and then the so-called Confessions Program begun in 1956, in which the government offered immunity to people who confessed to immigration irregularities, regularized the status of tens of thousands of people, resulted in the deportation of some, and closed thousands of "slots" for future migration. The combined effect of these policies was to bring more than a half century of paper immigration to a close.

As paper immigration disappeared, however, parentage testing expanded. Most of those listed in the government's "Digest of Blood Tests of White Persons" were from Poland, Czechoslovakia, Hungary, Russia, and the Ukraine, suggesting that the test lost its association with the Chinese but not its Cold War logic. As the procedure was redirected toward immigrants from behind the Iron Curtain, its essential function as a tool to defend the nation persisted. By the late 1950s and 1960s, the ideas and technologies of modern paternity were firmly entrenched in immigration and citizenship proceedings. Blood testing to prove parentage had become routine and uncontroversial.[108]

The Lee case sits at the crux of momentous changes in the histories of race and immigration. In addition to the fact that it coincided with the historic *Brown v. Board* decision, the Lees' lawyer, Benjamin Gim, was the first Asian American to argue a case before the Supreme Court. The children were among the very last immigrants to pass through Ellis Island as well as the first to be subjected to a new genetic method that, by the conclusion of their case, had become permanently incorporated into the government's tool kit. The case also augured important continuities. While the Chinese exclusion laws had been abrogated, anti-Asian racial quotas lived on in the new Immigration Act promulgated months after the Lees' arrival. When their case presented an opportunity to make a categorical ruling against racial discrimination in immigration proceedings, the Supreme Court opted instead for a pragmatic sort of nondetermination. Rather than signaling the triumphant eclipse of discrimination, genetic testing modernized race-conscious policies. The method was purportedly objective but was applied in discriminatory ways. Blood group tests were understood as an indicator not only of biological compatibility but also of social credibility and truthfulness—qualities with their own long-standing racial associations in the case of the Chinese.

While the blood testing saga of the Lees emerged from the racialized history of Chinese American citizenship and the anti-Communist paranoia of the Cold War United States, it was not a peculiarly American story. Australia, the Philippines, and possibly Canada also began parentage testing of Chinese immigrants in the 1950s and early 1960s.[109] It is not by chance that the new technology was first applied to the Chinese. Global immigration history reveals many examples of exclusionary policies first applied to Asians and later extended to other immigrants.[110] The midcentury Chinese blood testing program turned out to be the first instance of what, in the age of DNA, has become a widespread global practice: testing kinship among refugees and immigrants seeking family reunification. The program also anticipated DNA's stratified application. In the present as in the past, states apply distinct definitions of kinship to different racial-national groups.

And what became of the Lee family? After their case ended in 1957, the family disappeared into the streets of Chinatown. At some point, they confided to their lawyer Benjamin Gim that, as the blood tests indicated, the three youngsters were not all biological children of Lee Ha and Wong Tew Hee.[111] What the blood tests did not reveal is that Kum Cherk was the couple's biological son, while Moon Wah was a niece and Kum Hoy a nephew; the children were kin, although not the kind U.S. citizenship law required.[112] If Benjamin Gim had successfully secured their citizenship, it was because he managed to shift the focus of the case from what the genetic method revealed to how and to whom it was applied in the first place.

PATERNITY IN THE AGE OF DNA

> [DNA] is one piece of evidence like others . . . for us it doesn't change anything.
>
> —Julio Canella, grandson of Giulio Canella, 2014

THE ENORMOUS TURQUOISE van trolls New York City neighborhoods, attracting the curious glances of passersby. It noses through midtown traffic and parks on commercial avenues and leafy side streets to await customers. On its side is emblazoned an improbable slogan: "Who's Your Daddy?" The vehicle is a mobile DNA unit, and since 2012, it has peddled an unusual product to doubting New Yorkers: paternity tests. A 2016 newspaper story describes the truck's owner, Jared Rosenthal. "He is not a priest or a psychologist. He is not a doctor or a lawyer." He is a self-described health entrepreneur, and he sells answers to questions that are both deeply personal and resolutely public: "who am I? Where do I come from?" The genetic testing business has made Rosenthal philosophical about such issues. "No one really knows who they are. . . . People have been asking questions about paternity for as long as couples have been making babies. Only now we have a test."[1]

The San Francisco doctor Albert Abrams once ludicrously asserted that he could read parentage in a single drop of blood. Today, an idea once peddled by visionaries and quacks has become so ordinary that it is hawked from a truck like soft-serve ice cream. The groundbreaking development of

The "Who's Your Daddy?" truck trolls the streets of New York City offering on-the-spot DNA testing to passersby.

Tommy Liggett/Shutterstock.com.

DNA technologies, coupled with the explosive rise of a global commercial market for paternity testing, marks a brave new era in the millennial search for the father. More powerful technologies are accessible to more people around the world than ever before. States as well as consumers have driven expanded applications of paternity science, and potent discourses reinforce the right to paternal knowledge.

Yet for all their novelty, these scientific, commercial, social, and political developments recapitulate rather than revolutionize the history of paternity in the twentieth century. The Who's Your Daddy truck is in a sense the apotheosis of modern paternity. The ideas it peddles—that paternity is a biological relationship, that it can and should be known, and that science is the way to know it—have been with us since the 1920s. What is more, the dramatic advances of the DNA era have not resolved the tensions and ambiguities that modern paternity introduced almost a century ago. The father is as ambiguous, as deeply contested, indeed as elusive, as ever.

REMARKABLE SCIENTIFIC ADVANCES have marked the science of paternity in recent decades. In the 1970s, a serological test based on genetic markers of white blood cells known as human leukocyte antigens (HLA) substantially enhanced the power of exclusion. When used with traditional red cell tests, HLA testing generated a probability of a given man's paternity of above 90 percent, although the expense and inaccessibility of certain tests meant it was rarely practical to use them all.[2] In any event, this technology was soon eclipsed by another, even more powerful one: the "gold standard for truth telling," DNA fingerprinting.[3] Pioneered in the early 1980s by British geneticist Alec Jeffreys, the technique reads strands of base pairs, combinations of four different nucleotides that form pairs with each other to create DNA's characteristic double helix. The precise sequence of these pairs is unique to every individual, hence the metaphor of the fingerprint. While DNA fingerprinting is perhaps most closely associated with forensic identification in criminal cases, its earliest application was to demonstrate not individual identity but parentage. Every individual's sequence of base pairs is unique, but because they are inherited, the DNA from two people can be compared to determine the probability that they are related. For the first time in human history, it became possible to establish paternity with a 99.99 percent degree of certainty. The impenetrable veil with which Nature once shrouded the father, the veil fetishized for millennia and tugged at insistently by scientists for more than half a century, was yanked aside for good.

Scientific discovery soon catalyzed commercialization, and with it the "truth machine" became faster, cheaper, and more accessible.[4] Today, some genetic testing companies promise a turnaround of just days and charge as little as $79 a test. Web sites, commercials, billboards, and colorful vans market scientific tests of parentage to the consumer the same as pregnancy tests, treatments for warts—or soft-serve ice cream. The miraculous has become mundane.

Dr. Albert Abrams, inventor of the oscillophore, would no doubt appreciate this scientific and commercial achievement, which echoes the mail-order blood testing business he established a hundred years ago. But he would probably be astonished by its scope. The chief U.S. lab accreditation agency reports that the number of paternity tests under its auspices increased by almost 400 percent between 1988 and 2010. The agency records somewhere around 400,000 tests a year, although that figure is a fraction of the total number because only half of all labs are accredited and 40 percent of those that are do not report their data. One industry expert estimates the true number at over one million tests annually.[5] Genetic testing is projected to be a $3.8 billion dollar industry by 2023.[6] In the age of DNA, modern paternity has become thoroughly commercialized thanks to a profit-seeking behemoth, Big Paternity.

Galloping commercial expansion is most conspicuous in the United States, but paternity testing is on the rise globally. In Spain, the price of the test has fallen by half in recent years, thanks to increased competition as clinics multiply. In Italy, a geneticist observes, "the more the information spreads, the more demand increases." Latin America has likewise witnessed a testing boom in the past two decades. In the wake of a 1998 reform of the Chilean civil code facilitating paternity actions, Santiago's Medico-Legal Institute rushed to import DNA sequencing machines from the United States. Requests rose from 120 tests a year to 10,000 a little over a decade later. Public labs in Ecuador and Honduras also report soaring demand.[7]

Today, geneticists at the Instituto Oscar Freire in São Paulo continue to assess paternity for the courts, although a cheek swab has dispensed with the elaborate examinations of faces and bodies that their predecessors conducted for much of the twentieth century. In 1988 Brazil became the first Latin American country to adopt DNA fingerprinting, just a few years after its initial development, and as in England it was first used for paternity testing. By the early 2000s, some 7 percent of all children born in Rio Grande do Sul, Brazil's fifth most populous state, were subjected to a test. When a popular magazine did a story in 2008 on forty inventions that had most transformed Brazilian life in the preceding forty years, it showcased disposable diapers, cell phones, and DNA paternity tests.[8]

Big Paternity has expanded well beyond modern paternity's transatlantic cradle. British and American companies specializing in paternity and relationship testing now operate on six continents. Because testing does not require proximity to a lab—customers can purchase kits online and mail in their samples—it can be marketed anywhere there is Internet access and mail service. California-based EasyDNA offers testing in thirty-nine countries, from Uganda to Ukraine. The Web site for United Kingdom–based Nimble Diagnostics explains paternity testing procedures in thirty-six languages.[9] Chinese and Indian biotech companies have also capitalized on the new opportunities. The owner of a private lab in Beijing reports a 20 percent increase in paternity testing requests since 2000. Delhi-based

Indian Biosciences ("indisputable answers to emotional questions") operates collection centers in 150 cities across India.[10]

AS IN THE PAST, states are avid consumers of these technologies. A significant portion of genetic testing continues to take place under the aegis of courts and other state agencies, for child support, custody, or immigration cases. In the United States, "legal forensic services" outsourced to private labs account for about 45 percent of the market.[11]

Scientific advances make new applications possible, but it is social and political exigencies that make them necessary. The explosion of state-sponsored paternity testing in the United States dates from the 1970s—a decade before DNA—and coincides with the rollback of welfare and rising rates of extramarital birth. Beginning with the Social Security Act of the 1930s, the federal government's welfare programs created incentives for the states to find fathers. By the late 1970s, HLA tests were powerful enough to positively identify progenitors rather than just exclude impossible ones. Meanwhile, extramarital birth rates quadrupled between 1950 and 1980, reaching unprecedented levels.[12] Courts and legislatures proved eager to find fathers for poor children who would otherwise fall on the welfare rolls, and by the early 1980s blood testing had skyrocketed.[13]

The new scientific tests presented an expedient means to resolve claims. A simple blood draw dispensed with the protracted, messy, and expensive legal saga of the sort in which Charlie Chaplin was once enmeshed. Paternity was increasingly a matter for the lab rather than the courtroom, so much so that some states held that because the scientific evidence was incontrovertible, indigent fathers in paternity cases had no right to counsel. The momentous 1996 reform that famously dismantled the American system of "welfare as we know it," further institutionalized the use of genetic testing. It required that mothers cooperate to establish fathers' identities as a condition of aid, ordered the states to use genetic tests to do so, and invested those tests with conclusive power to fix paternity.[14] By 2000, the states were spending more than $32 million on lab tests annually.[15]

Latin American states' enthusiastic embrace of testing in the last two decades reflects an even more remarkable historical watershed. In the nineteenth century, law defined paternity in social and volitional terms: the father was the man who publicly and willingly performed this role before the community. In the twentieth century, modern paternity redefined fatherhood as a biological and obligatory relationship. Recent public policies have capitalized on this idea, as countries across the region have pursued "responsible paternity" campaigns to systematically identify fathers and compel them to take economic responsibility for their children. The task appears especially urgent in a region where marriage rates are low, extramarital birth rates high (in several countries, in excess of 70 percent), and many poor children are reared by mothers on their own—trends that policy makers believe contribute to female and child poverty.[16]

Responsible paternity schemes draft DNA into these efforts. At least a dozen countries have passed laws making testing available to poor people paid for by the state. In Brazil, which boasts an especially ambitious campaign to find errant fathers, mobile DNA testing vans fan out to hundreds of poor, urban neighborhoods and hardscrabble provincial towns. Here is a very different sort of Who's Your Daddy truck, one that circulates under the aegis not of a health entrepreneur but of the state, not to resolve personal questions about identity but to establish legal paternity and secure child support. The task of locating genetic fathers is imbued with vast and dubious powers to ameliorate social problems. In communities grappling with poverty, crime, police violence, and inadequate housing, health, and education, DNA testing promises a quick, apolitical techno-fix.[17]

Such policies have attracted remarkably widespread political support. Feminists and child rights advocates regard responsible paternity as a boon to gender equality and child welfare. Conservatives tout how these measures strengthen the family and promote individual responsibility. Meanwhile, it is no coincidence that paternity science has gained ascendance in an era of downsized states and social disinvestment. In Latin America as in the United States, paternity testing is an expedient tool for privatizing responsibility for poor children. If modern paternity was in part the creation of early-twentieth-century welfare states, ironically the retrenchment of those states at the end of the century has further galvanized efforts to find the father.

ANOTHER STATE PRIORITY, national security, has also spurred parentage testing. In 1983, thirteen-year-old Andrew Sarbah, a British teenager of

Ghanaian descent, flew from Ghana to Britain. At Heathrow airport, immigration officials examined his passport, which showed him to be a British citizen born in Britain, and declared it fraudulent. His mother, Christiana, hired a lawyer who began amassing evidence—photographs, documents, and blood tests—to prove Andrew's identity as her son and by extension his status as a British citizen. The Home Office rejected the evidence, including the blood tests, which in this pre-DNA era could not determine whether Christiana was Andrew's mother or his aunt. Then the Sarbahs' lawyer read in *The Guardian* about the new technique developed by geneticist Alec Jeffreys. Andrew Sarbah's case became the first ever application of DNA fingerprinting. The test demonstrated that Andrew was indeed Christiana Sarbah's son, and the family won their case. Jeffreys was soon inundated by requests for testing in other immigration cases: "we had enquiries by the thousand—box file after box file after box file of enquiries."[18]

While Jeffreys's technique was of course new, testing kinship at the borders of the nation was not. The Cold War–era U.S. immigration service pioneered this use of parentage testing when it began drawing the blood of Chinese migrants. Today more than twenty countries in Europe, North America, and Oceania routinely use genetic analysis to verify family relations among immigrants and refugees seeking to reunite with family members.[19] (Apparently the only place outside of North America, Europe, and Oceania to do so is Hong Kong, which tests Chinese applicants—a legacy of the U.S. consulate's policy in the 1950s?) Receiving countries typically require DNA evidence when applicants' documentary proof of family relation is unavailable or considered unreliable. In practice, the subjects of testing are almost always nonwhite immigrants from the global south. "The genetic tie is inherently paradoxical," notes legal scholar Dorothy Roberts. "It is at once a means of connection and a means of separation. It links individuals together while it preserves social boundaries." The boundaries that paternity testing preserves in this context are simultaneously national and racial.[20]

Such boundary making is not uncontested. In 2006, France passed a controversial law expanding the uses of DNA testing in family reunification, a policy that implicitly targeted African immigrants. Echoing the Chinese American community of the 1950s, antiracism advocates denounced testing as "detestable," and the policy was scrapped a few years later.[21] The creeping use of DNA testing in the United States, in contrast, has received less public scrutiny. The application of DNA testing to "national security" expanded exponentially after September 11, 2001. In 2008, officials suspended a program to reunite Somali refugees with family members residing in the United States after a pilot DNA program revealed that many refugees were not biologically related to their sponsors.[22] In an argument that could have been lifted directly from the INS inspectors of the 1950s, a government memo warned that "leaving open loopholes that allow potential terrorists, or simply fraudulent applicants to insert themselves in legitimate family groups not only threatens our security" but wastes "vast resources."[23] (What officials interpreted as widespread fraud, refugee advocates argued was a product of the conditions that produce refugees in the first place: conflict often separates families, and adults frequently take in orphaned children.) When the Somali program was reinstated four years later, it featured a beefed-up DNA requirement. More recently, the Department of Homeland Security has committed to introducing portable "rapid DNA" machines into refugee camps.[24] A compact device that can verify biological kinship with near certainty in less than two hours makes Albert Abrams's oscillophore look remarkably prescient. Incongruously, this state-of-the-art technology may soon appear amid the misery of the world's displaced persons camps.

A more recent, and potentially even more disruptive, development concerns the use of DNA to denaturalize citizens. In one recent case, a Yemeni-American lost his U.S. citizenship twenty years after obtaining it when a DNA test showed he was not biologically related to his citizen-father. It turned out he had been adopted as a baby (a fact Department of Justice officials declared "entirely irrelevant.")[25] In the 1950s, immigration authorities contemplated the mass denaturalization of Chinese Americans through retroactive blood group testing. A government lawyer rejected the policy, warning that naturalized citizens could not be forced to take blood tests and that, in any event, "the courts are not friendly to the invention of new ways to derogate already established rights."[26] It remains to be seen whether such caution will prevail or be swallowed up by the gaping maw of immigration enforcement. For its part, the biotech industry forecasts an annual growth rate of "DNA analysis in Homeland Security" of close to 14 percent as well as opportunities for future expansion in Asia and Europe.[27]

If many contemporary uses of parentage testing have precedents in the past, some appear genuinely novel. In Argentina, genetic technologies have been mobilized to address human rights violations committed during the country's Cold War military dictatorship. In a now notorious episode, authorities kidnapped some 500 children from parents considered to be political dissidents and placed them in irregular adoptions. Decades later, the Abuelas, an organization of the children's grandmothers, led the charge to locate the kidnapped children and restore their natal identities. As a result of their efforts, in 1987 Argentina established a national genetic database to assist with reunifications. Argentina was the cradle of the world's first "Mendelian" paternity test, Lehmann Nitsche's somatic analysis of the Arcardini family. Seventy years later, another genetic innovation was born here: the *índice de abuelidad,* or "index of grandpaternity," a technique to establish the genetic link between two people separated by a generation.[28] (The Abuelas joked that the fact the technique used mitochondrial DNA, inherited through the mother, was proof that God was female.) It is yet another example of how politics—in this case, the quest for justice in the aftermath of state terror—fuels genetic innovation. The politically motivated removal of children from their parents is hardly an Argentine peculiarity. In Spain, dictator Francisco Franco kidnapped some 30,000 children from their dissident parents. For several weeks in the spring of 2018, the Trump administration systematically seized thousands of Central American migrant children at the U.S.–Mexico border. In the wake of these episodes of state-sponsored child abduction, DNA has been used to facilitate reunifications.

THE EXPANDING DEMAND FOR kinship science is fueled not just by state policies or abuses but also by private demand. The enduring popular fascination with origins has long spurred such technologies; Big Paternity has commercialized that allure. Today the curious, the uncertain, and the aggrieved can purchase what the industry calls "discretionary" or, more colorfully, "peace of mind" tests for their own personal reasons. Within a decade of DNA fingerprinting, Identigene began marketing paternity tests directly to American consumers and four years later sold the first test on the Internet.[29] By 2007, test kits became available in pharmacies. Easy, accessible, and increasingly inexpensive, home testing today accounts for more than 36 percent of the American DNA market and is also available to consumers from Australia to South Africa, Italy to Malaysia. As industry analysts point out, the more commercially accessible DNA tests are, the more lucrative for the industry.[30]

Commercialization has in turn reshaped who has access to the truths paternity science tells. Modern paternity sought to transplant paternal truth from the courtroom to the laboratory. Now Big Paternity has uprooted it from both, propagating it on the Internet, on the pharmacy aisle, and in the mobile van. Modern paternity promoted the categorical advantages of knowing the father, but in practice that prerogative was always tempered by concerns about the destabilizing effects of unchecked knowledge. Even its avatars warned of such dangers, as when doctors and lawyers in Catholic countries insisted that protecting marriage, public propriety, and the child's legitimacy trumped any inalienable right to biogenetic truth. Paternity science had to be applied in judicious, socially responsible ways, with the courts and scientists themselves as the gatekeepers.

Today such ideas are to some extent still with us. Some observers express deep reservations about the prospect of consumers testing without the consent of all parties, particularly when the well-being of minors is at stake.[31] In France—where doctors once warned that including ABO blood types on identity cards might unwittingly expose adulterous children—private parentage testing is today illegal: tests can be obtained only with a court order. The state asserts the prerogative to protect the parent–child relationship and to manage potentially harmful information about kinship. The child's right to a father—even a nonbiological one—trumps the parent's or child's right to genetic knowledge.

But to many, such policies appear quaint, paternalistic, and, in any event, futile. Evasion of the French law appears to be widespread. Labs in Spain and Switzerland reporting a booming business testing surreptitious tissue samples sent by French citizens, who risk steep fines and even jail time for skirting the law. In any event, France's restriction of paternity testing is striking mostly because it is exceptional. Elsewhere, private testing tends to be regulated loosely, if at all. Some countries prohibit testing without the consent of both parents, but in practice it is often possible to submit a toothbrush or hair sample without the knowledge of its owner. Almost as if they were speaking directly to the circumspection of earlier testers, industry analysts boast that what is "revolutionary" about paternity testing today is that it does "not require involvement of doctors or lawyers."[32] Genetic knowledge has become a de facto prerogative, conferred not by deliberate legislation but by the vast commercial reach of the testing industry itself. Big Paternity has achieved what modern paternity did not.

Access to genetic knowledge is the product of political developments as well as commercial ones. The past half century has witnessed an increasingly powerful discourse surrounding the individual's right to know his or her origins, defined implicitly or explicitly in biogenetic terms. This right is codified in international child rights treaties, most notably the 1989 United Nations Convention on the Rights of the Child, which holds that "children have the right to an identity" (the article's inclusion is a result of lobbying efforts by Argentina's Abuelas). It is a rallying cry of adoptee rights organizations like Bastard Nation, a North American group that lobbies to open sealed records containing adoptees' "historical, genetic and legal identities." Advocates assert that donor-conceived people have such a right in the context of assisted reproductive technologies. In responsible paternity campaigns in Latin America, the right to a father is often framed in terms of the child's right to an identity. If the erasure of natal identity in the twentieth century has often resulted from political forces—state violence, repressive social norms, patriarchal prerogatives, racism, war—then so too must its recuperation. "The right to know one's identity," asserts Bastard Nation, "is a political issue."[33] Today a right that modern paternity championed, albeit unevenly, has become significantly more robust.

IDENTITY IS NOT JUST a political issue, however; it also makes for a good story. Chronicles of identity lost and found have long been the narrative expression of modern paternity. Today the media continues to tell these stories, often with new twists thanks to modern reproductive technologies. Babies are no longer likely to be swapped in hospitals, for example, but laboratory mix-ups of gametes and embryos can produce similar dramas. DNA technology also supplies fodder for melodrama. A hundred years after the Sconosciuto di Collegno was found wandering in a cemetery, the story of the mysterious amnesiac is the subject of recent books, a television miniseries, and commemorations by the town of Collegno. In 2014, the popular Italian television show *Chi L'ha Visto?* (Who Has Seen Him?), which explores unsolved mysteries, revisited the famous case. The show hired a geneticist to conduct a DNA test on the Canella family descendants. The genetic technology used was different from the odontological assessment of Luiz Silva eight decades earlier, but its logic was exactly the same: to elucidate the identity of the man by way of his putative offspring. The analysis sought to compare the DNA of Julio, the grandson of one of Giulia and Giulio Canella's children born before the war, to Camillo, a son born to Giulia and the Sconosciuto after the war.

Yet the revelations of DNA are as likely to create mysteries as to solve them. Recreational ancestry testing has led some people to discover their family trees were not what they thought, like the biologist who gave 23andMe kits to his parents as gifts and stumbled on a half brother, opening a Pandora's box of secrets that led to his parents' divorce, or the Irish Catholic woman whose test inexplicably revealed her to be half Jewish. Her quest to explain the test result led to the extraordinary discovery (extraordinary less because it happened than because she could document it) that after his birth in a Bronx Hospital in 1913 her father had been accidentally swapped with another baby.[34] There is now a Facebook group NPE—"not parent expected"—for people whose DNA dalliances reveal shocking secrets about their identities. It has more than 1,000 members.[35]

Popular coverage of DNA testing endlessly repeats this storyline: a person takes a DNA test and discovers a startling truth about his or her identity or family. At the heart of this narrative is the central idea of Euro-American kinship: that it exists in nature, that it is sometimes hidden whether deliberately or accidentally, and that truth consists of uncovering the natural facts. Some observers argue that even as the trope of identity lost and found continues to fascinate, the moral of contemporary stories about DNA, which often dwell on the primacy of rearing and emotion over birth or blood, is new: "technology has brought love to the foreground" writes one journalist.[36] In fact, rather than asserting the triumph of the social over the biological and nurture over nature, these stories tend to reify these distinctions. In doing so, they recapitulate rather than challenge deep cultural beliefs about what kinship is and how it works.

Meanwhile, in another unmistakable continuity with the past, the truth uncovered in these accounts often has to do with race. Racial revelation is a recurring plotline: the Irish woman turned half Jewish, the self-described "Southern white girl" whose DNA test revealed her to be "part-African," the African-American woman found to be 98 percent non-African, the García who discovers she has no discernible Mexican ancestry.[37] Such narratives emerge seamlessly from the long history of modern paternity. The scientific blood tests of the 1920s, which sought racial truths in vibrating electrons, blood types, and chromatic alchemy, may have relied on very different technologies and a distinct idea of race. Yet they share with contemporary DNA narratives the axiom that testing parentage is always also about testing race and ethnicity. This is not only because, as the biologist who now regrets giving his parents 23andMe kits observes, ancestry testing services are "essentially really advanced paternity tests."[38] It is also because of the abiding cultural logic in which ethnicity remains the sap of the family tree.

In keeping with this logic, contemporary media reports about swapped gametes and embryos home in with special fascination on stories involving white parents who mistakenly wind up with nonwhite children. In the age of modern biotech, Natus Ethiopus, the trope of the white mother and her tawny infant, is alive and well.[39] It is further telling that, while the dramatic tales of in vitro fertilization gone awry could just as easily involve eggs or embryos, it is usually wayward sperm that are the culprit.[40] This is a story not just about parentage but about paternity—about how, in the fertile imaginary surrounding paternal uncertainty, obsessions with racial ambiguity, mixing, and substitution continue to flourish.

IN THE TWENTIETH CENTURY, the press produced new ways of thinking about paternity and its technologies. In the twenty-first century, it continues to play this generative role. Charlie Chaplin was the protagonist of perhaps the greatest paternity scandal of the twentieth century, but in succeeding decades hundreds of other celebrities, statesmen, and tycoons have become embroiled in similar disputes. Alongside the time-honored paternity travails of the rich and famous, new media genres have emerged. In the late 1990s, flamboyant media personality Maury Povich pioneered the Who's Your

Daddy talk show, in which troubled couples recount their tales of love and woe to a studio audience, with suspense building up to the moment when the host announces the results of a DNA test. The formula—since adopted by numerous other shows—has proven a boon to ratings, so much so that paternity dominates storylines during sweeps months, when television advertising rates are set.[41] The genre is hardly an American peculiarity: paternity revelation has become a staple of talk shows from Britain to Brazil, Germany to Mexico, to, perhaps most recently, Uganda.[42] The uncertain father is on the move again. Having seamlessly passed from the nineteenth-century novel to the early-twentieth-century tabloid, at the turn of the new millennium he has resurfaced on television and the Internet.

Nowadays, however, his job is not merely to entertain: he is also a salesman. It is not by chance that the rise of the Who's Your Daddy show in the 1990s coincided with the emergence of direct-to-consumer marketing of paternity tests. DNA companies develop media partnerships to showcase their wares. Ohio-based DNA Diagnostics Center touts a media portfolio that includes *Maury* and twenty other television shows. One of them is *Lauren Lake's Paternity Court,* in which the judge listens to people recount their kinship conundrum and then dramatically produces the test results from an envelope emblazoned with the company logo. The business manager of one lab credits daytime television as "the major reason" for the growth in private testing.[43] The revelations of the paternity test have been a source of popular fascination since news of the oscillophore's miraculous exploits first hurtled out across the international telegraph wires. Now paternity drama sells not just newspapers whose audiences can follow others' dramas but home kits that permit them to resolve—or create?—their own.

Indeed, Big Paternity is in the business of selling DNA tests, and it therefore must peddle not just entertainment but doubt. After all, people will purchase a test only if they have reason to believe it will reveal something they did not already know. Marketing slogans help sow doubt, urging consumers to "Discover your truth"; "Don't let uncertainty tear you apart."[44] They warn that countless men are unknowingly rearing other men's children. Identigene, which pioneered direct-to-consumer DNA tests, campaigns to raise awareness of paternity fraud.

Big Paternity is not alone in creating and perpetuating narratives of rampant "paternal misattribution." Across the Internet, men's rights groups decry what they believe is an epidemic of duped dads. They, too, articulate a right to biogenetic identity, but one that is less about the right of children to know their progenitors than the right of men to disavow children to whom they are not biologically related. Some groups refer to "paternity rape"—the violence that women commit against men (and children) through their sexual treachery. According to one statistic endlessly recycled by men's rights groups, commercial DNA testing Web sites, popular media, and even scholarly sources, paternal misattribution rates are as high as 30 percent.[45] That is, a third of people are not the biological offspring of the man they hold to be their father, and by extension, lots of men are unknowingly rearing other men's children. It is an extraordinary statistic. It is also patently false. The source of the figure is debated, but it appears to derive from an American Association of Blood Banks finding about rates of negative test results in their labs. The problem with the statistic is that people who seek out a paternity test are hardly a representative sample of the general population.[46]

Self-identified men's rights groups have been decrying female betrayal and paternal misattribution since the 1920s, when the Vienna-based Rights for Men League called for the expanded use of blood group tests. Today the Internet and the synergies with Big Paternity breathe new power into these propositions. In recent years, legislators in Tennessee, Kansas, and New Jersey have introduced bills to make DNA testing obligatory for all newborns. "I've heard different stories about [duped] fathers who are raising children and paying child support," a New Jersey assemblyman explained.[47] None of these bills has come to fruition, but one can guess where he heard those stories. As in the past, the technologies of modern paternity are as much about women and sex as about men and parentage.

TODAY THE EXISTENCE of powerful new genetic technologies, their large-scale use by states and consumers, and the embrace of these technologies in the media and popular culture would seem to suggest an unmistakable watershed. At the turn of the millennium, the biological father appears triumphant. Modern paternity's promise that biological kinship can and should be known has, almost a century after its emergence, come to full fruition. Science has definitively vanquished social and legal (mis)understandings of paternity, kinship, and identity. Commercialization has provided unfettered access to testing. The will to biological truth has displaced other social values once and for all.

In fact, the latest chapter in the history of paternity turns out to be more complicated. While these developments have reinforced the biogenetic father, he is by no means hegemonic. His older social personas endure, not least in the realm of family law. Take the presumption of marital paternity, which holds that the husband is always the father of a child born in marriage. In 1949, an Italian court affirmed this principle when it declared Remo Cipolli the father of the "mulattino" Antonio. Four decades later, a case from California upheld the same principle. A married woman named Carole gave birth to a daughter, Victoria. DNA evidence showed Carole's boyfriend, not her husband, was the child's biological father. Who was Victoria's legal father? *Michael H. v. Gerald D.*—the case was named for the two possible fathers— was appealed up to the Supreme Court, where in 1989 Justice Antonin Scalia authored the majority opinion. In an elliptical turn of phrase, he declared: "under California law, Victoria is not illegitimate." Because she had been born to a married mother, the presumption of legitimacy determined her paternity. Regardless of genetic evidence, Victoria's father—like Antonio Cipolli's—was her mother's husband.

In contemporary U.S. law, the presumption of marital legitimacy is considered "rebuttable," but the extent to which it may be rebutted varies by state. If a man has played the role of the father to a child over a period of time, he may not be able to disavow paternity despite the finding of an eventual DNA test.[48] The enduring significance of marital paternity is just one instance of a broader principle. Almost a century after the genesis of modern paternity, biology does not necessarily define the legal father. The reason France outlaws discretionary DNA testing is that French civil law defines paternity as a social relationship. This principle—at first glance a French oddity—is in fact widely shared. Brazil has witnessed a DNA testing boom, but recent Brazilian law has repeatedly reaffirmed the primacy of "socio-affective paternity."[49] In the U.S., courts have ruled that a man's social behavior, his intent, contract, the nature of his relationship with the mother, and the importance of stability in a preexisting parent–child relationship may all define legal paternity. The man named on a birth certificate is presumed to be the father, as is he who, in the phrasing of the California Family Code, "receives the child into his home and openly holds out the child as his natural child."[50] Even unmarried men who act as fathers under the assumption that they are biological progenitors may continue to have support obligations even if the biological tie is later shown to be absent. This is precisely what Roque Arcardini's heirs discovered in Buenos Aires 100 years ago. Then as now, social performance may define the father.

If men with no biological link can be considered legal fathers, U.S. courts have also repeatedly found that "the mere existence of a biological link" does not make a man a father with paternal rights. In order for paternity to come into being, he must show emotional or economic commitment to the child.[51] Paternity is not born, it is made, and the principle of paternal intention therefore remains central to its definition. The principle of intent likewise governs the definition of parentage in assisted reproductive technologies. Take gestational surrogacy, in which a woman agrees to carry a child conceived using the gametes of a commissioning couple. In some U.S. states, the surrogate's husband is considered the legal father of the child even though he (and often she) is not genetically related to it. At the same time, most state laws governing sperm donors hold that a man bears no responsibility for a child conceived with his anonymous sperm because he clearly does not plan to act as a father. In both surrogacy and sperm donor scenarios, the law assigns paternity (as well as maternity) based on the principle of intent (who intended to make and rear the child) as opposed to genetic connection.[52]

The idea that paternity must be intended and performed also applies to citizenship law. A child born on foreign soil to an unmarried U.S. citizen mother can receive American citizenship almost automatically. But for an unmarried father to transmit his nationality, a mere biological link does not suffice. The man must perform his paternity, by providing proof of it, legally recognizing the child, and pledging to provide support for her. The heightened requirement for men to transmit their citizenship reflects the enduring desire to prevent the unrecognized (and often nonwhite) offspring of American soldiers born abroad from claiming U.S. citizenship. Biology is a necessary but insufficient condition for transmitting this valuable asset.[53] In the age of DNA, as in modern paternity's earlier eras, the logic of the social persists. So too does the gender asymmetry of maternity and paternity, as well as paternity's implicit racial dimension.

DNA'S TRUTH MACHINE has clearly not spawned the wholesale biologization of kinship. Nor has it resolved the tensions between the social and the biological, the performative and the natural, the volitional and the obligatory, that modern paternity intensified. If anything, these tensions have been thrown into even sharper relief by a scientific test that promises to reveal the genetic tie with power and perfection.

For some critics, the failure of biogenetic paternity to triumph in light of definitive physical proof is illogical and unjust. How can a man be required to support a child that science can show is not biologically his? This is the question that observers asked of the Chaplin verdict more than three-quarters of a century ago. Today, such scenarios still routinely arise.[54] Men's rights groups are especially vocal in decrying them as evidence of law's baffling and oppressive unfairness. But even feminist legal scholars find the law's inconsistent recourse to biological and social paternities to be noteworthy and perhaps problematic.[55]

This is just in the realm of family law. Moving across arenas of state practice, such inconsistencies become even more apparent. Where family law often recognizes social elements as defining the parent–child relation, the child support bureaucracies that arose in association with modern paternity have tended to emphasize biogenetic filiation. So, too, do immigration policies that define descent for migrants and refugees. In citizenship law, meanwhile, children born to unmarried fathers abroad must demonstrate both biological ties and social ones in order to gain nationality. To the question of what filiation is, a social or biological relation, the answer is: it depends.

Perhaps the problem is not inconsistency per se but stratification, that is, the fact that paternity's definition varies not just according to circumstance but depending on whose paternity is in question. Such social variability has long characterized the quest for the father. In colonial contexts paternity was often defined as distinct from that in the metropole. Law and custom strategically obfuscated the fatherhood of certain categories of men—the slave owner, the priest, the colonizer, the soldier. In many of these contexts, race in particular was determinant: paternity was deemed especially "unknowable" when white men fathered children with nonwhite women.

To some extent, modern paternity tempered such stratification. The impulse to know the father, all fathers, weakened the traditional idea that some progenitors enjoyed the prerogative of anonymity. Likewise, modern paternity eroded discrimination against illegitimate children; such a prejudice became harder to justify if biology, rather than marriage, was what defined the father. But modern paternity did not obliterate stratification. In the 1950s, U.S. immigration officials regarded the paternity of Chinese American fathers as inscrutable where that of other expatriates was considered obvious; indeed, this "fact" justified the race-based application of blood testing. In the age of DNA, too, stratification persists. As critics have noted, defining paternity in one way for citizens and another for immigrants is patently discriminatory. Kinship practices like adoption, stepparenthood, and relationships based on socio-affective bonds are considered perfectly legitimate when practiced by natives but are vilified as fraudulent and criminal when practiced by foreigners.[56] If the quest for the father has always been political—shaped by law, structures of inequality, public priorities, and social and racial norms—it should come as no surprise that it continues to be so, even in an era of newfound certainty.

At the same time, distinct ways of defining paternity have no necessary politics. Biological essentialism is not inherently "conservative." Genetic kinship can be mobilized as part of a state's murderous racial ideology, in the name of welfare privatization, or for national security. It can be a rallying cry of men's rights groups that wear their misogyny on their sleeve. But it can also be championed by human rights activists like the Abuelas of Argentina and by advocates of adoptee and children's rights. Nor is there anything inherently "progressive" about a social constructivist vision of kinship. Napoleonic tradition defined paternity in terms of male volition, social performance, and contract—a socio-affective definition if there ever was one. It did so to the benefit of patriarchal privilege, privacy, and property and at the expense of unmarried women and their children. In 1998, as Chilean lawmakers prepared to pass a reform allowing mothers and children to bring suits to establish the identity of illegitimate fathers for the first time since the nineteenth century, some conservatives expressed opposition. Should a child with no relationship to its progenitor be allowed to show up one day, make financial and moral claims on him, and disrupt his legitimate family, merely because of a genetic link? They argued that allowing such actions privileged a narrowly biological paternity over a more socially meaningful one based on affection, responsibility, and paternal will.[57] In *Michael H. v. Gerald D.*, conservative justice Anton Scalia likewise marshaled a social constructivist kinship when he assigned paternity to the husband and not the biological father of the child in question.

A global perspective reveals paternity's political contingency with particular clarity. The "truth machine" of DNA has become a routine part of judicial practice, public policy, media culture, and social life across an ever greater swathe of the planet. There is an unprecedented consensus among scientists, legal experts, and the public that the technology "works." But as in the past, the work it does continues to be dramatically shaped by local circumstance. In Latin America, many feminists have embraced DNA testing to combat irresponsible paternity, which they characterize as "a form of economic and emotional violence against women." In India, women's rights advocates have emphatically opposed its growing use as a cynical weapon used to control wives.[58] Paternity may be inherently political, but it has no preordained politics. Context is everything.

GIVEN ITS COMPLEX, contested, and sometimes sordid associations in the present, it is easy to forget that paternity was once a subject of formidable intellectual energy. Friedrich Engels and Sigmund Freud, Victorian novelists and early anthropologists, feminists and sociobiologists all put paternity at the center of their accounts of human society. The quest for the father was an abiding motor of human history, advancing alongside—indeed helping to define—the march from primitivism to modernity. For some thinkers, the problem of paternity explained the emergence of monogamous marriage and was linked to the transmission of private property. For others, it explained patriarchy or, as an evolutionary challenge, determined human behavior and morphology. As anthropologist Bronislaw Malinowski once put it, the problem of paternity is "the most exciting and controversial issue in the comparative science of man."[59]

Influenced by such accounts, many observers over the course of the twentieth century predicted that the perfection of paternity science would herald a watershed in the millennial search for the father. "With a scientifically accurate determination of paternity," a Baltimore newspaper prophesied in 1922, "a world of misery, suspicion, lies and faithlessness would be abolished from the earth." Science fiction writer Arthur C. Clarke predicted such a technology would bring about the end of marriage. In the 1960s, a Scottish jurist mused that it would be as explosive as nuclear physics. At the dawn of the DNA era, an American geneticist speculated that while motherhood had always been certain, now fatherhood would be as well: "We will have come one step closer to equality of the sexes."[60]

Not one of these spectacular predictions has come to pass. The quest for the father—deeply politicized, culturally fetishized, and now thoroughly commercialized—did not end with the advent of DNA. On the contrary, at the very moment that science has promised paternal certainty for the first time in human history, uncertainty as a cultural force appears as powerful as ever. This is in part because a global industry commercialized doubt. It is in part because, as recreational testing shows, DNA can just as easily create mysteries as solve them. But above all it is because science was never capable of finding the father in the first place. It was not a lack of knowledge that produced the quest for the father; the quest was always a social and political one. The truly significant question about paternity is thus not an empirical one—who is the father?—but a normative one—what do we want him to be? Which criteria, whose interests, intentions, or desires, should define paternity?

Over the course of the last century, paternity testing has morphed from a chimera to a "truth machine." But the truth that machine produces is indeterminate. Its meanings vary according to what it is being used for, the context of its use, and who is using it. This is true not only for states but also for ordinary people and their understandings of family and identity.

For Dunja Cipolli, the daughter of Antonio Cipolli, the Italian "mulattino" ostracized by his family and his country, science holds the key to a lost identity. Dunja has spent twenty years wondering about the identity of the man who fathered her father. Several years ago, after reading about DNA testing, she ordered a test kit. It matched her with second cousins in the United States, who explained the history of the Buffalo Soldiers. It was then that she understood the basic contours of the story: that her grandfather had been an African American GI stationed with the occupying troops. The man's identity remains a mystery, but she feels a deep and abiding connection with him. If she finds him, she says, she might move to the United States to be with him if he is still alive or else to be with his descendants.

And then there is Julio Canella, grandson of Giulio Canella, the mysterious amnesiac. After he and his uncle underwent a DNA test on the Italian television show a few years ago, the host handed him an envelope with the results. The analysis showed that the children of Giulia Canella born before the war (including Julio's father) and those born after (his uncle) had been

fathered by two different men. The amnesiac found in the cemetery therefore could not have been Giulia's long-lost husband; he must have been the
imposter Mario Bruneri. For Julio Canella, however, the DNA finding elicited a shrug. "It's not the result I expected," he conceded. "But this is one
proof like all the others, there are many in favor and many against, for us
this doesn't change anything."[61] The family's quest to prove the identity of
Giulio Canella and vindicate the family name continues. Whatever the
truths of DNA, for Julio Canella as for Dunja Cipolli the question of the
father, now the grandfather, remains unanswered.

They are still looking.

ABBREVIATIONS

IOF	Instituto Oscar Freire, Universidade de São Paulo
JAMA	*Journal of the American Medical Association*
NARA	National Archives and Records Administration
RAC	Rockefeller Archives Center
RCPML	*Revista de Criminología, Psiquiatría y Medicina Legal*
RDMIRF	*Rivista del Diritto Matrimoniale Italiano e dei Rapporti di Famiglia*
RLNA	Robert Lehmann Nitsche Archives, Ibero-Amerikanisches Institut

NOTES

PROLOGUE

Epigraph: Mary O'Brien, *The Politics of Reproduction* (London: Routledge & Kegan Paul, 1981), 29.

1. Tom Caton, "Chaplin Assailed in Scott Argument," *Los Angeles Times,* December 30, 1944, A1.

2. "Judge Refuses Chaplin Plea," *Baltimore Sun,* March 9, 1944, 1.

3. Marcia Winn, "Rips Chaplin as 'Liar, Cad, and Buzzard,'" *Chicago Daily Tribune,* December 30, 1944, 1.

4. "Chaplin Jolted Twice in Court Pleas; Selection of Joan Berry Jury Begins," *Los Angeles Times,* December 15, 1944, A1.

5. Bronislaw Malinowski, "Foreword," in Ashley Montagu, *Coming into Being among the Australian Aborigines: The Procreative Beliefs of the Australian Aborigines* (1937; repr., Abingdon: Routledge, 2004), xvi.

6. "Film Comedian Called 'Menace' and 'Crucified,'" *Los Angeles Times,* January 3, 1945, 2.

7. Marcia Winn, "Physicians Say Chaplin Is Not Father of Baby," *Chicago Daily Tribune,* December 28, 1944, 7.

8. "Science May Turn to 'Mother Instinct' in Puzzle to Establish Parentage in Baffling Case," *New York Tribune,* August 17, 1930, C2.

9. "Chaplin Case to Jury Today," *Baltimore Sun,* January 3, 1945, 11.

10. Winn, "Physicians Say Chaplin Is Not Father of Baby."

11. "Film Comedian Called 'Menace' and 'Crucified.'"

12. Both from Sidney Schatkin, *Disputed Paternity Proceedings,* 3rd ed. (New York: Matthew Bender and Company, 1953), 255–256. The California quote is from *The Boston Herald.*

13. "Film Comedian Called 'Menace' and 'Crucified.'"

14. "Hints Chaplin Test in Baby Case Is Phony," *Chicago Daily Tribune,* March 2, 1944, 1.

1. LOOKING FOR THE FATHER

Epigraph: Cited in Rachel G. Fuchs, *Contested Paternity: Constructing Families in Modern France* (Baltimore: Johns Hopkins University Press, 2008), 2.

1. One version of the story is found in Emmanuele de Azevedo, *Vita di Sant' Antonio di Padova* (Bologna: Lelio dalla Volpe, 1790), 90–92.

2. Tom MacFaul, *Poetry and Paternity in Renaissance England* (New York: Cambridge University Press, 2010); MacFaul, *Problem Fathers in Shakespeare and Renaissance Drama* (Cambridge: Cambridge University Press, 2012).

3. Jenny Davidson, *Breeding: A Partial History of the Eighteenth Century* (New York: Columbia University Press, 2009).

4. Ivy Pinchbeck and Margaret Hewitt, *Children in English Society* (London: Routledge & K. Paul, 1969), chapter 8; Michael Grossberg, "Duped Dads and Discarded Children," in *Genetic Ties and the Family: The Impact of Paternity Testing on Parents and Children*, ed. Mark A. Rothstein (Baltimore: Johns Hopkins University Press, 2005), 97–131; U. R. Q. Henriques, "Bastardy and the New Poor Law," *Past & Present* 37 (1967): 103–129; Laura Gowing, *Common Bodies: Women, Touch and Power in Seventeenth-Century England* (New Haven, CT: Yale University Press, 2003); Lisa Forman Cody, "The Politics of Illegitimacy in an Age of Reform: Women, Reproduction, and Political Economy in England's New Poor Law of 1834," *Journal of Women's History* 11, no. 4 (2000): 131–156.

5. Magistrate's quote: Henriques, "Bastardy and the New Poor Law," 106.

6. "Impenetrable veil": see Fuchs, *Contested Paternity*, 52; on the mystery of nature and maternity: Nara Milanich, *Children of Fate: Childhood, Class, and the State in Chile, 1850–1930* (Durham, NC: Duke University Press, 2009), 54. Also, Camille Robcis, *The Law of Kinship: Anthropology, Psychoanalysis, and the Family in France* (Ithaca, NY: Cornell University Press, 2013).

7. Staffan Müller-Wille and Hans-Jörg Rheinberger, *A Cultural History of Heredity* (Chicago: University of Chicago Press, 2012).

8. Suzanne Desan, *The Family on Trial in Revolutionary France* (Berkeley: University of California Press, 2006); Isabel V. Hull, *Sexuality, State, and Civil Society in Germany, 1700–1815* (Ithaca, NY: Cornell University Press, 1997); Milanich, *Children of Fate.*

9. Balzac, *Old Goriot* (1835); Thomas Hardy, "The Imaginative Woman" (1894); August Strindberg, "The Father" (1887); Machado de Assis, *Dom Casmurro* (1899); Guy de Maupassant, "Useless Beauty" (1926).

10. Jenny Bourne Taylor, "Nobody's Secret: Illegitimate Inheritance and the Uncertainties of Memory," *Nineteenth-Century Contexts* 21, no. 4 (2000): 565–592; Goldie Morgentaler, *Dickens and Heredity: When Like Begets Like* (New York: St Martin's Press, 2000); Ross Shideler, *Questioning the Father: From Darwin to Zola, Ibsen, Strindberg, and Hardy* (Stanford, CA: Stanford University Press, 1999).

11. Friedrich Engels, *The Origin of the Family, Private Property and the State* (Harmondsworth: Penguin Books, 1985); "only certain parents": 42; "final result": 98; "most decisive": 86; Johann Jakob Bachofen, *Myth, Religion, and Mother Right: Selected Writings of J. J. Bachofen* (Princeton, NJ: Princeton University Press, 1992); Rosalind Coward, *Patriarchal Precedents: Sexuality and Social Relations* (London: Routledge & Kegan Paul, 1983); Carol Delaney, "The Meaning of Paternity and the Virgin Birth Debate," *Man* 21, no. 3 (1986): 494–513.

12. *Family Romance* (1909); *Moses and Monotheism* (1939); "Notes upon a Case of Obsessional Neurosis" (1909).

13. On "historical process . . . [as] a psychic one," see Ann Taylor Allen, "Feminism, Social Science, and the Meanings of Modernity: The Debate on the Origin of the Family in Europe and the United States, 1860–1914," *American Historical Review* 104, no. 4 (1999): 1110.

14. Delaney, "The Meaning of Paternity."

15. Foreword to Ashley Montagu, *Coming into Being among the Australian Aborigines: The Procreative Beliefs of the Australian Aborigines* (1937; repr., Abingdon: Routledge, 2004), xvii.

16. Aldous Huxley, *Brave New World* (New York: HarperCollins, 2010), 44–45.

17. Arthur C. Clarke, *Childhood's End* (New York: Ballantine, 1953), 73.

18. Allen, "Feminism, Social Science, and the Meanings of Modernity"; Gloria Steinem, "Wonder Woman," in *The Superhero Reader,* ed. Charles Hatfield, Jeet Heer, and Kent Worcester (Jackson: University Press of Mississippi, 2013), 209; see also Cynthia Eller, *The Myth of Matriarchal Prehistory: Why an Invented Past Won't Give Women a Future* (Boston: Beacon Press, 2000).

19. Mary O'Brien, *The Politics of Reproduction* (London: Routledge & Kegan Paul, 1981), 8.

20. O'Brien, *The Politics of Reproduction*: "male seed": 29; "mud hut": 56; "huge and oppressive" and "intransigent reality": 60–61.

21. Sarah Blaffer Hrdy, *Mother Nature: A History of Mothers, Infants, and Natural Selection* (New York: Pantheon Books, 1999), 217. The literature on paternal confidence in diverse species, including humans, is enormous. Steven J. C. Gaulin and Alice Schlegel, "Paternal Confidence and Paternal Investment: A Cross Cultural Test of a Sociobiological Hypothesis," *Evolution and Human Behavior* 1, no. 4 (1980): 301–309; Steven M. Platek and Todd K. Shackelford, *Female Infidelity and Paternal Uncertainty: Evolutionary Perspectives on Male Anti-Cuckoldry Tactics* (Cambridge: Cambridge University Press, 2006); Kermyt G. Anderson, Hillard Kaplan, and Jane B. Lancaster, "Confidence of Paternity, Divorce, and Investment in Children by Albuquerque Men," *Evolution and Human Behavior* 28, no. 1 (2007): 1–10.

22. Annette Gordon-Reed, *Thomas Jefferson and Sally Hemings: An American Controversy* (Charlottesville: University of Virginia Press, 1999).

23. Giulia Barrera, "Patrilinearità, razza e identità: L'educazione degli italo-eritrei durante il colonialismo italiano (1885–1934)," *Quaderni Storici* 37, no. 109 (2002): 26.

24. Emmanuelle Saada, *Empire's Children: Race, Filiation, and Citizenship in the French Colonies* (Chicago: University of Chicago Press, 2011).

25. Although there is remarkably little historical or ethnographic work on paternity and its science, several excellent studies deserve mention: Heide Castañeda, "Paternity for Sale: Anxieties over 'Demographic Theft' and Undocumented Migrant Reproduction in Germany," *Medical Anthropology Quarterly* 22, no. 4 (2008): 340–359; Sueann Caulfield, "The Right to a Father's Name: A Historical Perspective on State Efforts to Combat the Stigma of Illegitimate Birth in Brazil," *Law and History Review* 30, no. 1 (2012): 1–36; Sueann Caulfield and Alexandra Minna Stern, "Shadows of Doubt: The Uneasy Incorporation of Identification Science into Legal Determination of Paternity in Brazil," *Cadernos de Saúde Pública* 33 (2017): 1–14; Sabrina Finamori, "Os sentidos da paternidade: Dos 'pais desconhecidos' ao exame de DNA" (PhD diss., Universidade Estadual de Campinas, 2012); Claudia Fonseca, "Following the Path of DNA Paternity Tests in Brazil," in *Reproduction, Globalization, and the State: New Theoretical and Ethnographic Perspectives*, ed. Carolyn Browner and Carolyn Fishel Sargent (Durham, NC: Duke University Press, 2011); Claudia Fonseca, "A certeza que pariu a dúvida: paternidade e DNA," *Estudos Feministas* 12, no. 2 (2004): 13–34; Giulia Galeotti, *In cerca del padre: Storia dell'identità paterna in età contemporanea* (Bari: Laterza, 2009); Shari Rudavsky, "Blood Will Tell: The Role of Science and Culture in Twentieth-Century Paternity Disputes" (PhD diss., University of Pennsylvania, 1996).

26. A Google Ngram analysis demonstrates this shift. The term "paternity test" remains high and rising through the twentieth century, with another discrete jump in the 1980s, coinciding with the emergence of DNA techniques of kinship assessment. The analysis is in English because the phrase for paternity test in romance languages (*prueba*/*prova*/*preuve*) means both "proof" and "test"—and hence encompasses both sociolegal methods and scientific ones.

27. The term is Gérard Noiriel and Ilsen About's, *L'identification: Genèse d'un travail d'état* (Paris: Belin, 2007).

28. Simon Szreter and Keith Breckenridge, *Registration and Recognition: Documenting the Person in World History* (Oxford: Oxford University Press, 2012); Jane Caplan and John C. Torpey, *Documenting Individual Identity: The Development of State Practices in the Modern World* (Princeton, NJ: Princeton University Press, 2001); Mercedes García Ferrari, *Ladrones conocidos, sospechosos reservados: Identificación policial en Buenos Aires, 1880–1905* (Buenos Aires: Prometeo Libros, 2010).

29. Francis Galton, "Personal Identification and Description," *Journal of the Anthropological Institute of Great Britain and Ireland* 18 (1889): 191.

30. Here I draw on wording from Sharrona Pearl's study of nineteenth-century physiognomy, *About Faces: Physiognomy in Nineteenth-Century Britain* (Cambridge, MA: Harvard University Press, 2010).

31. "Charlie Chaplin to Speak on Tyranny of U.S. Women," *St Petersburg Times,* May 26, 1928. The quote is from a Vienna-based men's rights group. On the modern girl and the anxieties she elicited, see Alys Eve Weinbaum, The Modern Girl around the World Research Group, Lynn M. Thomas, Priti Ramamurthy, Uta G. Poiger, Madeleine Yue Dong, and Tani E. Barlow, eds., *The Modern Girl around the World: Consumption, Modernity, and Globalization* (Durham, NC: Duke University Press, 2008); Susan K. Besse, *Restructuring Patriarchy: The Modernization of Gender Inequality in Brazil, 1914–1940* (Chapel Hill: University of North Carolina Press, 1996); Atina Grossmann, *Reforming Sex: The German Movement for Birth Control and Abortion Reform, 1920–1950* (New York: Oxford University Press, 1995).

32. Fuchs, *Contested Paternity;* Allen, "Feminism, Social Science, and the Meanings of Modernity"; Asunción Lavrín, *Women, Feminism, and Social Change in Argentina, Chile, and Uruguay, 1890–1940* (Lincoln: University of Nebraska Press, 1998); Marcela M. A. Nari, *Políticas de maternidad y maternalismo político: Buenos Aires, 1890–1940* (Buenos Aires: Editorial Biblos, 2004).

33. Elinor Ann Accampo, Rachel Ginnis Fuchs, and Mary Lynn Stewart, *Gender and the Politics of Social Reform in France, 1870–1914* (Baltimore: Johns Hopkins University Press, 1995); Joshua Cole, *The Power of Large Numbers: Population, Politics, and Gender in Nineteenth-Century France* (Ithaca, NY: Cornell University Press, 2000); Nancy Stepan, *The Hour of Eugenics: Race, Gender, and Nation in Latin America* (Ithaca, NY: Cornell University Press, 1991); Alexandra Minna Stern, *Eugenic Nation: Faults and Frontiers of Better Breeding in Modern America* (Berkeley: University of California Press, 2016); Alexandra Minna Stern, "Responsible Mothers and Normal Children: Eugenics, Nationalism, and Welfare in Post-Revolutionary Mexico, 1920–1940," *Journal of Historical Sociology* 12, no. 4 (1999): 369–397.

34. Fuchs, *Contested Paternity;* Anne-Emanuelle Birn, "Uruguay's Child Rights Approach to Health: What Role for Civil Registration?" in *Registration and Recognition: Documenting the Person in World History,* ed. Simon Szreter and Keith Breckenridge (Oxford: Oxford University Press, 2012), 415–447; Nara Milanich, "To Make All Children Equal Is a Change in the Power Structures of Society: The Politics of Family Law in Twentieth Century Chile and Latin America," *Law and History Review* 33, no. 4 (2015): 767–802.

35. Cited in "Feminist Movements Are Different Abroad," *New York Times Magazine,* October 10, 1915, 11–12.

36. Helena Bergman and Barbara Hobson, "Compulsory Fatherhood: The Coding of Fatherhood in the Swedish Welfare State," in *Making Men into Fathers: Men, Masculinities and the Social Politics of Fatherhood,* ed. Barbara Hobson (New York: Cambridge University Press, 2002), 92–124.

37. Antônio Ferreira de Almeida Júnior, *As provas genéticas da filiação* (São Paulo: Revista dos Tribunais, 1941), 6–8.

38. Werner Sollors, *Neither Black nor White yet Both: Thematic Explorations of Interracial Literature* (Cambridge, MA: Harvard University Press, 1999); for a contemporary case, see Maïa de la Baume, "In France, a Baby Switch and a Lesson in Maternal Love," *New York Times,* February 24, 2015, https://www .nytimes.com/2015/02/25/world/europe/in-france-a-baby-switch-and-a-test-of -a-mothers-love.html?rref=world/europe&module=Ribbon&version =context&region=Header&action=click&contentCollection=Europe&pgtype =article&_r=0; Helen Weathers, "Why Am I Dark, Daddy? The White Couple Who Had Mixed Race Children after IVF Blunder," *Daily Mail,* June 13, 2009, https://www.dailymail.co.uk/news/article-1192717/Why-I-dark-daddy-The -white-couple-mixed-race-children-IVF-blunder.html.

39. Alphonse Bertillon, "The Bertillon System of Identification," *Forum* 11 (1890–91): 330–341.

40. Mary Ann Glendon, *The Transformation of Family Law: State, Law, and Family in the United States and Western Europe* (Chicago: University of Chicago Press, 1996); Harry D. Krause, "Bastards Abroad: Foreign Approaches to Illegitimacy," *American Journal of Comparative Law* 15, no. 4 (1967): 726–751; Milanich, "To Make All Children Equal."

41. On maternity and paternity, see Coward, *Patriarchal Precedents.* On revelation, see David M. Schneider and Richard Handler, *Schneider on Schneider: The Conversion of the Jews and Other Anthropological Stories* (Durham, NC: Duke University Press, 1995), 222. On nature and culture as defining Euro-American kinship, see Marilyn Strathern, *After Nature: English Kinship in the Late Twentieth Century* (Cambridge: Cambridge University Press, 1992); Janet Carsten, *After Kinship* (Cambridge: Cambridge University Press, 2004); Sarah Franklin, "Anthropology of Biomedicine and Bioscience," in *The SAGE Handbook of Social Anthropology,* 2 vols. (London: SAGE Publications Ltd, 2012), 42–55.

42. Extramarital children in Latin American countries accounted for between 20 percent and more than 60 percent of births, whereas in most of Europe, the statistic was in the low single digits. Göran Therborn, *Between Sex and Power: Family in the World, 1900–2000* (London: Psychology Press, 2004); J. A. Bauzá, "Importancia del factor ilegitimidad en la mortalidad infantil," *Boletín del Instituto Internacional Americano de Protección a la Infancia* 14, no. 3 (1941): 397; Béla Tomka, *A Social History of Twentieth-Century Europe* (London: Routledge, 2013).

43. The phrase is from Edwin Sidney Hartland, *Primitive Paternity: The Myth of Supernatural Birth in Relation to the History of the Family* (London: D. Nutt, 1910).

44. Fritz Schiff, *Blood Groups and Their Areas of Application,* Selected Contributions to the Literature of Blood Groups and Immunology, vol. 4, part 2 (1933; repr., Fort Knox, Kentucky: U.S. Army Medical Research Laboratory, 1971), 311–313;

Fritz Schiff, "Abstammungsproben in alter Zeit," *Deutsche Medizinische Wochenschrift* 27 (1929): 1141–1143; Alexander S. Wiener, "Determining Parentage," *Scientific American* 40, no. 4 (1935): 323–331; A. Almeida Júnior, *Paternidade* (São Paulo: Companhia Nacional, 1940); Arnaldo Amado Ferreira, *Determinação médico-legal da paternidade* (São Paulo: Comp. Melhoramentos de São Paulo, 1939).

45. "Modern Solomons": "Can Scientists Tell Babies Apart?" *Science News-Letter* 18, no. 488 (August 16, 1930), 98; "obscured paternity": "How the Blood Tests Made a Love Test," *Atlanta Constitution,* April 24, 1921, G4.

2. THE CHARLATAN AND THE OSCILLOPHORE

Epigraph: "Scientists Doubt Findings of Russian Savant Claiming Blood of Jews and Gentiles Differs," *Jewish News Service,* October 17, 1930.

1. "Paternity of Child up to Experts / Baby Not Mine, Declares Italian," *San Francisco Chronicle,* January 29, 1921, 9.

2. "How the Blood Test Made a Love Test," *Atlanta Constitution,* April 24, 1921, G4. The same article appeared in the *San Francisco Chronicle,* May 1, 1921, SM6.

3. "Absolutely conclusive": "Court Establishes Parentage of Baby by Electric Blood Test," *New York Tribune,* February 14, 1921, 1.

4. "Easy to Determine Child's Parentage," *Los Angeles Times,* February 13, 1921, IV11.

5. Prior to the Vittori case, Abrams was known mostly to a circle of followers. He and his work were largely absent from the mainstream press, so Nolan could not have known of his work. One follower wrote to Abrams's journal: "a little over one year ago the name of Abrams was unknown to me. I recall seeing his name in a newspaper which stated that he had been asked to determine the parentage of a child from a few drops of blood." Robert Rosen, "Preliminary Report on the Electronic Reactions of Abrams," *Physico-Clinical Medicine* 7, no. 11 (January 1924): 2.

6. In fact, the doctor and the judge knew each other prior to the Vittori case. Several years before, Graham had presided over a dispute between Abrams and his in-laws over Abrams's deceased wife's estate. "Long Contest over Will of Mrs. Blanche Abrams Settled," *San Francisco Chronicle,* December 23, 1916, 12.

7. Ohms list: "Racial Rates," *Physico-Clinical Medicine* 4, no. 2 (December 1919): 219; "racial rates of vibration are transmitted," "a child through generations": "The Physics of Human Phenomena," *Physico-Clinical Medicine* 4, no. 2 (December 1919): 197–201.

8. "Court Establishes Parentage of Baby by Electric Blood Test."

9. "A Blood Polemic," *Physico-Clinical Medicine* 6, no. 1 (September 1921): 3.

10. "Preposterous": "Blood Test Bunk, Says Dean Cabot," *Los Angeles Times,* September 30, 1922, I16; "dean": James Harvey Young, *The Medical Messiahs: A Social History of Health Quackery in 20th Century America* (Princeton, NJ:

Princeton University Press, 2015), 137. Asked to examine one of his gadgets for a fraud investigation, physicist Robert Millikan characterized it as "the kind of device a ten-year-old boy would build to fool an eight-year-old." Young, *The Medical Messiahs*, 140.

11. Minutes of the Society of Forensic Medicine, New York City, October 1921, *Medico-Legal Journal* 38 (1921): 80.

12. Minutes of the Society of Forensic Medicine, 94.

13. Major symbol: Porqueres i Gené, "Kinship Language and the Dynamics of Race," in *Race, Ethnicity and Nation: Perspectives from Kinship and Genetics,* ed. Peter Wade (New York: Berghahn Books, 2007), 127; allegorical meaning: Douglas Starr, *Blood: An Epic History of Medicine and Commerce* (New York: Random House, 2012), 72.

14. Bruno J. Strasser, "Laboratories, Museums, and the Comparative Perspective: Alan A. Boyden's Quest for Objectivity in Serological Taxonomy, 1924–1962," *Historical Studies in the Natural Sciences* 40, no. 2 (2010): 149–182.

15. A retrospective on Reichert's contributions is available at http://www .historyofinformation.com/expanded.php?id=2681. Certain phenomena: Arthur St. George Joyce, "Blood Will Tell," *Technical World Magazine* 20, no. 2 (October 1913): 188–191 and 294. Much of this article was reprinted as "Blood Will Tell: According to European Scientists They Will Make Identification of Criminal Certain," *Washington Post,* November 30, 1913, MS4. Within a year: "Blood Test for Parentage. University Professor Can Determine Ancestry by Character of Blood Crystals," *Washington Post,* June 22, 1913, R2.

16. Arthur Benjamin Reeve, *The Ear in the Wall* (New York: Harper, 1916), 311, 312.

17. "Making Blood Tell: Science's Newest and Surest," *San Francisco Chronicle,* December 29, 1918, SM2. The article did not mention parentage testing but cited an article in *Popular Science Monthly* (from which it was largely drawn), which did discuss this application: Anna Heberton Ewing, "Blood Will Tell," *Popular Science Monthly,* December 1918, 72–73. The lawyer's confusion may have stemmed from the fact there was another Reichert—Frederick L.— affiliated with Johns Hopkins around this time, who wrote on the use of blood groups in paternity determination: Frederick L. Reichert, "On the Present Status of the Inheritance of the Blood Groups," *Eugenical News,* June 1922, 65–67.

18. England: "Machine to Determine Paternity," *Manchester Guardian,* February 17, 1921, 7; "Paternity Proved by Mechanism," *Manchester Guardian,* February 18, 1921, 7; "Positive Paternity," *Observer,* August 7, 1921, 13; France: Tristan Le-Roux, "Séducteurs, Gare a l'Hematogramme!" *La Presse,* April 3, 1922, 1; the invention was also reported in *Le Journal* (Paris), *L'Ouest-Eclair, La Griffe. Journal hebdomadaire de critique, politique, et satirique,* and probably elsewhere; Italy: "Un apparecchio per l'accertamento della paternità," *La Stampa,* February 18, 1921, 2; Australia: "The Parent Finder. Invention to Save Honor. New

Yankee Theory," *Horsham Times* (Victoria), July 8, 1921, 3; "Find-the-Father Test," *Daily News* (Perth), April 4, 1921, 3; Brazil: "As maravilhas da sciencia," *A Lucta,* May 25, 1921, 2 (Sobral, Ceará); Argentina: "Invención de un notable aparato," *Caras y Caretas,* no. 1236, June 10, 1922, 124.

19. "Blood Vibrations Determine Parentage," *Popular Mechanics,* July 1921, 9–10. Readership: Clifton Fadiman, "What Does America Read?" in *America as Americans See It,* ed. Frederick Julius Ringel (New York: Harcourt, Brace, 1932), 77.

20. "Use Ouija Board in Baby Tangle, Suggests Lawyer," *Atlanta Constitution,* May 10, 1921, 1.

21. Boston: "No Blood Test to Pick Boy's Mother," *Boston Daily Globe,* October 18, 1922, 1; Chicago: "Mothers Bare Souls in Fight for Illicit Baby," *Chicago Daily Tribune,* June 12, 1923, 6.

22. "No Blood Test for Bull," *New York Tribune,* October 10, 1922, 8; "Farmers' Counsel in Illinois Cites Decisions Favoring Innocent Offspring," *New York Times,* October 7, 1922, 3.

23. Lucy Bland, "'Hunnish Scenes' and a 'Virgin Birth': A 1920s Case of Sexual and Bodily Ignorance," *History Workshop Journal* 73, no. 1 (2012): 118–143.

24. "Sex trials": "Expert Explains How Blood Test Proves Paternity," *Atlanta Constitution,* October 1, 1922; an ad for Abrams's lecture appears in *New York Tribune,* October 7, 1922, 6; "Dr. Abrams Talks on His Medical Theory," *Boston Daily Globe,* October 9, 1922, 16; Russell divorce: "Dream Babies that Would Perplex a Modern Solomon," *San Francisco Chronicle,* October 8, 1922, SM3.

25. "Father by Blood Test," *New York Times,* June 19, 1922, 30. The doctor in question, Dr. C. L. Thudicum, appears in Abrams's publications as one of the doctors who had studied his method. Thudicum also carried on an "extended correspondence" with John Tiernan about blood testing: "Professor Invokes Blood Test to Fix Infant's Paternity," *New York Tribune,* September 6, 1922, 7.

26. Letter from Mrs. C. W. Womack to Dr. Albert Abrams, Page Collection, University of California San Francisco (UCSF) Archives.

27. "Has Science Found Answer to Question of Parentage?," *Baltimore Sun,* October 29, 1922, part 6, 2.

28. "How the Blood Test Made a Love Test," *Atlanta Constitution.*

29. Frederick Lewis Allen, *Only Yesterday: An Informal History of the 1920s* (1931; repr., New York: John Wiley & Sons, 1997), 74.

30. "How the Blood Test Made a Love Test," *Atlanta Constitution.*

31. Letter from Mrs. C. W. Womack to Dr. Albert Abrams.

32. "The Parent Finder."

33. Minutes of the Society of Forensic Medicine, 94: "Dr. Herzog stated that he had been informed by a member of the Society, residing in San Francisco, that Judge Graham, who it was claimed, had decided a case in accordance with Dr. Abrams' testimony, had stated that it was not because of the Medical Testimony, but because [of] the legal presumption" of marital legitimacy.

34. Michael Schudson, *Discovering the News: A Social History of American Newspapers* (New York: Basic Books, 1978), 129.

35. "How the Blood Test Made a Love Test," *Atlanta Constitution* and *San Francisco Chronicle*.

36. "Vittori Blood Test Case Made Famous," *San Francisco Chronicle*, May 3, 1921, 5.

37. The papers also reproduced the correspondence from the case. The *San Francisco Chronicle* published the letter from the parents' attorney to Graham, and the *Atlanta Constitution* published on its front page Abrams's expert report and Judge Graham's assessment of the report, which had been sent to the Atlanta lawyer. "Vittori Blood Test Case Made Famous"; "Use Ouija Board in Baby Tangle."

38. "Blood Test to Fix Parentage Stirs Doctors," *San Francisco Chronicle*, February 19, 1921, 10; purchase reported in *Physico-Clinical Medicine* 6, no. 1 (September 1921): 30.

39. Minutes of the Society of Forensic Medicine, 94.

40. Statistics taken from the back matter of *Physico-Clinical Medicine* 6 (September 1921): 30–34.

41. Prices given in his journal *Physico-Clinical Medicine,* for 1919, 1921; advertisements for Abrams courses also appear in Brazilian newspapers in the 1920s.

42. 3,000: "Man to Dine, He Says, On Electric Vibrations," *New York Tribune,* October 2, 1923, 1.

43. The interview was conducted with an *Evening News* journalist, probably in London, and was reproduced in an Australian newspaper. "Solomon Outdone," *Cairns Post* (Queensland), June 24, 1922, 12. Follower Dr. C. L. Thudichum also performed paternity tests and sparred with critics of the Vittori verdict. Letter from C. L. Thudichum to Mr. Sam H. Clark (editor of *Jim Jam Jems*), no date, criticizing the publication for "your attack on Dr. Abrams re the Vittori paternity case," Page Collection, UCSF Archives.

44. "As theorias scientificas do dr. Abrams," *O Jornal,* March 18, 1923, 11; "La Maison du Miracle," *Bulletin de l'Ordre de l'Étoile d'Orient,* no. 3 (July 1923): 22–40.

45. A series of advertisements for classes appear in the classified section of *Correio da Manhã,* February 1921.

46. Díaz's doctor is mentioned in *Physico-Clinical Medicine* 6, no. 1 (September 1921): 42; "Dr. Abrams Arrives," *Los Angeles Times,* July 2, 1923, II2, discusses President Obregon's invitation; Sophie Treadwell, "Calles Pledges Further Aid to Mexican Laborers," *New York Herald Tribune,* November 23, 1924, 6, mentions Abrams's treatment of president-elect Plutarco Calles. Abrams also appeared in the Mexican press: "Un doctor que se las trae," *El Siglo de Torreon* (Coahuila), March 27, 1923, 3; and more critically, V. Salado Alvarez, "Un brujo moderno," *El Siglo de Torreon,* April 8, 1923, 3.

47. "Blood Test Bunk, Says Dean Cabot," September 30, 1922, *Los Angeles Times,* II6; "Scoff at Blood Tests," *Los Angeles Times,* February 21, 1921, I5; also see

"Paternity Blood Test Finds Paris Skeptical," *New York Times,* February 19, 1921, 3.

48. "No Blood Test to Pick Boy's Mother," *Boston Daily Globe,* October 18, 1922, 1.

49. American Medical Association, *Minutes of the House of Delegates Proceedings, Annual Sessions,* 1923, 13; 1924, 13; 1925, 12.

50. Morris Fishbein, *The Medical Follies* (New York: Boni and Liveright, 1925); and Fishbein, *The New Medical Follies* (New York: Boni and Liveright, 1927); "The Name of Albert Abrams," *Hygeia,* May 1938, 462. The full sequence of *JAMA*'s coverage of Abrams is reconstructed by Ken Raines, http://www.petitioneurope.com/radionique/index.php/abrams-dr.

51. "Ridicules": "Our Abrams Investigation-I," *Scientific American,* October 1923, 230.

52. "Is Parentage Determinable by Blood Tests?" *JAMA* 79, no. 15, October 7, 1922, 1246–1247.

53. This assessment, from the *Atlanta Constitution,* appeared the same week as the *JAMA* editorial. "Expert Explains How Blood Test Proves Paternity," *Atlanta Constitution,* October 1, 1922, A14. An ad for an Abrams lecture in Carnegie Hall also appeared that same day. *New York Tribune,* October 7, 1921, 6.

54. In all its Abrams coverage, *JAMA* never mentioned his paternity test. Moreover, the journal only began to denounce Abrams a full year into the media furor surrounding his parentage test, which began with the Vittori case. Paternity testing launched Abrams's star, but you would never know this from the coverage of his most powerful critic. This probably explains why histories of Abrams have overlooked his paternity test.

55. "Blood Test to Fix Parentage Stirs Doctors," *San Francisco Chronicle,* February 19, 1921, 10.

56. "Albert Abrams: A Defense by Upton Sinclair," *JAMA* 78 (April 29, 1922): 1334.

57. The lengthy interview was published in an Australian paper but appears to have been conducted by an English interviewer. "The Parent Finder."

58. Austin C. Lescarboura, "Our Abrams Verdict," *Scientific American,* September 1924, 158.

59. Upton Sinclair, "The House of Wonder," *Pearson's Magazine,* June 1922, 14.

60. $2,000,000: Fishbein, *Medical Follies,* 116. Perhaps predictably, this success also became fodder for anti-Semitic critique. "Abrams was an American Jew, and naturally, almost instinctively, he made a commercial success of his discoveries," noted a British observer. A. J. Clark, "Universal Cures, Ancient and Modern," *British Medical Journal* 2, no. 3329 (October 18, 1924): 733.

61. Lescarboura, "Our Abrams Verdict," 160.

62. Identigene claims to be the first company to market DNA tests directly to consumers. "Identigene Turns 20! Identigene DNA Laboratory Celebrates 20 Year Anniversary," http://www.identigene.com/news/identigene-turns-20-identigene-dna-laboratory-celebrates-20-year-anniversary.

63. Editorial, "Solomon in Cleveland," *New York Herald Tribune,* September 26, 1927, 20.

3. BLOOD WORK

Epigraph: Leone Lattes, "I gruppi sanguigni e la ricerca della paternità," *Atti della Società lombarda di scienze mediche e biologiche* 16 (1927): 319.

1. Fritz Schiff and Lucie Adelsberger, "Die Blutgruppendiagnose als forensische Methode," *Aerztliche Sachverständigen-Zeitung* 11 (1924): 101–103; Fritz Schiff, "Wie häufig läßt sich die Blutgruppendiagnose in Paternitätsfragen heranziehen?," *Aerztliche Sachverständigen-Zeitung* 24 (1924): 231–233. The lecture is described in Mathias Okroi and Peter Voswinckel, "'Obviously Impossible'—The Application of the Inheritance of Blood Groups as a Forensic Method. The Beginning of Paternity Tests in Germany, Europe and the USA," *International Congress Series* 1239 (2003): 711–714. See also Gunther Geserick and Ingo Wirth, "Genetic Kinship Investigation from Blood Groups to DNA Markers," *Transfusion Medicine and Hemotherapy* 39 (2012): 163–175.

2. "The facts rest on so secure a basis that the test is sure to be used in evidence sooner or later." "Blood Groups and Paternity," *Lancet,* August 5, 1922, 285.

3. "Mystique": Fritz Schiff, *Die Blutgruppen und ihre Anwendungsgebiete: Indikation und Technik der Bluttransfusion* (Berlin: Verlag von Julius Springer, 1933), published in English as *The Blood Groups and Their Areas of Application,* Selected Contributions to the Literature of Blood Groups and Immunology, vol. 4, part 2 (Fort Knox, Kentucky: U.S. Army Medical Research Laboratory, 1971), 335. On blood, race, and politics in this period in Europe and the United States, see Rachel E. Boaz, *In Search of "Aryan Blood": Serology in Interwar and National Socialist Germany* (Budapest: Central European University Press, 2012); Jonathan Marks, "The Origins of Anthropological Genetics," *Current Anthropology* 53, Suppl. 5 (2012): 161–172; Pauline M. H. Mazumdar, "Blood and Soil: The Serology of the Aryan Racial State," *Bulletin of the History of Medicine* 64, no. 2 (1990): 187–219; William H. Schneider, "Chance and Social Setting in the Application of the Discovery of Blood Groups," *Bulletin of the History of Medicine* 57, no. 4 (1983): 545–562; William H. Schneider, "Blood Group Research in Great Britain, France, and the United States between the World Wars," *Yearbook of Physical Anthropology* 38 (1995): 87–114; Myriam Spörri, *Reines und gemischtes Blut: Zur Kulturgeschichte der Blutgruppenforschung, 1900–1933* (Bielefeld: Transcript Verlag, 2013).

4. Schneider, "Chance and Social Setting."

5. Emil von Dungern, and Ludwik Hirschfeld [*sic*], "Concerning Heredity of Group Specific Structures of Blood," *Transfusion* 2, no. 1 (1962): 70–74 [originally published in 1910 as Emil von Dungern, "Ueber Nachweis und Vererbung biochemischer Strukturen und ihre forensische Bedeutung," *Münchener Medizinische Wochenschrift* 57 (1910): 293–295]; cuckoo's egg: Ludwik Hirszfeld, *Ludwik Hirszfeld: The Story of One Life* (Rochester: University Rochester Press, 2010), 19.

6. The American doctor Reuben Ottenberg had reached the same conclusion in an earlier study, but given that study's small size, von Dungern and Hirszfeld are usually considered the first researchers to work out the Mendelian inheritance of blood groups.

7. "Ideal marker": Pauline M. H. Mazumdar, *Species and Specificity: An Interpretation of the History of Immunology* (Cambridge: Cambridge University Press, 2002), 301. On the importance of blood groups to modern genetics, see Schneider, "Blood Group Research."

8. Laurence H. Snyder, "Human Blood Groups: Their Inheritance and Racial Significance," *American Journal of Physical Anthropology* 9, no. 2 (1926): 233–263, provides a lengthy but incomplete bibliography of family studies between 1920 and 1926. Leone Lattes, *Aspetti biologici della ricerca della paternità* (Modena: Facoltà di Giurisprudenza della R. Università di Modena, 1927) provides a longer list, dividing the studies by publication dates (1910–1925 and 1926) but does not cite the actual studies. Presumably most of them overlapped with those cited by Snyder and occurred after 1920.

9. Schneider, "Chance and Social Setting"; Mazumdar, "Blood and Soil"; Boaz, *In Search of "Aryan Blood."*

10. Snyder, "Human Blood Groups," 236. His comment links a discussion of the Salonika study and related research with a consideration of von Dungern and Hirszfeld's experiment.

11. E. O. Manoiloff, "Discernment of Human Races by Blood. Particularly of Russians from Jews," *American Journal of Physical Anthropology* 10, no. 1 (1927): 15–16. On Manoiloff's research, see Charles E. Abromavich Jr. and W. Gardner Lynn, "Sex, Species, and Race Discrimination by Manoilov's Methods," *Quarterly Review of Biology* 5, no. 1 (March 1930): 68–78.

12. The five reagents were methyl blue and cresyl violet (staining agents), silver nitrate (used in early photography and explosives), hydrochloric acid (used in chemical products), and potassium permanganate (an oxidant, disinfectant, and bleaching agent).

13. Manoiloff cited Landsteiner's work and the Salonika study as well as other research on the racial and geographic distribution of blood groups. Neither he nor Poliakowa cite Schiff's parentage work, which was probably too new.

14. Leone Lattes, *Individuality of the Blood in Biology and in Clinical and Forensic Medicine* (Oxford: Oxford University Press, 1932), cites Manoiloff and Poliakowa in his extensive bibliography. Other researchers attempted to replicate his results but were unsuccessful: Boaz, *In Search of "Aryan Blood,"* 129–30.

15. "Blood Unalike": *Science News-Letter* 18, no. 493 (September 20, 1930): 180. The *Science News-Letter* was a nonprofit publication devoted to providing the American press with high-quality information on the latest scientific developments. "Startling accuracy": "Have You a Double?" *Popular Science Monthly*, April 1926, 16.

16. "Scientists Doubt Findings of Russian Savant Claiming Blood of Jews and Gentiles Differs," *Jewish News Service,* October 17, 1930. Unpopular with whites: Algernon Jackson, "Chemical Test May Determine Race Identity," *Philadelphia Tribune,* December 24, 1931, 15. A number of African American papers covered the story. Other critiques by anthropologists include Earnest Hooton, cited in Jonathan Marks, "Blood Will Tell (Won't It?): A Century of Molecular Discourse in Anthropological Systematics," *American Journal of Physical Anthropology* 94, no. 1 (1994): 61.

17. Anna Poliakowa, "Manoiloff's 'Race' Reaction and Its Application to the Determination of Paternity," *American Journal of Physical Anthropology* 10, no. 1 (1927): 23–29.

18. Lattes, *Individuality of the Blood;* and Schiff, *Blood Groups and Their Areas of Application.* Serologists who wrote about both racial serology and the serology of kinship include Hirszfeld, of course, as well as Reuben Ottenberg, Laurence Snyder, Fritz Schiff, and Otto Reche.

19. Fritz Schiff, "Die Blutgruppenverteilung in der Berliner Bevölkerung," *Klinische Wochenschrift* 36 (1926): 1660–1661; and Fritz Schiff, "Die sogenannten Blutgruppen des Menschen und ihr Vorkommen bei den Juden," *Jüdische Familienforschung* 4 (1926): 178–180. He concluded that there was no significant difference between the Jewish and non-Jewish inhabitants of the city. Boaz, *In Search of "Aryan Blood."*

20. Leone Lattes, "La dimostrazione biologica," *La Riforma Medica* 39, no. 8 (1923): 172; "Is Parentage Determinable by Blood Tests?" *JAMA* 79, no. 15, October 7, 1922: 1247.

21. Lattes, *Aspetti biologici.*

22. Lattes, "I gruppi sanguigni e la ricerca," 310; "much misinformation": "Blood Tests for Paternity," *JAMA* 87, no. 14 (October 2, 1926): 1130. One wonders if this was a reference to the lingering fallout from the Abrams affair two years earlier.

23. Peter Schiff (Fritz Schiff's youngest son), telephone interview with author, May 12, 2016.

24. Leone Lattes, "Necrologia. Dr. Fritz Schiff," *La Prensa Médica Argentina* 27, no. 2 (1940): 2134.

25. Georg's father, Fritz Strassmann, was also a well-known forensic specialist who had been called to perform somatic analysis in a parentage dispute as far back as 1903.

26. On Strassmann and their early collaboration, see Schiff, "Wie häufig läßt sich die Blutgruppendiagnose in Paternitätsfragen heranziehen?"; Georg Strassmann, "Über individuelle Blutdiagnose," *Deutsche Zeitschrift für die gesamte gerichtliche Medizin* 5 (1925): 184–192.

27. On inflated expectations in the press, see Schiff, "Wie häufig läßt sich die Blutgruppendiagnose in Paternitätsfragen heranziehen?"; Fritz Schiff, "Kann eine Blutuntersuchung in Vaterschaftsfragen herangezogen werden?," *Juristische*

Wochenschrift 54 (1925): 343–344; Fritz Schiff, "Die Blutuntersuchung bei strittiger Vaterschaft in Theorie und Praxis," *Deutsche Zeitschrift für die gesamte gerichtliche Medizin* 7 (1926): 360–375. See also Spörri, *Reines und gemischtes Blut,* chapter 7.

28. The case was described in Paul Moritsch, "Über den Wert der Blutgruppen-bestimmung in der Paternitätsfrage," *Wiener Klininische Wochenschrift* 39 (1926): 961–962.

29. The controversy is discussed in Spörri, *Reines und gemischtes Blut,* chapter 7.

30. Julius Schwalbe, "Die praktische Bedeutung der Blutgruppenuntersuchung, insbesondere für die Gerichtliche Medizin: Eine Umfrage," *Deutsche Medizinische Wochenschrift* 30 (1928): 1240–1244.

31. 5,000: Schiff, *Blood Groups and Their Areas of Application,* 331. The statistics are from 1929; "surprising": Fritz Schiff, "The Medico-Legal Significance of Blood Groups," *Lancet* 214 (1929): 921; 700 Austrian cases: Magdalene Schoch, "Determination of Paternity by Blood-Grouping Tests: The European Experience," *Southern California Law Review* 16 (1942): 177–192; Schiff reproduced a series of decrees from ministries of justice around Germany dating from the late 1920s and early 1930s; Fritz Schiff, "Anhang. Veröffentlichungen von Justizbehörden," in *Die Technik der Blutgruppenuntersuchung für Kliniker und Gerichtsärzte: Neben Berücksichtigung ihrer Anwendung in der Anthropologie und der Vererbungs- und Konstitutionsforschung* (Berlin: Springer, 1932), 74–80.

32. Soviet Union: Lattes, *I gruppi sanguigni e la ricerca,* 312–313; Czechoslovakia and Poland: Schoch, "Determination of Paternity by Blood-Grouping Tests," 182; University of Copenhagen: Louis Christiaens, *La recherche de la paternité par les groupes sanguins: Étude technique et juridique* (Paris: Masson et cie, 1939), 86.

33. Associated Press wire: "Accepts Blood Test in Paternity Case," *Boston Daily Globe,* March 11, 1926, A10; and "A investigação de maternidade pelo exame do sangue," *A Manhã,* April 7, 1926, 1; Australia: "Blood Test. Paternity of a Child. Court Case," *Sydney Morning Herald,* February 21, 1928, 11; Cuba: Antonio Barreras y Fernandez and Manuel Barroso y Mensaque, "Informe sobre investigación de la paternidad," *Revista de Medicina Legal de Cuba* 7, no. 12 (December 1929): 375–386; Colombia: Guillermo Uribe Cualla, "Investigación de la paternidad por los grupos sanguíneos," *Revista de la Facultad de Medicina* 4, no. 1 (1935): 32–33; Peru: Leonidas and Jorge Avendaño, "Investigación de la paternidad y grupos sanguíneos," *La Crónica Médica* 52, no. 864 (1935): 198–205 (the case actually took place in 1932).

34. United States: "Chicagoan Asks Blood Tests to Determine Child's Father," *Atlanta Constitution,* April 8, 1926, 4; "Blood Test for Paternity," *Times of India,* December 27, 1926, 7.

35. France: Schoch, "Determination of Paternity by Blood Grouping Tests," 183; England: Christiaens, *La recherche de la paternité,* 82.

36. The Hirszfelds were Polish but had trained in Germany; Ludwik had, of course, earlier worked in Heidelberg. Before the war, the couple transferred to Zurich.

37. The discovery of MN was the work of Landsteiner and his colleague Phillip Levine; the Rh discovery, that of Landsteiner and Alexander Weiner. Both Weiner and Levine became important public spokesmen for parentage testing. Other early American authorities on blood group testing include Reuben Ottenberg and Laurence Snyder.

38. Malcolm McDermott, "The Proof of Paternity and the Progress of Science," *Howard Law Journal* 1 (1955): 41.

39. Schoch, "Determination of Paternity by Blood-Grouping Tests"; Morris Ploscowe, "The Expert Witness in Criminal Cases in France, Germany, and Italy," *Law and Contemporary Problems* 2, no. 4 (1935): 504–509.

40. In some German states (Bavaria, Saxony, Württemberg), institutes of forensic medicine added blood testing to their list of duties. Elsewhere, such as Berlin, Frankfurt, and Nuremberg, individual specialists in serology or heredity became certified as court experts for the task of analyzing blood group evidence. Schiff, "The Medico-Legal Significance of Blood Groups," 922.

41. Béla Tomka, *A Social History of Twentieth-Century Europe* (London: Routledge, 2013), 88. Austria was followed by Sweden (15.8 percent) and then Germany. The statistics are for the year 1930.

42. Michelle Mouton, *From Nurturing the Nation to Purifying the Volk: Weimar and Nazi Family Policy, 1918–1945* (Cambridge: Cambridge University Press, 2007); Georg Lilienthal, "The Illegitimacy Question in Germany, 1900–1945: Areas of Tension in Social and Population Policy," *Continuity and Change* 5, no. 2 (August 1990): 249–281.

43. 150,000 births: Fritz Schiff, "Die sogenannte Blutprobe und ihre sozial Bedeutung," *Fortschritte der Gesundheitsfürsorge: Organ der deutschen Gesund-heitsfürsorgeschule* 9 (1928): 356, cited in Spörri, *Reines und gemischtes Blut,* 278.

44. Atina Grossmann, *Reforming Sex: The German Movement for Birth Control and Abortion Reform, 1920–1950* (Oxford: Oxford University Press, 1995).

45. Mouton, *From Nurturing the Nation to Purifying the Volk,* 204. These statistics are from Munster and Westphalia, respectively.

46. Maria Teschler-Nicola, "The Diagnostic Eye—On the History of Genetic and Racial Assessment in Pre-1938 Austria," *Collegium Antropologicum* 28 (2004): 11.

47. Mouton, *From Nurturing the Nation to Purifying the Volk,* 206.

48. Spörri, *Reines und gemischtes Blut,* chapter 7.

49. Schiff, *Blood Groups and Their Areas of Application,* 332.

50. "Hundreds of false judgments": Otto Reche, "Anthropologische Beweisführung in Vaterschaftsprozessen," *Österreichische Richterzeitung* 19 (1926): 157; "capri-ciousness": Emil Blank, "Pater semper incertus?," *Österreichische Richterzeitung* 20 (1927): 136–138. Similar comments are made in Albert Harrasser, "Zur prozessualen Bedeutung des naturwissenschaftlichen Vaterschaftsbeweises," *Österreichische Richterzeitung* 25 (1932): 125–126; and Albert Hellwig, "Meineidsverhütung durch Blutgruppenprobe," *Kriminalistische Monatshefte* 3 (1929): 75–77.

51. "Blood Tests to Establish Paternity," *JAMA* 100, no. 7, February 18, 1933, 510.

52. "Paternity Suit Blood Test," *Irish Times,* December 3, 1927, 7; "Paternity by Blood Test," *Times of India,* December 5, 1927, 9; "El examen de sangre como determinante de la paternidad," *El Orden* [Santa Fe], December 28, 1927, http://www.santafe.gov.ar/hemerotecadigital/diario/35/?page=1&zl=2&xp=-252&yp=-150. It appeared in multiple U.S. newspapers as well.

53. Strassmann reports a perjury case in which he intervened as early as September 1924, although it was apparently inconclusive. Strassmann, "Über individuelle Blutdiagnose." Hellwig reports a case from Potsdam: Hellwig, "Meineidsverhütung durch Blutgruppenprobe."

54. The cost of blood tests is quoted in Hans Mayser, "Erfahrungen mit gerichtlichen Blutgruppenuntersuchungen," *Deutsche Zeitschrift für die gesamte gerichtliche Medizin* 10 (1927): 638–651; and Georg Strassmann, "Die forensische Bedeutung der Blutgruppenfrage," *Zeitschrift für Medizinal-Beamte und Krankenhausärzte* 40 (1927): 327–336. Costs varied somewhat by location.

55. Monthly child support awards were set in some places according to the socioeconomic status of the mother and in others in relation to the cost of living. These data, for Westphalia, are cited in Mouton, *From Nurturing the Nation to Purifying the Volk,* 207–209.

56. Strassmann, "Ein Beitrag zur Vaterschaftsbestimmung," *Deutsche Zeitschrift für die gesamte gerichtliche Medizin* 10 (1927): 343.

57. The most complete biography of Fritz Schiff is Mathias Okroi, "Der Blutgruppenforscher Fritz Schiff (1889–1940): Leben, Werk und Wirkung eines jüdischen Deutschen" (PhD diss., Universität zu Lübeck, 2004). Biographical information is also drawn from an interview with Peter Schiff.

58. Nielsen, cited in Okroi, "Der Blutgruppenforscher Fritz Schiff," 75.

59. Wilhelm Zangemeister, "A New Paternity Test," *Medico-Legal Journal* 46 (1929): 5.

60. "Blood Serum Test Proves Parentage: New Method Developed by Professor at German University Is Reported Successful," *Baltimore Sun,* December 15, 1929, AT4.

61. In addition to German scientific journals, Zangemeister's research was reported in *JAMA,* the *British Medical Journal,* the *Journal of Heredity,* the *Medico-Legal Journal,* the *American Journal of Police Science,* and *Actualidad médica mundial* (Argentina) as well as popular venues such as *Popular Science* and *Popular Science Monthly* in the United States and *Careta* and *Jornal Pequeno* (Recife) in Brazil.

62. John MacCormac, "Americans Answer Men's Rights Call," *New York Times,* August 5, 1929, 5; "Fair Deal to Men from Women Is League's Aim," *The Milwaukee Journal,* August 10, 1929; "Charlie Chaplin to Speak on Tyranny of U.S. Women," *St. Petersburg Times,* May 26, 1928.

63. "Here's News! Folks in Far-off Vienna Are Taking Pity on the Men in Spokane," *Spokane Daily Chronicle,* August 9, 1929.

64. "Bitter feelings": Blank, "Pater semper incertus?," 136. See also Reche, "Anthropologische Beweisführung in Vaterschaftsprozessen," 157–159. On the

group and its founder, see Elisabeth Malleier, "Der 'Bund für Männerrechte.' Die Bewegung der 'Männerrechtler' im Wien der Zwischenkriegszeit," *Wiener Geschichtsblätter* 58, no. 3 (2003): 208–233; and Kerstin Christin Wrussnig, "Wollen Sie ein Mann sein oder ein Weiberknecht?" (Diploma diss., University of Vienna, 2009). The group is well known to contemporary "men's rights" Web sites and activists.

65. "League for Rights of Men Abandoned," *Montreal Gazette,* October 6, 1930, 13; "Alas! Poor Yorick," *Christian Science Monitor,* March 12, 1931, 16.

66. European and U.S. blood group scientists, as well as the historians who have written about them, have paid little attention to developments in Latin America. An exception is the Italian Leone Lattes, who was aware of the relevant literature in Brazil, Argentina, and Cuba.

67. Flamínio Favero and Arnaldo A. Ferreira, "Determinação da paternidade pelos grupos sanguineos," *Archivos da Sociedade de Medicina Legal e Criminologia de São Paulo* 2, no. 1 (November 1927): 53–72, cite several Brazilian studies that report relative proportions of blood types as well as data on race (or nationality) but did not attempt to correlate the two. Possibly the first study specifically concerned with the racial and geographic distribution of blood types in Brazil was Octávio Torres's "Estudo geral sobre os grupos sanguineos," *Archivos Brasileiros de Medicina* 21 (1931): 189–214. Later such studies would become common: Sueann Caulfield and Alexandra Minna Stern, "Shadows of Doubt: The Uneasy Incorporation of Identification Science into Legal Determination of Paternity in Brazil," *Cadernos de Saúde Pública* 33 (2017): 1–14.

68. The request must have been in 1925 or 1926.

69. Favero and Ferreira, "Determinação da paternidade pelos grupos sanguineos."

70. The Instituto Vital in Niteroi produced vaccines, pharmaceuticals, and other medical products before also providing serum. The first IOF thesis on blood tests, produced under the direction of Favero and featuring a multilingual bibliography, is José Augusto Lefevre, *Da hereditariedade dos grupos sanguineos e sua applicação na investigação da paternidade* (São Paulo: Irmãos Ferraz, 1928).

71. Arnaldo Amado Ferreira, *Determinação médico-legal da paternidade* (São Paulo: Companhia Melhoramentos de São Paulo, 1939), 133.

72. Virginity examinations were common at the IOF into the 1960s. On Brazilian hymenology, see Sueann Caulfield, *In Defense of Honor: Sexual Morality, Modernity, and Nation in Early-Twentieth-Century Brazil* (Durham, NC: Duke University Press, 2000).

73. Flamínio Favero, *Classificação de Oscar Freire para as fórmas hymenaes* (São Paulo: Editora Limitada, 1930); Arnaldo Amado Ferreira, "Conceito de defloramento na legislação penal brasileira," *Revista Judiciária* 1, no. 2 (1935).

74. Exames e Pareceres Medico-Legais, IOF, vol. 2, 1925–1927, L. 149, F. 154. The accused's professional positions are reconstructed from Diário Oficial do Estado de São Paulo. The case was reported in the scientific literature (Favero and

Ferreira, "Determinação da paternidade pelos grupos sanguineos") and the local press ("Sociedade de Medicina Legal e Criminologia," *Correio Paulistano,* September 16, 1927, 6; also published in *Diario Nacional,* September 16, 1927, 4). Later, it was lauded in histories of the IOF: Amar Ayush, "Revisão da Experiência do Instituto Oscar Freire na Perícia Hematológica de Paternidade e Maternidade" (Thesis, Faculdade de Medicina, Universidade de São Paulo, 1966).

75. Modell: "Chicagoan Asks Blood Tests to Determine Child's Father," *Atlanta Constitution,* April 8, 1926, 4; Vanderbilt Whitney: "Dancer Asks Blood Test of Paternity," *Boston Daily Globe,* August 16, 1922, 8, "Dancer to Again Sue Whitney," *Los Angeles Times,* July 25, 1930, 3; Nebraska: "Blood Test for Paternity," *Times of India,* December 27, 1926, 7; superintendent: "N.Y. Father Plans Blood Test of 'Sonny'," *Baltimore Sun,* September 24, 1927, 22; Iowa women: "Unwed Mother Pleads in Court for Babe," *Des Moines Register,* October 2, 1927, 61.

76. "How the Blood Test Is Made a Love Test," *Atlanta Constitution,* May 1, 1921, G4.

77. "Vittori Blood Test Case Made Famous," *San Francisco Chronicle,* May 3, 1921, 5. The letter was written to Albert Abrams.

78. "Science to Decide Parentage of Baby," *Baltimore Sun,* September 16, 1927, 1; "Blood Tests May Decide Parentage," *Boston Daily Globe,* September 19, 1927, 8.

79. "Baby Test Due Tomorrow: Mother in Cleveland Eagerly Awaits Verdict to Be Given by Experts in Asserted Shuffle," *Los Angeles Times,* September 18, 1927, 10.

80. "Family Examined in 'Baby Puzzle,'" *Boston Daily Globe,* September 22, 1927, 32.

81. Letters: "See No Way to End Tangle over Baby: Doctors Doubt They Can Determine Whether Cleveland Woman Is Girl's Mother," *New York Times,* September 19, 1927, 27.

82. "Solomon in Cleveland," *New York Herald Tribune,* September 26, 1927, 20.

83. "Blood Tests Fail to Solve 'Baby Shuffle,'" *Atlanta Constitution,* September 20, 1927, 1; "Blood Will Not Tell, Say Doctors," *Boston Daily Globe,* September 20, 1927, 32.

84. "Girl Awarded: Judge Rules on Birth Tangle," *Los Angeles Times,* September 23, 1927, 1.

85. Richard W. Wertz and Dorothy C. Wertz, *Lying-in: A History of Childbirth in America* (New Haven, CT: Yale University Press, 1989), 133; see also Judith Walzer Leavitt, *Brought to Bed: Childbearing in America, 1750–1950* (Oxford: Oxford University Press, 1988). This transition was most marked in cities. Hospital births in rural areas also increased but at a much slower rate.

86. "George or Georgia?" *Washington Post,* September 24, 1927, 6.

87. Leavitt, *Brought to Bed.*

88. "Age of efficiency" and "lot of bundles": Marjorie MacDill, "If Hospitals Mix up Babies," *Science News-Letter,* 13, no. 359 (1928): 115.

89. "Bogy" and "well organized": "Hospital Babies Always Labeled," *New York Times,* September 25, 1927, XX13.

90. Morris Fishbein, "Mixed Babies," *New York Herald Tribune,* October 9, 1927, SM15.

91. "Bamberger Flees with 'Watkins' Baby," *Chicago Daily Tribune,* July 25, 1930, 1.

92. "Shuffled Babes Tossed on Lap of Mothers' Jury," *Chicago Daily Tribune,* July 27, 1930, 3.

93. "Hospital Adds New Defense in Mixed Baby Case," *Chicago Daily Tribune,* August 9, 1930, 1.

94. Antônio Ferreira de Almeida Júnior, *As provas genéticas da filiação* (São Paulo: Revista dos Tribunais, 1941), 223–224; Arnaldo Amado Ferreira, *A perícia técnica em criminologia e medicina legal* (São Paulo: n.p., 1948), 391.

95. Cuba: Barreras y Barroso, "Informe sobre Investigación de la Paternidad"; France: Wiener, "Blood Tests for Paternity," *JAMA* 95, no. 9 (1930): 681.

96. Schiff, *Blood Groups and Their Areas of Application,* 328; Alexander Wiener, "On the Usefulness of Blood-Grouping in Medicolegal Cases Involving Blood Relationship," *Journal of Immunology* 24 (1933): 454.

97. Wiener, "Blood Tests for Paternity," 681.

98. "Baby Shuffle Still Puckers Sages' Brows," *Chicago Daily Tribune,* July 22, 1930, 1.

99. The blood test was first reported in the first full article on the case but was treated as inconclusive. "Baby Mix-up Stumps Sages; Try New Tests," *Chicago Daily Tribune,* July 20, 1930, 1. A day later the press reported that the "various tests to establish their identity have so far failed": "Hope to Settle Baby Mixup in Hospital Today" *Chicago Daily Tribune,* July 21, 1930, 5.

100. "Moving pictures" and "Ooch": "Shuffled Babies Howl as Science Toils on Puzzle," *Chicago Daily Tribune,* July 23, 1930, 3.

101. "Smears": "Bamberger Flees with 'Watkins" Baby," 1, 8.

102. "Human testimony": "Shuffled Babies Howl as Science Toils on Puzzle," *Chicago Daily Tribune,* July 23, 1930, 3.

103. "Science May Turn to 'Mother Instinct' in Puzzle to Establish Parentage in Baffling Cases," *NY Herald Tribune,* August 17, 1930, C2.

104. "Bamberger Flees with 'Watkins' Baby," 1, 8.

105. Grouping business: cited in Shari Rudavsky, "Blood Will Tell: The Role of Science and Culture in Twentieth-Century Paternity Disputes" (PhD diss., University of Pennsylvania, 1996), 83; against expert say-so: "Five Years after the Famous Baby Mix-up," *Philadelphia Inquirer,* Magazine Section, 1935, available from http://fultonhistory.com/Newspapers%2023/Philadelphia%20PA%20Inquirer/Philadelphia%20PA%20Inquirer%201935/Philadelphia%20PA%20Inquirer%201935%20-%204115.pdf. See also "Whose Baby Am I?" *Liberty Magazine,* October 11, 1930: 36–44.

106. Afrânio Peixoto, *Novos rumos da medicina legal* (Rio de Janeiro: Editora Guanabara, 1932), 98.

107. "Topics of the Times. Blood Will Not Tell Paternity," *New York Times,* October 9, 1922, 10; "Blood Tests May Decide Parents of Little Child," *Atlanta Constitution,* May 4, 1921, 1.

108. John E. Lodge, "Can You Prove Who You Are?" *Popular Science Monthly* 127, no. 6 (December 1935): 15.

109. Lattes, *Individuality of the Blood.*

4. CITY OF STRANGERS

Epigraph: Judge Adolfo Casabal verdict, 1a Instancia of Arcardini v. Arcardini (suc.), July 10, 1917, *Jurisprudencia Argentina* 2 (1918): 780.

1. Lehmann Nitsche's expert report appears in "Las leyes de herencia en caso de filiación natural," *Revista de la Universidad Nacional de Córdoba* 6, no. 9–10 (November / December 1919): 52–81. He also wrote an explanation of the theory and method behind his analysis, which appeared in four different publications between 1917 and 1919. One is Roberto Lehmann Nitsche, "Peritaje somático en casos de filiación natural," *RCPML* 6 (1919): 80–93. The case is referenced in Leone Lattes, "Dimostrazione biologica della paternità," *La Riforma Médica* 39 (February 19, 1923): 169–172; Alejandro Raitzin, *La investigación médico-forense de la paternidad, la filiación y el parentesco* (Buenos Aires: Imprenta de E. Spinelli, 1934); Leone Lattes, "Demonstración biológica de la paternidad," *El Día Médico* (October 9, 1939): n.p.; Lorenzo Carnelli, *Los caracteres grupales, el derecho, y la ley* (Montevideo: Claudio García y Cia, 1940); Giuseppe Sotgiu, *La ricerca della paternità* (Rome: Croce Editore, 1951).

2. Francis Galton, "Personal Identification and Description," *Journal of the Anthropological Institute of Great Britain and Ireland* 18 (1889): 191.

3. That village was Domodossola; its population is cited in Amato Amati, *Dizionario corografico dell'Italia,* vol. 3 (Milan: Dottor Francesco Vallardi, 1868), 456; Jose C. Moya, *Cousins and Strangers: Spanish Immigrants in Buenos Aires, 1850–1930* (Berkeley: University of California Press, 1998), 149.

4. The following narrative is reconstructed from Argentine censuses and vital records (marriage and baptismal records). Information about the family's real estate transactions comes to light primarily in legal suits concerning their properties, such as litigation concerning deposits and rents, found in jurisprudence publications.

5. Although this particular company, the Italo-Platense, folded several years later, it was nevertheless prescient, anticipating the tide of immigrants that would soon cross the Atlantic in ever greater numbers. Giuseppe Moricola, *Il viaggio degli emigranti in America Latina tra Ottocento e Novecento. Gli aspetti economici, sociali, culturali* (Naples: Guida Editori, 2008).

6. He appears as a concessioner in the province of Neuquén and also in Chaco Austral, further north. Melitón González, *El Gran Chaco Argentino* (Buenos Aires: Comp. Sud-Americana de Billetes de Banco, 1890).

7. April 1914 list; $4 million: Testamentaria Arcardini, *La Nación,* August 23, 1919, 7.

8. Casabal verdict, 1a Instancia, 770.

9. A tenant's suit two years after the birth (in 1905) provides details of the property, at Calle Tacuarí 1242, which was owned by María Luisa Arcardini and administered by Roque. Notably, the suit was written up by Ernesto Quesada, prosecutor in the litigation concerning the three children, and the lawyer was Roque Humberto's godfather. Causa 166, *Fallos y disposiciones de la exma. Cámara de Apelaciones de la Capital,* vol. 175 (Buenos Aires: Adolfo Grau, 1909).

10. These and other details are reconstructed from the transcripts of the suits in the first and second instance: Casabal verdict, 1a Instancia, 781 and Arcardini v. Arcardini (suc.), November 9, 1918, Cámara Civil, 2da Instancia, *Jurisprudencia Argentina* 2 (1918). Parts of the voluminous record were also published by the Arcardini lawyers in Drs. Rivarola, *Hijos artificiales. Algunas piezas del juicio sobre nulidad del reconocimiento de tres menores como hijos de D. Roque Arcardini* (Buenos Aires: Imprenta Tragant, 1917).

11. Arcardini v. Arcardini (suc.), 2da Instancia, 777.

12. Casabal verdict, 1a Instancia, 769.

13. Arcardini v. Arcardini (suc.), 2da Instancia, 776.

14. Lehmann Nitsche, "Peritaje somático," 80–81.

15. Casabal verdict, 1a Instancia, 770.

16. Augusto Carette, *Diccionario de la jurisprudencia argentina* (Buenos Aires: J. Lajouane & Cia, 1908); Agustín Pestalardo, "La posesión de estado de hijos naturales," *Revista Jurídica y de Ciencias Sociales* 32 (1915): 260–275.

17. Cámara de Representantes de Uruguay, *Diario de sesiones* (1915), 236.

18. Illegitimacy in Buenos Aires: Isabella Cosse, "Estado y orden doméstico," unpublished paper, courtesy of the author, 3; comparative European statistics: Victor von Borosini, "The Problem of Illegitimacy in Europe," *Journal of the American Institute of Criminal Law and Criminology* 4, no. 2 (1913): 212–236. On economic and social change in Argentina in this period, see Leandro Losada, *La alta sociedad en la Buenos Aires de la Belle Époque: Sociabilidad, estilos de vida e identidades* (Buenos Aires: Siglo Veintiuno Editores, 2008); Sandra Gayol, *Sociabilidad en Buenos Aires. Hombres, honor y cafés: 1862–1910* (Buenos Aires: Ediciones del Signo, 2000); Moya, *Cousins and Strangers;* Fernando Devoto and Marta Madero, *Historia de la vida privada en la Argentina: La Argentina plural, 1870–1930* (Buenos Aires: Aguilar, 1999).

19. Lehmann Nitsche, "Peritaje somático," 81.

20. Between 1895 and 1914 (when transnational migration was temporarily interrupted by World War I), the population of Argentina nearly doubled.

21. 1914: Richard Walter, *Politics and Urban Growth in Buenos Aires* (Cambridge: Cambridge University Press, 2003), 7.

22. In Argentina, the unusually close alliance between scientific experts and statesmen has led historians to speak of Argentina's "medicolegal state." Ricardo

Salvatore, "Positivist Criminology and State Formation in Argentina," in *Criminals and Their Scientists: The History of Criminology in International Perspective,* ed. Peter Becker and Richard F. Wetzell (Cambridge: Cambridge University Press, 2006), 254.

23. Lattes, "Dimostrazione biologica."

24. Silvia De Renzi, "Resemblance, Paternity, and Imagination in Early Modern Courts," in *Heredity Produced: At the Crossroads of Biology, Politics, and Culture, 1500–1870,* ed. Staffan Müller-Wille and Hans-Jörg Rheinberger (Cambridge, MA: MIT Press, 2007), 61–83. There is a large literature on maternal imagination; see the citations in Jenny Davidson, *Breeding: A Partial History of the Eighteenth Century* (New York: Columbia University Press, 2009), 213, footnote 19.

25. De Renzi, "Resemblance, Paternity and Imagination."

26. Ayman Shabana, "Paternity between Law and Biology: The Reconstruction of the Islamic Law of Paternity in the Wake of DNA Testing," *Zygon* 47, no. 1 (2012): 214–239, especially 224–225; Ron Shaham, *The Expert Witness in Islamic Courts: Medicine and Crafts in the Service of Law* (Chicago: University of Chicago Press, 2010), chapter 6.

27. Davidson, *Breeding: A Partial History of the Eighteenth Century,* especially chapter 1.

28. Courtney Kenny, "Physical Resemblance as Evidence of Consanguinity," *Law Quarterly Review* 39 (1923): 298.

29. Regulations concerning this type of evidence varied widely across American states, but only a handful of states flat out rejected it. E. Donald Shapiro, Stewart Reifler, and Claudia L. Psome, "The DNA Paternity Test: Legislating the Future Paternity Action," *Journal of Law and Health* 7 (1992–93): 1–47. It was accepted in England but rejected in Scotland.

30. My study of hundreds of such suits in the nineteenth and early twentieth century in neighboring Chile, for example, turned up almost no references to physical similarity. Nara Milanich, *Children of Fate: Childhood, Class, and the State in Chile, 1850–1930* (Durham, NC: Duke University Press, 2009). In Argentina, judges occasionally asked for physical assessments of paternity, and the jurisprudence mentions resemblance (*parecido*), but possession of status clearly remained central. See, for example, I. Torino, "El parecido," *Revista Argentina de Ciencias Médicas* 6 (1886): 23–32.

31. "Boy Wins the Slingsby Suit," *Manchester Guardian,* February 4, 1915, 12.

32. "Boy Wins the Slingsby Suit," 12.

33. On the early modern practice of citing painters to assess resemblance, see Lattes, "Dimostrazione biologica," 169. On sculptors and painters in early twentieth-century Anglo-American courts, see Frances Newboldt, "Evidence of Resemblance in Paternity Cases," *Transactions of the Medico-Legal Society* 18, no. 31 (1923–1924): 39; "Evidence of Paternity," *The Lancet,* October 13, 1923, 840; "Likeness Not Proof of Paternity, Says Artist," *San Francisco Chronicle,* June 11,

1921, 4; "Art and Science to Fix Parentage of Boy in Court," *San Francisco Chronicle,* May 21, 1921, 6.

34. "Family Resemblances," *The Lancet,* February 13, 1915: 337. The article was also partially reproduced in the press; "Medical Comment on Slingsby Case," *New York Times,* May 6, 1915, 13. For another favorable comment on sculptors' ability to discern likeness, made in the context of the Russell divorce suit, see "An Infant's Evidence," *Times of India,* August 5, 1922, 12.

35. Fritz Strassmann, the father, wrote an analysis of the case for an Italian journal, "La rassommiglianza fisica in tribunale," *Archivio di psichiatria, neuropatologia, antropologia criminale, e medicina legale* 25 (1904): 109–115. See also discussions in Annes Dias, Ulysses Nonohay, and Luiz Guedes, "Investigação de paternidade," *Archivos de Medicina Legal e Identificação* 5, no. 11 (1935): 111–129.

36. "Likeness Not Proof of Parentage, Says Artist."

37. "Bust of Wendel Stirs Laughter in Will Contest," *New York Times,* November 17, 1932, 3.

38. "Evidence-Resemblance on the Issue of Paternity," *Medico-Legal Journal* 42 (1925): 171–172; Kenny, "Physical Resemblance as Evidence of Consanguinity."

39. "Ampthill Heir Denies Paternity of Child," *New York Times,* July 19, 1922, 11; see also "Russell Baby before the Jury. A Resemblance Test. Judge's Reference to All Parents' Experience," *Manchester Guardian,* July 19, 1922, 3.

40. "Child as exhibit": Roswell H. Johnson, "The Determination of Disputed Parentage as a Factor in Reducing Infant Mortality," *Journal of Heredity* 10, no. 3 (March 1, 1919): 121–124.

41. See the cases cited by Shari Rudavsky, "Blood Will Tell: The Role of Science and Culture in Twentieth-Century Paternity Disputes" (PhD diss., University of Pennsylvania, 1996), 245, 308. Shapiro, Reifler, and Psome, "The DNA Paternity Test," provides an overview of this evidence in English and American courts. In 1972, a Wisconsin court rejected the practice: "Paternity Decision Is Upheld," *Milwaukee Journal,* March 4, 1972, 12.

42. Kenny, "Physical Resemblance as Evidence of Consanguinity," 302.

43. "An Infant's Evidence."

44. Kenny, "Physical Resemblance as Evidence of Consanguinity, 299; "Physical Resemblance as Evidence in Cases of Disputed Paternity," *The Lancet,* December 26, 1925, 1349; Stanley B. Atkinson, "Heredity and Affiliation," *Medico-Legal Journal* 28 (1910–1911): 17–21; "Russell Baby before the Jury"; and several of the cases in Shapiro, Reifler, and Psome, "The DNA Paternity Test."

45. Lehmann Nitsche, "Leyes de Herencia," 68; italics are in original.

46. Lehmann Nitsche, "Peritaje somático," 90.

47. Diego Alberto Ballestero, "Los espacios de la antropología en la obra de Roberto Lehmann-Nitsche, 1894–1938" (PhD diss., Universidad Nacional de la Plata, 2013).

48. Various correspondence, Robert Lehmann Nitsche Archives, Ibero-Amerikanisches Institut, Berlin (hereafter RLNA).

49. Eye color: Gertrude C. Davenport and Charles B. Davenport. "Heredity of Eye-Color in Man," *Science* 26, no. 670 (November 1, 1907): 589–592.

50. Lehmann Nitsche, "Peritaje somático," 85–86.

51. Lehmann Nitsche, "Leyes de Herencia": Piedmont: 69; central Europe: 61; children's racial origins, 62–63.

52. Lehmann Nitsche, "Peritaje somático," 91.

53. Lehmann Nitsche, "Peritaje somático," 88.

54. Lehmann Nitsche, "Peritaje somático," 89.

55. Letter from Camilo Rivarola to Roberto Lehmann Nitsche, December 22, 1914, RLNA.

56. Method of attack: Johnson, "Determination of Disputed Parentage," 124.

57. Charles Davenport, "Research in Eugenics," in *Eugenics, Genetics and the Family,* Scientific Papers of the Second International Congress of Eugenics, vol. 1 (Baltimore, MD: Williams & Wilkins, 1923), 23.

58. "Tracing Parentage by Eugenic Tests," *New York Times,* September 23, 1921, 8.

59. The case in Rio was reported in the newspaper, "A investigação de maternidade pelo exame do sangue," *A Manhã,* April 7, 1926, 1, more than a decade after it occurred, and was presented in contrast to reports of new blood group tests in Vienna. The gist of the article is that the Brazilian somatic test was conclusive where the Viennese blood test was not. Norway: Maria Teschler-Nicola, "The Diagnostic Eye. On the History of Genetic and Racial Assessment in Pre-1938 Austria," *Collegium Antropologicum* 28, Suppl. 2 (2004): 7–29; Soviet Union: Albert Harrasser, "Zur prozessualen Bedeutung des naturwissenschaftlichen Vaterschaftsbeweises," *Österreichische Richterzeitung* 25 (1932): 125–126; Anna Poliakowa, "Manoiloff's 'Race' Reaction and Its Application to the Determination of Paternity," *American Journal of Physical Anthropology* 10, no. 1 (1927): 23–29; Poland and Hungary: Otto Reche and Anton Rolleder, "Zur Entstehungsgeschichte der ersten exakt wissenschaftlichen erbbiologischanthropologischen Abstammungsgutachten," *Zeitschrift für Morphologie und Anthropologie* 55, no. 2 (1964): 283–293; Portugal: Azevedo Neves, "Do valor do retrato na investigação da paternidade," *Archivo de Medicina Legal* 3 (1930): 19–31; and comment by Victor Delfino, "Valor del retrato en la investigación de la paternidad," *Medicina Latina* 7, no. 28 (1930): 343–344.

60. The two exchanged correspondence in the late 1920s: Letters from Reche to Lehmann Nitsche, January 16, 1928, and November 28, 1927, RLNA.

61. Lehmann Nitsche reported his paternity work in Spanish in Argentine journals and does not appear to have written about it in German publications.

62. The paternity of first-born María Mafalda, whom Arcardini had not formally recognized, was not technically under consideration in this particular suit, although the filiation of all three children was obviously understood as linked. The question of María Mafalda's paternity would be taken up in later litigation.

63. Letter from Larguia to Roberto Lehmann Nitsche, September 1917, RLNA.

64. The request was made via the Arcardini lawyers. Rodolfo Rivarola to Roberto Lehmann Nitsche, January 21, 1918, RLNA. Quesada and Lehmann Nitsche maintained friendly correspondence on unrelated topics around the time of the Arcardini case.

65. Quesada's report appeared as "La prueba científica de la filiación natural. Aplicación del mendelismo a los casos forenses," *RCPML* 6 (1919): 595–612; and as a book: *La prueba científica de la filiación natural. El mendelismo* (Córdoba: Bautista Cubas, 1919).

66. "Absolute or exclusive": Quesada, "La prueba científica,", 604; "restless/false": Lehmann Nitsche, "Leyes de herencia," 61.

67. Quesada, "La prueba científica," 471.

68. Quesada, "La prueba científica," 612. The phrase actually quotes the judge's verdict, which was added as a footnote.

69. Casabal verdict, 1a Instancia, 778–779.

70. Casabal verdict, 1a Instancia. The verdict was appealed by the Arcardinis but was confirmed by an appellate court in November 1918.

71. A Buenos Aires court was impressed by the litigation's "singular magnitude." *María Luisa Arcardini de Costadoat contra Luis Jacobé, Gaceta del Foro* 162 (January 3, 1943): 21.

72. This was Emilio Benjamín Passini Costadoat, Roque's grand-nephew, grandson of his sister María Luisa. Passini Costadoat became a public figure as did his brother, Carlos Alberto, also a lawyer (and Arcardini heir), who wrote about inheritance law.

73. María Carmen, the youngest child (born 1905) appears to have died by the mid-1920s, married but childless. Her portion passed to her husband, who in turn ceded his rights to one Luis Jacobé. Roque Humberto (born 1904) died at some point in the 1930s, having designated as his universal heir Enrique Rottger—possibly the well-known Peronist military official of the same name. This narrative is pieced together from litigation in the decades after the civil suit.

74. Lehmann Nitsche participated in Martín Ramón Arana and Niceforo Castellano, "Filiación-Juicio criminal," *Archivos de Medicina Legal* 6 (1936): 293–306 (the case dated from 1923); his method was a model for other scientists, such as Nerio Rojas y Eliseo Ortiz, "Informes medicolegales. Filiación natural," *Revista Argentina de Neurología, Psiquiatría y Medicina Legal* 1 (1927): 317–325. The method was cited two decades later by Carnelli, *Caracteres grupales* and Lattes, "Aspetti biologici"; on skepticism, see Carnelli, *Caracteres grupales,* 15.

75. Leone Lattes, "Processi giudiziari in tema di filiazione naturale," *Minerva Medicolegale* 70 (1950): 92.

76. "Precise and complete": Lattes, "Processi giudiziari in tema di filiazione naturale," 91. The expert report is reproduced in Andrés Sein, José T. Erenchun, and Alfredo Ferrer Zanchi, "Los grupos sanguíneos en los juicios de filiación. Fallo judicial," *Archivos de Medicina Legal* 18 (1948): 71–102; Lattes offered a

commentary for the court: "Valor de las pruebas biológicas en la investigación de la paternidad," *Archivos de Medicina Legal* 18 (1948): 210–237.

77. Contrary to the system: M. Dalloz, *Recueil périodique et critique de jurisprudence* (1936): 41. See also Rachel G. Fuchs, *Contested Paternity: Constructing Families in Modern France* (Baltimore: Johns Hopkins University Press, 2008); and Emmanuelle Saada, *Empire's Children: Race, Filiation, and Citizenship in the French Colonies* (Chicago: University of Chicago Press, 2011). Contemporary French law: Solenn Briand, Claire Delarbre, and Anne-Cécile Krygiel, "Filiation, Genetics and the Judge: Conciliating Family Stability and the Right to Biological Truth" (Themis Competition, European Judicial Training Network, 2016), retrieved from http://www.ejtn.eu/PageFiles/14775/Written%20paper_France.pdf.

5. BODIES OF EVIDENCE

Epigraph: Luís Reyna Almandos, "El drama de la familia del Profesor Canella," *Revista de Identificación y Ciencias Penales* 11 (September 1933–December 1934): 136.

1. The scene of the couple's encounter would become iconic, reproduced over and over in print, visual images, theater, and later film.

2. Background on the story draws from Lisa Roscioni's *Lo smemorato di Collegno: Storia italiana di un'identità contesa* (Turin, Italy: G. Einaudi, 2007).

3. At one point, a reporter from Rio arrived to interview her for a front-page feature story. He delivered a cache of sympathetic letters addressed to her from Rio's elite: "O enigma que ha cinco anos apaixona a opinião italiana," *O Jornal,* October 16, 1931, 1.

4. Even as it was happening, the Canella–Bruneri affair was the subject of a 1930 play by Pirandello, made two years later into a film featuring Greta Garbo. It was also the subject of a 1962 film featuring Totò, a number of television shows beginning in 1970, and a 1981 book by Leonardo Sciascia, among many other works. The best scholarly work on the episode is Roscioni, *Lo smemorato di Collegno.* While Italian historians, writers, dramaturges, journalists, and television programs have thoroughly explored the first part of this story, its Brazilian sequel has been largely forgotten.

5. One was a catastrophic 1897 fire at a charity bazaar in Paris. The other was the arson of the German consulate in Santiago, Chile, a decade later.

6. "Teeth and Madness": *A Tribuna,* Santos, November 6, 1926; "Pacifiers": *A Tribuna* (Santos), January 13, 1927; "Accidents" and "National Security": *Diario de São Paulo,* various dates, September–October 1936; "Portrait": *O Globo,* October 30, 1943. A long (but incomplete) list of his publications in the press and scientific journals can be found in *Diario Oficial do Estado de São Paulo* (1977): 98–100.

7. Silva describes his impressions of the Sconosciuto in "Odontologia Legal e Psiquiatria 'Italiana,'" *Arquivos de Polícia e Identificação* 2 (1938–1940):

443–488. The two other experts were Argentine fingerprint specialist Luís Reyna Almandos and Ricardo Gumbleton Daunt, director of the Identification Service of the State of São Paulo. Reyna Almandos gave the inaugural speech at the conference; after reading about it, the Marquês and his famous son-in-law invited the scientists to meet with them.

8. João Maringoni, Emilio Viégas, and Raul Marques Negreiros, *Um caso de investigação de paternidade* (São Paulo: Revista dos Tribunais, 1940), 23.

9. "Most terrible tragedy": cited in Roscioni, *Lo smemorato di Collegno,* 10; "perpetual offense": Reyna Almandos, "El drama de la familia," 136.

10. "Revive o famoso caso do desmemoriado de Collegno. Conseguira a sciencia brasileira destruir as provas dos scientistas italianos?" *Folha da Noite,* September 11, 1934, 2–3. The coverage ran in the newspaper daily from September 11 through October 4, 1934.

11. In internal documents, the director of the Central Identification Service of the School of Scientific Police in Rome—the counterpart of the institution for which Silva worked in São Paulo—dismissed his results as "not deserving of any consideration, whether from the scientific or practical point of view." The informer and director are cited in Roscioni, *Lo smemorato di Collegno,* 250.

12. "Vae ser revisto o processo do Desmemorado de Collegno," *Diario da Noite,* July 4, 1939, 2. "Lone voice": "E o professor Canella!" *Diario da Noite,* July 5, 1939, 1; "personally and in homage": "Quer interessar-se junto a Embaixada italiana no Brasil para a revisão do famoso processo do prof. Julio Canella," *Diario da Noite,* July 6, 1939, 2.

13. "Noticias," *Boletim do Instituto Oscar Freire* 2, no. 2 (April 1935): 26.

14. The ability to taste phenylthiocarbamide (PTC) was discovered to be hereditary and proposed as a method of parentage determination in the early 1930s. Arnaldo Amado Ferreira, *A perícia técnica em criminologia e medicina legal* (São Paulo: n.p., 1948), 392.

15. Arnaldo Amado Ferreira, *Determinação médico-legal da paternidade* (São Paulo: Companhia Melhoramentos de São Paulo, 1939), 131. This phrase and variations of it were repeated constantly in the IOF's expert reports.

16. Francis Galton, *Finger Prints* (London: Macmillan, 1892), 192–193. See also Daniel Asen, "'Dermatoglyphics' and Race after the Second World War" in *Global Transformations in the Life Sciences, 1945–1980,* ed. Patrick Manning and Mat Savelli (Pittsburgh, PA: University of Pittsburgh Press, 2018), 61–77.

17. "On the Skin-Furrows of the Fingers," *Nature,* October 28, 1880, 605. Simon Cole argues that fingerprinting advocates deliberately suppressed these early hereditary inquiries because of their unsavory eugenic associations and because they threatened the bedrock principle of unique individual identification: *Suspect Identities: A History of Fingerprinting and Criminal Identification* (Cambridge, MA: Harvard University Press, 2009).

18. Mercedes García Ferrari, *Marcas de identidad: Juan Vucetich y el surgimiento transnacional de la dactiloscopia (1888–1913)* (Rosario, Argentina: Prohistoria Ediciones, 2015).

19. Alejandro Raitzin, *Investigación medico-forense de la paternidad* (Buenos Aires: Imprenta de E. Spinelli, 1934), 52. Among these protégés was Luís Reyna Almandos, who accompanied his friend Luiz Silva to the Canellas' home in Rio to meet the famous amnesiac.

20. The researcher was Rodolfo Senet, cited in Alejandro Raitzin, "Hereditariedad dactiloscópica," *RCPML* 22 (1935): 223–257, on 228. In 1906, he presented his findings, one of the first studies of the topic, at the Eleventh Congress of Criminal Anthropology in Turin, Italy.

21. In the first half of the twentieth century, they included researchers from Argentina, Brazil, Italy, Norway, Germany, France, Portugal, Egypt, and no doubt elsewhere. On post–World War II East Asian developments, see Daniel Asen, "Fingerprints and Paternity Testing: A Study of Genetics and Probability in Pre-DNA Forensic Science," unpublished paper, courtesy of the author.

22. Raitzin, "Hereditariedad dactiloscópica," 225.

23. A. Lauer and H. Poll, "Tracing Paternity by Finger Prints," *American Journal of Police Science* 1 (1930): 92.

24. Hieroglyphics: IOF, vol. 54, 1947, L. 6177; all great mysteries: Raitzin, "Hereditariedad dactiloscópica," 224.

25. 1953 conference: Ilse Schwidetzky, "Forensic Anthropology in Germany," *Human Biology* 26, no. 1 (1954): 6; an example of this insistence is Tansella, "Lo studio delle impronte digitali e del padiglione auricolare," *Minerva Medica* 49 (1958): 3110–3116; Hungary: "Kleine Nachrichten: Fingerabdrücke als Vaterschaftsnachweis," *Passauer Neue Presse,* November 7 and 8, 1959.

26. The Austrian test was developed by Otto Reche, discussed in Chapter 6. On the early Russian test, see Anna Poliakowa, "Manoiloff's 'Race' Reaction and Its Application to the Determination of Paternity," *American Journal of Physical Anthropology* 10, no. 1 (1927): 23.

27. The following believed Bertillon's method was useful: Leonídio Ribeiro, "Perícia da investigação da paternidade," *Arquivos de Medicina Legal e Identificação* 4, no. 8 (1934): 144–153; Luiz Silva, "Investigação odonto-legal do 'Desconhecido de Collegno,'" *Revista de Identificación y Ciencias Penales* 14, no. 52–54 (1936): 40–96; Annes Dias, Ulysses Nonohay, and Luiz Guedes, "Investigação de paternidade," *Archivos de Medicina Legal e Identificação* 5, no. 11 (1935): 111–129; Juan Caride, Alberto Rodríguez Egaña, and Alberto Bonhour, *Filiación natural* (Buenos Aires: Imprenta de E. Spinelli, 1939); and the Cuban Jorge Castroverde, cited in Ismael Castellanos, *La odontología legal en la investigación de la paternidad* (Havana: Cultural, S.A., n.d.). Afrânio Peixoto and Leonídio Ribeiro disputed its utility in a statement

in a legal suit: Virgilio Barbosa, Luiz Novaes, and Gastão Neves, *Investigação de paternidade ilegítima* (Rio de Janeiro: Typ. do Jornal do Commercio, 1936), 106–109, as did the Cuban race scientist Castellanos, *La odontología legal.*

28. "Decisive conclusions": Ribeiro, "Perícia da investigação da paternidade," 146; "spoken portrait": Peixoto and Ribeiro, cited in Barbosa, Novaes, and Neves, *Investigação de paternidade ilegítima,* 108 (apparently, Ribeiro, a noted medicolegal expert, had changed his mind).

29. Ferreira, *Determinação médico-legal da paternidade,* 17.

30. "In conformity": Ferreira, *A perícia técnica,* 365.

31. On biotypology and its appeal among Latin American scientists, see Alexandra Minna Stern, "What Kind of a Morph Are You? Biotypology in Transit, 1920s–1960s," *REMEDIA* (blog), February 10, 2016, https://remedianetwork .net/2016/02/10/what-kind-of-a-morph-are-you-biotypology-in-transit-1920s -1960s/; in Brazil, see Olívia Maria Gomes da Cunha, *Intenção e gesto: Pessoa, cor e a produção cotidiana da (in)diferença no Rio de Janeiro, 1927–1942* (Rio de Janeiro: Arquivo Nacional, 2002).

32. On Roquette Pinto, see Vanderlei Sebastião de Souza, "Retratos da nação: Os 'tipos antropológicos' do Brasil nos estudos de Edgard Roquette-Pinto, 1910–1920," *Boletim do Museu Paraense Emílio Goeldi. Ciências Humanas* 7, no. 3 (2012): 645–669; da Cunha, *Intenção e gesto,* chapter 3.

33. *The Blood Groups and Their Areas of Application,* Selected Contributions to the Literature of Blood Groups and Immunology, vol. 4, part 2 (Fort Knox, Kentucky: U.S. Army Medical Research Laboratory, 1971), 327.

34. The phrase constantly recurs in their expert reports.

35. IOF, vol. 35, 1943, L. 4149. For other cases of biracial couples that reject a transparent reading of this fact, see IOF, vol. 29, 1942, L. 3497; vol. 11, 1952, L. 8615. On race and paternity science in the Brazilian milieu, Sueann Caulfield and Alexandra Minna Stern, "Shadows of Doubt: The Uneasy Incorporation of Identification Science into Legal Determination of Paternity in Brazil," *Cadernos de Saúde Pública* 33 (2017): 1–14.

36. Nara Milanich, *Children of Fate: Childhood, Class, and the State in Chile, 1850–1930* (Durham, NC: Duke University Press, 2009), 54.

37. IOF, vol. 34, 1943, L. 4022; vol. 29, 1942, L. 3477.

38. "Do repertorio dentario," *Careta,* August 6, 1937, 16; "Está em moda investigar a paternidade pelos dentes," *Careta,* May 28, 1938, 12; "Filhos Curiosos," *Beira-Mar,* June 11, 1938, 4. The dispute with the other dentist was covered in "Será feita a investigação da paternidade pelos dentes. Curioso e inedito episodio no Fôro," *A Noite,* May 3, 1938, 1; "A prova de paternidade pelos dentes," *A Noite,* July 19, 1938, 1; Frederico Eyer, "Pelos dentes não se pode investigar a paternidade," *A Noite,* August 5, 1938, 1; "Pelos dentes não se pode investigar a paternidade," *O Radical,* August 13, 1938, 1.

39. Maringoni, Viégas, and Negreiros, *Um caso de investigação,* 13.

40. Luiz Silva and José Ramos de Oliveira Júnior, "Investigação de paternidade pelos exames comparativo, prosopográfico e prosopométrico," *Arquivos de Polícia e Identificação* 2 (1938–40): 534.

41. Azevedo Neves, "Do valor do retrato na investigação da paternidade," *Archivo de Medicina Legal* 3 (1930): 19–26; Victor Delfino, "Valor del retrato en la investigación de la paternidad," *Medicina Latina* 3 (1930): 343–344; Albert Harrasser, "Die Laienphotographie als Hilfsmittel für erbbiologische Beobachtungen," *Mitteilungen der Anthropologischen Gesellschaft in Wien* 62 (1932): 338–342.

42. Mia Fineman, *Faking It: Manipulated Photography before Photoshop* (New York: Metropolitan Museum of Art, 2012), 106–115.

43. One Argentine specialist characterized as "exceptionally advantageous" a case in which the father was still alive and could be examined directly: Nerio Rojas, "Filiación y prueba médica," *Archivos de Medicina Legal* 1, no. 1 (1931): 196–203. See similar comments by the Cuban Castellanos, *La odontología legal.*

44. Here I draw on Mary Bouquet's contention that photography is a reproductive technology that "makes" kinship by posing people together in what looks like a family: "Making Kinship, with an Old Reproductive Technology," in *Relative Values: Reconfiguring Kinship Studies*, ed. Sarah Franklin and Susan McKinnon (Durham, NC: Duke University Press, 2002), 85–115.

45. Maringoni, Viégas, and Negreiros, *Um caso de investigação*, n.p.

46. Maringoni, Viégas, and Negreiros, *Um caso de investigação*, n.p.

47. Separation of the head: Maringoni, Viégas, and Negreiros, *Um caso de investigação*, 118.

48. Various cases of exhumation are discussed in Antônio Ferreira de Almeida Júnior, *As provas genéticas da filiação* (São Paulo: Revista dos Tribunais, 1941), 52. Silva's exhumation request: "A prova odonto-legal constitue por si só um elemento decisivo na investigação da paternidade," *Folha da Noite,* July 1, 1941, 10, 5; on the Argentine case, see Leone Lattes, "Processi giudiziari in tema di filiazione naturale," *Minerva Medicolegale* 70 (1950): 87.

49. The lawyers were so pleased with the outcome (which no doubt earned them a handsome cut of Alfredo Oliveira's estate) that they published highlights of the case. Maringoni, Viégas, and Negreiros, *Um caso de investigação*. Silva's analysis also appeared as Silva, "Investigação de paternidade."

50. Almeida Junior, *As provas genéticas*, 179, citing a 1938 study by Essen-Möller and Geyer.

51. Otto L. Mohr, "A Case of Hereditary Brachy-Phalangy Utilized as Evidence in Forensic Medicine," *Hereditas* 2, no. 2 (1921): 290–298.

52. Cited in Maria Teschler-Nicola, "The Diagnostic Eye—On the History of Genetic and Racial Assessment in Pre-1938 Austria," *Collegium antropologicum* 28 (2004): 22. The anthropologist was Otto Reche.

53. Arnaldo Amado Ferreira, "Investigação medico-legal da paternidade," *Revista Médica* 85, no. 4 (2006) [originally 1953]: 144. The phrase was repeated constantly in their reports.

54. Arnaldo Amado Ferreira, "Determinação medico-legal da paternidade," *Arquivos de Polícia Civil de São Paulo* 2 (1941): 53. Others made similar statements: Raitzin, "Hereditariedad dactiloscópica," 232; Almeida Júnior, *As provas genéticas.*

55. "Deceptive, dangerous": Ferreira, *A perícia técnica,* 366. See a similar argument in IOF, 1955, L. 10828, and many other expert reports.

56. Almeida Júnior, *As provas genéticas,* VII.

57. For example, the Oliveira sisters cited the IOF doctors to challenge Silva's method, and Silva published critiques of the institute's expert reports: Luiz Silva, "Um laudo odonto-legal de investigação de paternidade e tres 'pareceres' da medicina-legal," *Arquivos da Polícia Civil de São Paulo* 13 (1947): 441–460.

58. IOF, vol. 34, 1943, L. 4022.

59. Lattes, "Processi giudiziari," 97.

60. Schiff, *Blood Groups and Their Areas of Application,* 326.

61. Ilse Schwidetzky, "New Research in German Forensic Anthropology," in *Men and Cultures: Selected Papers of the Fifth International Congress of Anthropological and Ethnological Sciences,* ed. Anthony F. C. Wallace (Philadelphia: University of Pennsylvania Press, 1960), 709. Lattes makes a similar argument in "Processi giudiziari," 97.

62. Flamínio Favero, "Discurso de abertura da 1ª Semana Paulista de Medicina Legal," *Archivos da Sociedade de Medicina Legal e Criminologia de São Paulo* 8, suppl. (1937): 11.

63. da Cunha, *Intenção e gesto,* introduction.

64. Olívia Maria Gomes da Cunha, "The Stigmas of Dishonor: Criminal Records, Civil Rights, and Forensic Identification in Rio de Janeiro, 1903–1940," in *Honor, Status, and Law in Modern Latin America,* ed. Sueann Caulfield, Sarah C. Chambers, and Lara Putnam (Durham, NC: Duke University Press, 2005), 300.

65. Maria Teschler-Nicola, "Volksdeutsche and Racial Anthropology in Interwar Vienna," in *"Blood and Homeland": Eugenics and Racial Nationalism in Central and Southeast Europe, 1900–1940,* ed. Marius Turda and Paul Weindling (Budapest: Central European University Press, 2007), 20. See also Chapter 6.

66. A variation on the phrase appears in the first expert report in 1927, and versions of it reappear for two decades thereafter; for example, IOF, vol. 55, 1947, L. 6267.

67. Almeida Júnior, *As provas genéticas,* 8.

68. Sueann Caulfield, "Changing Politics of Freedom and Virginity in Rio de Janeiro, 1920–1940," in Caulfield, Chambers, and Putnam, *Honor, Status, and Law,* 226.

69. The timing seems to have coincided with the retirements of Favero and Ferreira, who were replaced by a new generation of forensic experts who were perhaps unwilling or unable to perform them.

70. Castellanos, *La odontología legal.*

71. Elisa Larenas, *Mais um crime do fascismo* (Rio de Janeiro: Casa do Estudante do Brasil, 1943). The Chilean-born Larenas had her own public profile as a distinguished collector of antiques and rare books.

72. Larenas, *Mais um crime do fascismo.* Brazil was the only Latin American country to join the Allies in combat.

73. Cited in "Giulio Canella, Mio Nonno," *L'Arena,* April 12, 2009, http://www .larena.it/permanent-link/1.2656926.

6. JEWISH FATHERS, ARYAN GENEALOGIES

Epigraph: Hanns Schwarz, *Jedes Leben ist ein Roman: Erinnerungen eines Arztes* (Berlin: Der Morgen, 1975).

1. Schwarz, *Jedes Leben ist ein Roman,* 157.

2. This discussion draws on Maria Teschler-Nicola, "The Diagnostic Eye: On the History of Genetic and Racial Assessment in Pre-1938 Austria," *Collegium antropologicum* 28 (2004): 7–29, which provides an overview of Reche's development of the test; Reche and Rolleder's retrospective essay, which recounted these events almost four decades later, Otto Reche and Anton Rolleder, "Zur Entstehungsgeschichte der ersten exakt wissenschaftlichen erbbiologischanthropologischen Abstammungsgutachten," *Zeitschrift für Morphologie und Anthropologie* 55, no. 2 (1964): 283–293; and the intellectual biography of Reche by Katja Geisenhainer, *"Rasse ist Schicksal." Otto Reche (1879–1966)—Ein Leben als Anthropologe und Völkerkundler* (Leipzig: Evangelische Verlagsanstalt, 2002).

3. Reche and Rolleder, "Zur Entstehungsgeschichte," 286–287.

4. Reche's biography draws on Andrew D. Evans, *Anthropology at War: World War I and the Science of Race in Germany* (Chicago: University of Chicago Press, 2010), 52; Reche and Rolleder, "Zur Entstehungsgeschichte," 286; Teschler-Nicola, "Diagnostic Eye," 10; and Geisenhainer, *"Rasse ist Schicksal."* Melanesian jaws: Annegret Ehmann, "From Colonial Racism to Nazi Population Policy," in *The Holocaust and History: The Known, the Unknown, the Disputed, and the Reexamined,* ed. Michael Berenbaum and Abraham J. Peck (Bloomington: Indiana University Press, 1998), 120–121.

5. The method and its merits are described in Otto Reche, "Anthropologische Beweisführung in Vaterschaftsprozessen," *Österreichische Richterzeitung* 19 (1926): 157–159.

6. Teschler-Nicola, "Diagnostic Eye," 13.

7. Emil Blank, "Pater semper incertus?," *Österreichische Richterzeitung* 20 (1927): 137.

8. The variety show was in Leipzig, where Reche would soon move. Teschler-Nicola, "Diagnostic Eye," 14; "ground-breaking" and "beautiful and detailed": Teschler-Nicola, "Diagnostic Eye," 12.

9. "Science of Resemblance Used in Austrian Courts for Proof of Paternity," *Schenectady Gazette,* March 11, 1927; "Tests Prove Satisfactory," *Lawrence Journal-World,* September 19, 1927.

10. Besides, Roberto Lehmann Nitsche's analysis in Argentina predated Reche's by more than a decade. Meanwhile, Fritz Strassmann, a forensic doctor and Georg's

father, describes an analysis he performed for a court in Berlin around the same time as Reche's, after a blood test by Schiff was inconclusive: see "Ein Beitrag zur Vaterschaftsbestimmung," *Deutsche Zeitschrift für die gesamte gerichtliche Medizin* 10 (1927): 341–345.

11. Reche, "Anthropologische Beweisführung," 159.

12. Teschler-Nicola, "Diagnostic Eye," 9. Reche made this statement in the context of a 1938 grant application seeking funds for the journal published by the German Society for Blood Group Research.

13. His collaborator, Judge Rolleder, became a Party member in 1931.

14. H. Schwarz, *Jedes Leben ist ein Roman,* 12.

15. Schwarz, *Jedes Leben ist ein Roman,* 110.

16. Schwarz, *Jedes Leben ist ein Roman,* 132.

17. Teschler-Nicola, "Diagnostic Eye," 13.

18. Weninger's skepticism: Teschler-Nicola, "Diagnostic Eye," 14; thrilled judge: Reche and Rolleder, "Zur Entstehungsgeschichte," 285–286; greater demand: Maria Teschler-Nicola, "*Volksdeutsche* and Racial Anthropology in Interwar Vienna: The 'Marienfeld Project,'" in *"Blood and Homeland": Eugenics and Racial Nationalism in Central and Southeast Europe 1900–1940,* ed. Marius Turda and Paul K. Weindling (Budapest: Central European University Press, 2007), 58; on the six examinations and photography equipment, see Teschler-Nicola, "Diagnostic Eye," 18. Analysis of photographic representations of bodies was a routine practice of Viennese anthropology.

19. Albert Harrasser, "Ergebnisse der anthropologisch-erbbiologischen Vaterschaftsprobe in der österreichischen Justiz," *Mitteilungen der Anthropologischen Gesellschaft in Wien* 65 (1935): 204–232.

20. Teschler-Nicola, "*Volksdeutsche* and Racial Anthropology," 63.

21. "Far from serving": Teschler-Nicola, "*Volksdeutsche* and Racial Anthropology," 63; funding agencies: Teschler-Nicola, "*Volksdeutsche* and Racial Anthropology," 70; Rolleder remembered Weninger's Working Group in Reche and Rolleder, "Zur Entstehungsgeschichte," 285–286.

22. Teschler-Nicola, "*Volksdeutsche* and Racial Anthropology," 69.

23. Teschler-Nicola, "Diagnostic Eye," 16.

24. In 1926, Reche proposed a study on blood groups in Württemberg. The Ministries of Justice and the Interior expressed support for the research based on its application to paternity and other judicial cases, although ultimately the project was not funded. Paul Weindling, *Health, Race and German Politics between National Unification and Nazism, 1870–1945* (Cambridge: Cambridge University Press, 1993), 466.

25. Teschler-Nicola, "*Volksdeutsche* and Racial Anthropology," 59.

26. Teschler-Nicola, "Diagnostic Eye," 18.

27. The University of Munich and the KWIA both began somatic paternity assessments in 1928.

28. H. Schwarz, *Jedes Leben ist ein Roman,* 146.

29. Different contexts demanded different degrees of "Aryanness." For more routine purposes, ancestral reckoning might include only parents and grandparents, but the highest ranking Nazi leaders might be required to demonstrate a lineage free of "racially alien" elements as far back as 1750. Eric Ehrenreich, *The Nazi Ancestral Proof: Genealogy, Racial Science, and the Final Solution* (Bloomington: Indiana University Press, 2007), chapter 4.

30. H. Schwarz, *Jedes Leben ist ein Roman,* 155–158.

31. Schwarz, *Jedes Leben ist ein Roman,* 187.

32. That reviewer was none other than well-known racial anthropologist Eugen Fischer, discussed subsequently. Teschler-Nicola, "*Volksdeutsche* and Racial Anthropology," 70.

33. Alexandra Schwarz, "Hans Koopmann (1885–1959)–Leben und Werk eines Hamburger Gerichtsmediziners," (PhD diss., University of Hamburg, 2010), 66. Similar language can be found in Thomas Pegelow, "Determining 'People of German Blood,' 'Jews' and 'Mischlinge': The Reich Kinship Office and the Competing Discourses and Powers of Nazism, 1941–1943," *Contemporary European History* 15, no. 1 (2006): 56

34. Ehrenreich, *Nazi Ancestral Proof,* 125.

35. A. Schwarz, "Hans Koopmann," 74; Patricia Szobar, "Telling Sexual Stories in the Nazi Courts of Law: Race Defilement in Germany, 1933 to 1945," *Journal of the History of Sexuality* 11, no. 1/2 (2002): 131–163.

36. On child welfare in the 1920s, see Michelle Mouton, *From Nurturing the Nation to Purifying the Volk: Weimar and Nazi Family Policy, 1918–1945* (Cambridge: Cambridge University Press, 2007); Geisenhainer, *"Rasse ist Schicksal,"* 377–379.

37. An annual report for 1935–1936 documents the applied activities performed by the Institute: sixty expert opinions on race for the RKO and twenty-eight more for the superior courts, in addition to twenty sterilization assessments for the health courts. Hans-Walter Schmuhl, *The Kaiser Wilhelm Institute for Anthropology, Human Heredity and Eugenics, 1927–1945: Crossing Boundaries* (Dordrecht: Springer, 2008), 203–206.

38. Benno Müller-Hill, *Murderous Science: Elimination by Scientific Selection of Jews, Gypsies, and Others in Germany, 1933–1945* (Plainview, NY: Cold Spring Harbor Laboratory Press, 1998), 138; Sheila Faith Weiss, *The Nazi Symbiosis: Human Genetics and Politics in the Third Reich* (Chicago: University of Chicago Press, 2010), 102.

39. Thomas Pegelow Kaplan, "'In the Interest of the Volk . . .': Nazi-German Paternity Suits and Racial Recategorization in the Munich Superior Courts, 1938–1945," *Law and History Review* 29, no. 2 (2011): 534–535; Robert Proctor, *Racial Hygiene: Medicine under the Nazis* (Cambridge, MA: Harvard University Press, 1988).

40. Pegelow, "Determining 'People of German Blood,'" 50; A. Schwarz, "Hans Koopmann," 70; Geisenhainer, *"Rasse ist Schicksal,"* 238.

41. Schmuhl, *Kaiser Wilhelm Institute,* 203–204.

42. Interview with Abel in Müller-Hill, *Murderous Science,* 141.

43. Reche and Rolleder, "Zur Entstehungsgeschichte," 285.

44. A. Schwarz, "Hans Koopmann," 72.

45. *Wer gehoert zu Wem?* [Who belongs to whom?], Germany, 1944. Held by the U.S. Holocaust Memorial Museum, Film ID 2502B.

46. Kaplan, "'In the Interest of the Volk.'"

47. Ehrenreich, *Nazi Ancestral Proof,* 203; Madelene Schoch, "Determination of Paternity by Blood-Grouping Tests: The European Experience," *Southern California Law Review* 16 (1942–1943): 185.

48. "Begründung zu dem Gesetz über die Änderung und Ergänzung familienrechtlicher Vorschriften und über die Rechtsstellung der Staatenlosen vom 12. April 1938," *Allgemeines Suchblatt für Sippenforscher* 2 (1938): 67.

49. Willy Schumacher, "The Iso-Agglutination Test as Evidence in Judicial Proceedings in German Courts to Determine Parenthood," *St. John's Law Review* 8, no. 2 (1934): 276–284.

50. Kaplan, "'In the Interest of the Volk,'" discusses Munich; Evan Burr Bukey, *Jews and Intermarriage in Nazi Austria* (Cambridge: Cambridge University Press, 2010), discusses Vienna; and Beate Meyer, *"Jüdische Mischlinge": Rassenpolitik und Verfolgungserfahrung 1933–1945* (Hamburg: Dölling und Garlitz, 1999), discusses Hamburg. Whereas the RKO tended to come after people, as in Hanns Schwarz's case, when Jews challenged their own racial status, they tended to do so in civil courts.

51. Ernst Klee, *Deutsche Medizin im Dritten Reich: Karrieren vor und nach 1945* (Frankfurt am Main: S. Fischer, 2001), 44–45; Ernst Klee, *Das Personenlexikon zum Dritten Reich: Wer war was vor und nach 1945* (Frankfurt am Main: Fischer Taschenbuch Verlag, 2005), 303; and Ute Felbor, *Rassenbiologie und Vererbungswissenschaft in der Medizinischen Fakultät der Universität Würzburg 1937–1945* (Würzburg: Königshausen und Neumann, 1995), 100–102.

52. In a sample of sixty-six cases involving sixty-eight people, Meyer finds that fifty-four of them were able to improve their racial status, becoming either "full Germans" or at least *mischlinge.* Meyer, *"Jüdische Mischlinge,"* 113.

53. Ehrenreich, *Nazi Ancestral Proof,* 106.

54. Meyer finds this for the court in Hamburg. Meyer, *"Jüdische Mischlinge,"* 112.

55. The Gestapo did not look kindly on such requests, causing conflict with the judiciary. Sometimes they accompanied men to the court or took their statements within the camp. Meyer, *"Jüdische Mischlinge,"* 112–116.

56. The few historical studies of paternity proceedings are cited here. I have found no oral histories or memoirs that mention the practice.

57. Marta A. Balinska and William H. Schneider, "Introduction" to Ludwik Hirszfeld, *Ludwik Hirszfeld: The Story of One Life* (Rochester, NY: University Rochester Press, 2010), xvi.

58. Hirszfeld, *Story of One Life,* 20.

59. Highest priority: Schmuhl, *Kaiser Wilhelm Institute;* Weiss, *Nazi Symbiosis,* 102.

60. On the increasing burden of reports, see Abel interview in Müller-Hill, *Murderous Science,* 138; Schmuhl, *Kaiser Wilhelm Institute,* 206–207. Fischer complained about the workload and lobbied the genealogical authority for increased compensation: Ehrenreich, *Nazi Ancestral Proof,* 131; Kristie Macrakis, *Surviving the Swastika: Scientific Research in Nazi Germany* (Oxford: Oxford University Press, 1993), 128. Reche's complaints: Geisenhainer, *"Rasse ist Schicksal,"* 247.

61. Ehrenreich, *Nazi Ancestral Proof,* 131.

62. Fischer and Mollison quotes: Müller-Hill, *Murderous Science,* 37.

63. Abel interview in Müller-Hill, *Murderous Science,* 138.

64. Geisenhainer, *"Rasse ist Schicksal,"* 258.

65. Abel interview in Müller-Hill, *Murderous Science,* 138.

66. Macrakis, *Surviving the Swastika,* 128.

67. Geisenhainer, *"Rasse ist Schicksal,"* 257.

68. The letter, dated April 15, 1933—a week after the passage of the Law for the Civil Service—is reproduced in Mathias Okroi, "Der Blutgruppenforscher Fritz Schiff (1889–1940): Leben, Werk und Wirkung eines jüdischen Deutschen" (PhD diss., Universität zu Lübeck, 2004). Throughout the 1930s, Landsteiner would field numerous pleas from colleagues desperate to leave Germany, a role facilitated by the fact that the secretary of the Emergency Committee in Aid of Displaced Foreign Scholars, Alfred Cohn, was a colleague at the Rockefeller Institute.

69. Letter from Karl Landsteiner to Herman Nielsen, June 16, 1933, reproduced in Okroi, "Der Blutgruppenforscher Fritz Schiff."

70. "They believe there is no future for Dr. Schiff in the United States," wrote the secretary of the Emergency Committee of the Rockefeller officials. Letter from George Baehr to Dr. Gustav Bucky, June 26, 1934, box 26, folder 15, Karl Landsteiner Papers, RAC.

71. Lab personnel: Letter from Schiff to Landsteiner, February 24, 1935, box 26, folder 15, Karl Landsteiner Papers, RAC; university appointment: Okroi, "Der Blutgruppenforscher Fritz Schiff," 83. His record decreased from an average of more than ten publications annually during the years 1924–1932 to just over five after 1933.

72. Okroi, "Der Blutgruppenforscher Fritz Schiff," 85.

73. H. Schwarz, *Jedes Leben ist ein Roman,* 190–191.

74. Schwarz, *Jedes Leben ist ein Roman,* 196–198, 361–362; on the Polyclinic that examined Schwarz, see Jürgen Matthäus, "Evading Persecution: German-Jewish Behavior Patterns after 1933," in *Jewish Life in Nazi Germany: Dilemmas and Responses,* ed. Francis R. Nicosia and David Scrase (New York: Berghahn Books, 2013), 63.

75. Pegelow, "Determining 'People of German Blood,'" 48.

76. H. Schwarz, *Jedes Leben ist ein Roman,* 196–198.

77. Friedrich Herber, *Gerichtsmedizin unterm Hakenkreuz* (Leipzig: Militzke, 2002), 362–363; Hanns Schwarz, "Existenzkampf unter den Bedingungen faschistischen

Rassenwahns," in *Ärzte: Erinnerungen—Erlebnisse—Bekenntnisse,* ed. Günter Albrecht and Wolfgang Hartwig (Berlin: Der Morgen, 1982), 32–58.

78. H. Schwarz, "Existenzkampf," 54. Schwarz is cagey about the exchange with Dubitscher. In his memoir, he does not mention it, perhaps because Dubitscher, who after the war was dogged by allegations of his Nazi past, was still alive. Schwarz may have chosen to distance himself from the controversy by leaving unremarked the fact that the two knew each other. In a brief autobiographical essay published seven years later, Schwarz notes that he knew his examiner and recounts the loan offer but does not mention Dubitscher by name. By that time, Dubitscher was dead.

79. H. Schwarz, *Jedes Leben ist ein Roman,* 203–207.

80. For example, Hans Koopmann, a Hamburg scientist and seasoned ancestry tester, and Hans Weinert, a racial anthropologist at the University of Kiel, were frequently hired to perform exams on the same claimants only to reach completely opposite conclusions. A. Schwarz, "Hans Koopmann," 67–68; Meyer, *"Jüdische Mischlinge,"* 125–131.

81. Müller-Hill, *Murderous Science,* 139.

82. The scientist in question was Karl Tuppa. Pegelow, "Determining 'People of German Blood,'" 58.

83. A. Schwarz, "Hans Koopmann," 67.

84. Müller-Hill, *Murderous Science,* 139.

85. Müller-Hill, *Murderous Science,* 40.

86. Kaplan, "'In the Interest of the Volk,'" 535.

87. H. Schwarz, *Jedes Leben ist ein Roman,* 209–210.

88. Poll: James Braund and Douglas G. Sutton, "The Case of Heinrich Wilhelm Poll (1877–1939): A German-Jewish Geneticist, Eugenicist, Twin Researcher, and Victim of the Nazis," *Journal of the History of Biology* 41, no. 1 (2008): 1–35; W. Paul Strassmann, *The Strassmanns: Science, Politics and Migration in Turbulent Times (1793–1993)* (New York: Berghahn Books, 2008); Weningers: Paul Weindling, "A City Regenerated: Eugenics, Race, and Welfare in Vienna," in *Interwar Vienna: Culture between Tradition and Modernity,* ed. Deborah Holmes and Lisa Silverman (Rochester, NY: Camden House, 2009), 104; see also H. Strauch and I. Wirth, "Persecution of Jewish Forensic Pathologists," *Forensic Science International* 144, no. 2–3 (2004): 125–127.

89. Lattes obituary for Dr. Fritz Schiff, *Prensa Médica Argentina* 27 (1940): 2134.

90. While his passing left his family in serious economic straits all over again, all three sons would go on to become very successful professionals. Peter Schiff, telephone interview with author, May 12, 2016; letter from Philip Levine to Karl Landsteiner, August 5, 1940, box 3, folder 3, Karl Landsteiner Papers, RAC.

91. Hirszfeld, *Story of One Life,* 20.

92. Various sources cite between 2,500 and 3,000 cases a year: Ilse Schwidetzky, "Forensic Anthropology in Germany," *Human Biology* 26, no. 1 (1954): 1–20;

"Wenn die Sachverständigen versagen," *Die Zeit,* November 10, 1961; Reche and Rolleder, "Zur Entstehungsgeschichte." The papers covered such cases: see, for example, "Barbara bleibt ohne Eltern," *Die Zeit,* June 27, 1957. The case of Josette Phellipeau, a child contested by two mothers, became a cause célèbre in France and Germany and the plot for a 1955 film.

93. Teschler-Nicola, "Diagnostic Eye."

94. Susanne Heim, Carola Sachse, and Mark Walker, *The Kaiser Wilhelm Society under National Socialism* (Cambridge: Cambridge University Press, 2009), 378.

95. Friedrich Keiter, "Advances in Anthropological Paternity Testing," *American Journal of Physical Anthropology* 3 (1963): 82.

96. List of experts: Schwidetzky, "Forensic Anthropology in Germany," 2. By 1964, Reche claimed there were fifty officially recognized paternity experts. Press reports on paternity science include the following: "Barbara bleibt ohne Eltern," "Wenn die Sachverständigen versagen." Entirely unknown: Schwidetzky, "Forensic Anthropology in Germany," 3. Schwidetzky, who had performed ancestry examinations as an assistant to von Eickstedt, presented work on paternity at an anthropology conference in Philadelphia in 1956; Schwidetzky, "New Research in German Forensic Anthropology," in *Men and Cultures: Selected Papers of the Fifth International Congress of Anthropological and Ethnological Sciences,* ed. Anthony F. C. Wallace (Philadelphia: University of Pennsylvania Press, 1960), 703–708. Keiter, "Advances in Anthropological Paternity Testing," followed the author's research stay in the United States.

97. On scientific amnesia: Müller-Hill, *Murderous Science,* 90; international literature: Leone Lattes, "Processi giudiziari in tema di filiazione naturale," *Minerva Medicolegale* 70 (1950): 85–99, discusses claims of "certainty," referring to this as a "formula used for many years now by the German courts of justice" (85) with no reference to the political context of its use; Angelo Vincenti, *La ricerca della paternità e i gruppi sanguigni nel diritto civile e canonico* (Florence: Casa Editrice Dott. Carlo Cya, 1955), 204, gives an overview of paternity testing in Germany that skips from the 1920s to the mid-1950s; Stanley Schatkin, "A Challenge to Precedent in Paternity Exclusion Cases," *Journal of Criminal Law and Criminology* 36, no. 1 (1945): 42–44 comments on the widespread use of blood tests in continental Europe but mentions only Denmark and Sweden. The racial use of paternity was acknowledged at the time by Louis Christiaens, *La recherche de la paternité par les groupes sanguins: étude technique et juridique* (Paris: Masson et cie, 1939) and Antônio Ferreira de Almeida Júnior, *As provas genéticas da filiação* (São Paulo: Revista dos Tribunais, 1941). In an Argentine case, a court official disparaged biological evidence that Lattes presented, referring to the "eminently racial" uses of such evidence in Germany. Leon [sic] Lattes, "Valor de las pruebas biológicas en la investigación de la paternidad," *Archivos de Medicina Legal* 18 (1948): 212.

98. In 1958, Reche was made honorary member of the Deutsche Gesellschaft für Anthropologie and received the Austrian Honorary Cross for Science and Art

in 1965. Katja Geisenhainer, "Otto Reches Verhältnis zur sogenannten Rassenhygiene," *Anthropos* 91, no. 4/6 (1996): 509.

99. "Wenn die Sachverständigen versagen."

100. Ehrenreich, *Nazi Ancestral Proof,* 129.

101. Reche and Rolleder, "Zur Entstehungsgeschichte," 292.

102. H. Schwarz, *Jedes Leben ist ein Roman,* 194.

7. TO THE WHITE HUSBAND A BLACK BABY

1. "Inexpert eye" and "negroid": I. del Carpio, *In tema di disconoscimento di paternità* (Pisa: Arti Grafiche Pacini Mariotti, 1946), 5. The story is reconstructed from del Carpio, the judicial sources associated with the case, and interviews with Antonio Cipolli and Dunja Cipolli.

2. Corsanego, Constitutional Assembly, April 22, 1947, 1146.

3. Cevolotto, Constitutional Assembly, April 17, 1947, 983.

4. "Nowadays": Calamandrei, Constitutional Assembly, April 17, 1947, 970. Other references to adultery in the Constitutional Assembly discussions include: Delli Castelli, April 19, 1947, 1043–1044; Sardiello, April 21, 1947, 1066; Rossi, April 21, 1947, 1102; Corsanego, April 22, 1947, 1152.

5. It is unclear what became of his petition for conjugal separation, but the couple did split up.

6. "Per la riforma del Codice Civile in tema di disconoscimento di paternità. Un bianco può essere padre legittimo di un figlio negro?" *RDMIRF* 14 (1947): 128–131; "absurd": *RDMIRF*, 130; "rigorous logic": *RDMIRF,* 131. See also Giuseppe Lenzi, "Conclusioni del Pubblico Ministero," *RDMIRF,* 131–134.

7. The 1942 code revised its predecessor, the Civil Code of 1865.

8. "Sweet name": Corsanego, Constitutional Assembly, April 22, 1947, 1146.

9. Quoted in Maria Porzio, *Arrivano gli alleati! Amori e violenze nell'Italia liberata* (Rome: Laterza, 2011), 68.

10. Centro Furio Jesi, *La menzogna della razza: Documenti e immagini del razzismo e dell'antisemitismo fascista* (Bologna: Grafis, 1994), 202.

11. The stereotype of the oversexed, drunken black soldier is repeated in the journalist Aldo Santini's memories of the period, *Tombolo* (Milan: Rizzoli, 1990); Gigi Di Fiore's polemical and racist *Controstoria della Liberazione* (Milan: Rizzoli, 2012); and Carla Forti, *Dopoguerra in provincia. Miscrostorie pisane e lucchesi, 1944–48* (Milan: FrancoAngeli, 2007).

12. The incident, which took place in May 1944 in an area southeast of Rome, is immortalized in Alberto Moravia's 1957 novel *La Ciociara* and in a subsequent film by Vittorio de Sica of the same name (*Two Women* in its English release), starring Sophia Loren. On racism, colonialism, and sexuality, see Vincenza Perilli, "Relazioni pericolose. Asimmetrie dell'interrelazione tra 'razza' e genere e sessualità interrazziale" in *Il colore della nazione*, ed. Gaia Giuliani (Florence: Le Monnier, 2015).

13. Journalists, novelists, and neorealist cinematographers portrayed the zone for audiences on both sides of the Atlantic. Saverio Giovacchini, "John Kitzmiller, Euro-American Difference, and the Cinema of the West," *Black Camera* 6, no. 2 (2015): 17–41.

14. Santini, *Tombolo.*

15. The song appeared on the soundtrack of the film *The Bicycle Thieves* in November 1948, although it was written earlier. Moe, "Naples '44 / 'Tammurriata Nera' / *Ladri di Biciclette,"* in *Italy and America, 1943–1944: Italian, American and Italian American Experiences of the Liberation of the Italian Mezzogiorno,* ed. John A. Davis (Naples: La Città del Sole, 1997), 433–477.

16. Giulia Barrera, "Patrilinearità, razza, e identità: L'educazione degli italo-eritrei durante il colonialismo italiano (1885–1934)," *Quaderni Storici* 37, no. 109 (1) (2002): 21–53.

17. One hundred: Lilian Scott, "Inter-Racial Group to Aid Deserted Italian War Babies," *Chicago Defender,* June 25, 1947, 1; 11,000: Spallicci, Constitutional Assembly, April 21, 1947, 1095–1096. The speaker cited an unidentified 1945 census. The figure may include mixed-race peoples in the African colonies, although even in this case it appears wildly inflated. Silvana Patriarca, "Fear of Small Numbers: 'Brown Babies' in Postwar Italy," *Contemporanea* 18, no. 4 (2015): 537–567.

18. "Caffe-latte": review of the film *Il Mulatto* (discussed subsequently), *La Stampa,* May 17, 1950, 5.

19. Constitutional Assembly, April 21, 1947, 1095–1096. The delegate was Aldo Spallicci, member of the center-left, liberal-anti-fascist Partito Repubblicano Italiano and by profession a doctor.

20. David I. Kertzer, *Sacrificed for Honor: Italian Infant Abandonment and the Politics of Reproductive Control* (Boston: Beacon Press, 1994).

21. Only the Communist press ever referred to them as citizens, notes Patriarca, "Fear of Small Numbers," 557.

22. Antonio Cipolli, telephone interview with the author, July 11, 2017.

23. Luigi Gedda, Angelo Serio, and Adriana Mercuri, *Il meticciato di guerra e altri casi* (Rome: Edizioni dell'Istituto Gregorio Mendel, 1960); Francesco Cassata, *Building the New Man: Eugenics, Racial Science and Genetics in Twentieth-Century Italy* (Budapest: Central European University Press, 2011), describes these ties and places the "war hybrid" study in the broader context of postwar Italian and transnational racial thinking.

24. Patriarca, "Fear of Small Numbers." On U.S.-based adoption schemes, see Lilian Scott, "Inter-Racial Group to Aid Deserted Italian War Babies," *Chicago Defender,* June 25, 1947, 1; Di Fiore, *Controstoria della Liberazione,* 259. On Afro-German children, see Heide Fehrenbach, *Race after Hitler: Black Occupation Children in Postwar Germany and America* (Princeton, NJ: Princeton University Press, 2005).

25. Cipolli, interview.

26. The character Angelo was played by Angelo Maggio, the abandoned child of an Italian woman and a black American GI who had been adopted by Neapolitan actor Dante Maggio.

27. The film was released as *Angelo* in English. Analyses include Shelleen Greene, *Equivocal Subjects: Between Italy and Africa-Constructions of Racial and National Identity in the Italian Cinema* (New York: Continuum, 2012); and Grace Russo Bullaro, *From Terrone to Extracomunitario: New Manifestations of Racism in Contemporary Italian Cinema* (Leicester: Troubador Publishing, 2010).

28. Fausto Gullo, Constitutional Assembly, April 18, 1947, 997.

29. Survivors, prisoners: Rossi, Constitutional Assembly, April 21, 1947, 1102. Other comments in the Constitutional Assembly include the following: Gullo, April 18, 1947; Delli Castelli, April 19, 1947; Sardiello, April 21, 1947. Press: Pacifico Fiori, "Quanti 'figli della guerra' sono fra gli illegittimi?," *Corriere della Sera,* October 31 / November 1, 1946, 2; "Come sarebbe finito quel figlio cieco e non suo?" *Corriere della Sera,* February 4, 1947, 2; "Non sono più abbandonati i giovani che escono dal carcere," *Corriere della Sera,* July 30, 1947, 2; "Un bambino di cinque anni conteso da due padri," *Corriere della Sera,* October 7, 1947, 2; Felice Chilanti, "Il figlio 'disconosciuto' dal reduce di San Severo," *Corriere della Sera,* July 30–31, 1948, 3; "Triste vicenda coniugale di un reduce dalla prigionia," *Corriere della Sera,* September 15–16, 1950, 2.

30. The letter was addressed to the Presidenza del Consiglio dei Ministri. Cited in Porzio, *Arrivano gli alleati!,* 117.

31. "Diminished manhood": Ruth Ben-Ghiat, "Unmaking the Fascist Man: Masculinity, Film and the Transition from Dictatorship," *Journal of Modern Italian Studies* 10, no. 3 (September 1, 2005): 337, 339; on suffrage, see Molly Tambor, *The Lost Wave: Women and Democracy in Postwar Italy* (Oxford: Oxford University Press, 2014).

32. Gullo, Constitutional Assembly, April 18, 1947, 997.

33. Gullo, Constitutional Assembly, April 18, 1947, 997.

34. Chilanti, "Il figlio 'disconosciuto.'"

35. Gullo cites the former scenario; on the latter, see Fiori, "Quanti 'figli della guerra.'"

36. Civil Code of 1942, tit. 7, sec. 1, art. 235, nos. 1–4.

37. Luciano Tonni, "Rigorisimi legislative in tema de azione di disconoscimento di paternità," *Il Foro Padano* (1949): 872.

38. "Genital organs": Antonio Carrozza, "Sulla paternità del mulatto partorito da donna bianca maritata ad un bianco," *Il Foro Italiano* 75 (1950): 112. See also Antonio Carrozza, "Di una nuova specie del disconoscimento di paternità e dei relativi effetti," *RDMIRF* 14 (1947): 134–138. Carrozza later became a well-known professor of agrarian law.

39. Carrozza, "Di una nuova specie" 134.

40. Carroza, "Di una nuova specie," 134.

41. See prosecutor Mario Comucci's untitled commentary in *Il Foro Padano* 72 (1949): 1211–1216.

42. "Al padre bianco resta il figlio nero," *La Stampa,* July 16, 1949, 1. See also "Gli nasce un figlio mulatto e non può disconoscerlo," *Corriere d'Informazione,* July 16, 1949, 1.

43. Plinio Citti, "Quando le leggi sono ingiuste. Il bimbo moro di Pisa," *Il Tirreno,* July 16, 1949, 3. Coverage continued: "Una proposta di legge per l'estensione dei casi di disconoscimento della paternità," *Il Tirreno,* July 27, 1949, 1; and "Quando le leggi sono inique. Il 'bimbo moro' di Pisa," *Il Tirreno,* August 6, 1949, 1.

44. Carrozza cites this article but his citation (*Oggi,* n31, July 28, 1949) is incorrect.

45. "Quando le leggi sono inique."

46. Decision of Corte d'Appello di Firenze, *Il Foro Padano* (1949): 870–871.

47. Tonni, "Rigorisimi legislative," 869. An almost identical statement appears in "Quando le leggi sono inique."

48. Lenzi, "Conclusioni del Pubblico Ministero," 132–133.

49. That decision read: "The case of a colored child born to a white woman in marriage to a man who is also of the white race, does not appear among the circumstances in which the civil law gives the husband the right to propose a suit for disavowal of paternity." Decision of Corte d'Appello di Firenze, *Il Foro Padano* (1949), 870.

50. "Proposta di legge d'iniziativa del deputato Silvio Paolucci," *RDMIRF* (1949): 183.

51. *RDMIRF* (1949), 183.

52. *RDMIRF* (1949), 183.

53. A working group of the Union of Italian Catholic Jurists discussed the issue: "Riforma degli istituti familiari," *Iustitia* (1950): 106.

54. L'Unione Cinematografica Internazionale fra i Cattolici was a new group that oversaw the morality of films. Patriarca, "Fear of Small Numbers."

55. The very first published piece on the case, a lecture by the doctor called to examine the baby, mentions this possibility: del Carpio, *In tema di disconoscimento* (it appeared in July 1946).

56. Decision of Corte d'Appello di Firenze, *Il Foro Padano* (1949), 873. See also Antonio Emanuele Granelli, *L'azione di disconoscimento di paternità* (Milan: Dott. A. Giuffrè: 1966), 156.

57. As Antonio Carrozza put it, "Is it not impotence that characterizes a subject of blood type O who absolutely cannot conceive a child of type AB?" Carrozza, "Di una nuova specie," 136. Hematological impotence: Granelli, *L'azione di disconoscimento,* 156; serological races: Angelo Vincenti, *La ricerca della paternità e i gruppi sanguigni nel diritto civile e canonico* (Florence: Casa Editrice Dott. Carlo Cya, 1955), 198.

58. From a review of the Paolucci proposal by a working group of the Union of Italian Catholic Jurists, see "Riforma degli istituti familiari," 106.

59. The phrase comes from second-century rhetorician Calpurnius Flaccus. Werner Sollors, *Neither Black nor White yet Both: Thematic Explorations of Interracial Literature* (Cambridge, MA: Harvard University Press, 1999), chapter 4.

60. Wendy Doniger and Gregory Spinner, "Misconceptions: Female Imaginations and Male Fantasies in Parental Imprinting," *Daedalus* 127, no. 1 (1998): 115.

61. J. Michael Duvall and Julie Cary Nerad, "'Suddenly and Shockingly Black': The Atavistic Child in Turn-into-the-Twentieth-Century American Fiction," *African American Review* 41, no. 1 (2007): 51–66; Sollors, *Neither Black nor White.*

62. Charles Benedict Davenport, *Heredity of Skin Color in Negro-White Crosses* (Washington: Carnegie Institution of Washington, 1913), 29.

63. "Lab science": Carrozza, "Sulla paternità del mulatto," 110.

64. "Simplest case": Otto Reche, cited in Maria Teschler-Nicola, "The Diagnostic Eye. On the History of Genetic and Racial Assessment in Pre-1938 Austria," *Collegium Antropologicum* 28 (2004): 16; English jurist: Courtney Kenny, "Resemblance as Evidence of Consanguinity," *Law Quarterly Review* 39 (1923): 306; South Africa: "Disputed Paternity-Appearance of the Child as Corroboration," *African Law Journal* (1943): 195–196; Scotland: Kenny, "Resemblance as Evidence"; U.S. states: Shari Rudavsky, "Blood Will Tell: The Role of Science and Culture in Twentieth-Century Paternity Disputes" (PhD diss., University of Pennsylvania, 1996), 54. Similar claims appear in Fritz Schiff, *Blood Groups and their Areas of Application,* Selected Contributions to the Literature of Blood Groups and Immunology, vol. 4, part 2 (Fort Knox, Kentucky: U.S. Army Medical Research Laboratory, 1971), 327; Amedeo Dalla Volta, *Trattato di medicina legale* (Milan: Società Editrice Libraria, 1933), 571; Madelene Schoch, "Determination of Paternity by Blood-Grouping Tests: The European Experience," *Southern California Law Review* 16 (1942–1943): 179; and many others. Brazilian forensic scientists were the only ones who questioned this logic; see Chapter 5.

65. Rudavsky, "Blood Will Tell," 54.

66. "Colored adultery": Carrozza, "Sulla paternità del mulatto," 113.

67. Jorge Bocobo, "The Conclusive Presumption of Legitimacy of Child," *Philippine Law Journal* 12 (1932–1933): 161.

68. Nomi Maya Stolzenberg, "Anti-Anxiety Law: Winnicott and the Legal Fiction of Paternity," *American Imago* 64 (2007): 345.

69. The Four Seas principle was quoted perhaps most recently by a California court in 1989. David D. Meyer, "Parenthood in a Time of Transition: Tensions between Legal, Biological, and Social Conceptions of Parenthood," *American Journal of Comparative Law* 54 (2006): 125–144.

70. Ron Shaham, "Law versus Medical Science: Competition between Legal and Biological Paternity in an Egyptian Civil Court," *Islamic Law & Society* 18, no. 2 (2011): 219–249.

71. Florencio García Goyena, *Concordancias, motivos y comentarios del Código Civil español,* vol. 1 (Madrid: Sociedad Tipográfico-Editorial F. Abienzo, 1852), 111.

72. Cited in Rudavsky, "Blood Will Tell," 227.

73. Matter of Findlay, 253 N.Y. 1, 11, 170 N.E. 471 (1930).

74. Piers v. Piers (1849), *The Jurist* 13, part 1 (1850): 572.

75. California law, which had a hard and fast presumption of paternity, allowed for the exception of racial impossibility. Rudavsky, "Blood Will Tell," 249. Cardozo invoked Lord Campbell's dictum as one of the "extravagances" of the presumption. Matter of Findlay, 253 N.Y. 1, 11, 170 N.E. 471 (1930).

76. The case is [1931] N.Z.L.R. 559, cited in Law Commission, *Blood Tests and the Proof of Paternity* (London: Her Majesty's Stationery Office, 1968); also discussed in "Blood Test Powers for Courts Urged," *Guardian,* November 1, 1968, 5.

77. Schoch, "Determination of Paternity," 185.

78. Shaham, "Law versus Medical Science."

79. Rudavsky, "Blood Will Tell," especially chapter 4. Rudavsky found lower courts to be more resistant than appellate ones, and those in some states (such as California) more resistant than those in others (New York, Pennsylvania).

80. Hill v. Jackson, 226 P. 2d 656 (1951), cited in Rudavsky, "Blood Will Tell," 257.

81. Jackson v. Jackson, 430 P 2d 291 (1967), cited in Rudavsky, "Blood Will Tell," 220.

82. The original Italian Civil Code of 1865 likewise restricted challenges to marital paternity. The 1942 reform changed slightly the clauses governing these proceedings but if anything made it more difficult for a husband to dispute his paternity.

83. See, for example, the *RDMIRF* and the popular press.

84. In Germany, such suits were impossible into the mid-1950s because soldiers could not appear in German civil court. It is unclear if such a prohibition was in effect in Italy as well. In England, women had a somewhat easier time pursuing paternity claims. In all instances, women would have been prohibited from pursuing recourse in American courts because the child was not born on U.S. territory nor did the mother and child live there. Fehrenbach, *Race after Hitler,* 69.

85. Silvia Cassamagnaghi, *Operazione spose di guerra: Storie d'amore e di emigrazione* (Milan: Feltrinelli, 2014). On similar dynamics in Germany, see Fehrenbach, *Race after Hitler.* See also Brenda Gayle Plummer, "Brown Babies: Race, Gender and Policy after World War II," in *Window on Freedom: Race, Civil Rights, and Foreign Affairs, 1945–1988,* ed. Brenda Gayle Plummer (Chapel Hill: University of North Carolina Press, 2003), 67–91.

86. As Plummer notes, in erasing black GI fathers, the U.S. military "ironically replicated the slavery-era principle that guaranteed the anonymity of mulattoes' fathers." "Brown Babies," 77. Such a policy affected not only the children but also black fathers themselves, who were deprived of the patriarchal role of responsible father ("Brown Babies," 85).

87. The papal encyclical *Casti Connubii* (1930) condemned eugenics, sterilization, and birth control. In the 1940s, church authorities took a public stand against

artificial insemination. *Humanae Vitae* (1968) reaffirmed the church's opposition to birth control.

88. Ludwik Hirszfeld, *Les groupes sanguins. Leur application à la biologie, à la médecine et au droit* (Paris: Masson et cie, 1938), 86. The quote was widely cited by Latin American and southern European authors, for example, Guillermo Tell Villegas Pulido, *La inquisición de la paternidad por el examen de la sangre* (Caracas: Cecilio Acosta, 1940), 14; Osvaldo Stratta, *Los grupos sanguíneos y el problema medico-legal de la paternidad y la filiación* (Santa Fe: Imprenta de la Universidad Nacional del Litoral, 1944), 81; Louis Christiaens, *La recherche de la paternité par les groupes sanguins: Étude technique et juridique* (Paris: Masson et cie, 1939), 104.

89. On the press, see G. Tansella, "Lo studio delle impronte digitali," *Minerva Medica* 49 (1958): 3110.

90. Palmieri's thoughts were published in an Argentine medical journal: Vincenzo Mario Palmieri, "Consideraciones médicolegales sobre la investigación biológica de la paternidad a pedido de particulares," *Jornada Médica* 6, no. 71 (1952): 305. On Palmieri's Catholicism, see Luigia Melillo, *Katyn, una verità storica negata. La perizia di V. M. Palmieri* (Naples: Università degli Studi di Napoli, 2009).

91. See, for example, Tansella, "Lo studio delle impronte digitali"; and Giorgio Benassi, "Ricerca della paternità davanti alla legge e davanti alla," *Rivista Trimestrale di Diritto e Procedura Civile* 3 (1949): 918–926. Benassi was an affiliate of the University of Bologna, Tansella of the University of Bari.

92. Palmieri, "Consideraciones médicolegales," 305.

93. Palmieri, "Consideraciones médicolegales," 305.

94. Vincenti, "Ricerca della paternità," 215.

95. Antônio Ferreira de Almeida Júnior, *As provas genéticas da filiação* (São Paulo: Revista dos tribunais, 1941), 218.

96. Palmieri, "Consideraciones médicolegales," 305.

97. Flamínio Favero, "A ética e a prova dos tipos sanguíneos," Artigos para *Folha da Manhã*, vol. 4, 301–400, IOF. The article apparently appeared in the newspaper on May 4, 1952.

98. "Suspicion on the mother": G. B. Funaioli, *Diritto civile. La filiazione naturale* (Pisa: Arti Grafiche Tornar, 1949–50), 143; "grave dangers": Cazzaniga, cited in Folco Domenici, *Gruppi sanguigni e ricerca della paternità* (Milan: Gentile Editore, 1946), 146.

99. "Motives clear": Palmieri, "Consideraciones medicolegales"; "danger of a proof of exclusion": Favero, "A ética e a prova." One Italian doctor recounted how he reluctantly agreed to perform a paternity test for a fellow doctor with marital troubles: Benassi, "Ricerca della paternità," 924.

100. Domenici, *Gruppi sanguigni*, 145.

101. Other examples include Pierre Barbier, member of the Faculty of Law in Paris, cited in William H. Schneider, "Chance and Social Setting in the Application of

the Discovery of Blood Groups," *Bulletin of the History of Medicine* 57, no. 4 (1983): 555.

102. Vincenti, "Ricerca della paternità," 215.

103. Favero, "A ética e a prova"; Arnaldo Amado Ferreira, "Investigação medico-legal da paternidade," *Revista Médica* 85 no. 4 (2006) [originally 1953]: 142–156.

104. Christiaens, *La recherche de la paternité*, 104–105.

105. Tambor, *Lost Wave.*

106. "Celebrated case of the mulatto": Bianca Checchini, "Accertamento e attribuzione della paternità" (PhD diss., University of Padova, 2008), 127; Giorgio Collura, Leonardo Lenti, and Manuela Mantovani, *Trattato di diritto di famiglia* vol. 2 Filiazione (Milan: Giuffrè Editore, 2012), 157; Luigi Balestra, *Commentario del codice civile. Della famiglia* (Turin: UTET Giuridica, 2011), 450; Laura Di Bona, *Rapporti a contenuto non patrimoniale e vincolatività del consenso* (Pesaro: Edizioni Studio @lfa, 2005), 36.

107. "Can you believe this story from Florence? Can you believe a judge would sign a sentence (a sentence!) holding that the father of a black child inherited from the passage of soldiers belongs to the husband of the mother?" Gian Antonio Stella, *Il maestro magro* (Milan: Rizzoli, 2005), 109, cited in Giulia Galeotti, *In cerca del padre: Storia dell'identità paterna in età contemporanea* (Bari: GLF Editori Laterza, 2009).

8. CITIZEN FATHERS AND PAPER SONS

1. Some 7,500 people were given travel affidavits between January 1, 1950, and May 1, 1953. Thousands more were issued by the Canton and Hong Kong consulates in the years prior. Sidney Schatkin, Leon Sussman, and Dorris Yarbrough, "Chinese Immigration and Blood Tests," *Criminal Law Review* 2, no. 1 (1955): 46.

2. Mae M. Ngai, "Legacies of Exclusion: Illegal Chinese Immigration during the Cold War Years," *Journal of American Ethnic History* 18, no. 1 (1998): 4; Mae M. Ngai, *Impossible Subjects: Illegal Aliens and the Making of Modern America* (Princeton, NJ: Princeton University Press, 2014), chapter 6.

3. Reopened Board of Special Inquiry Hearing, November 5, 1953, in published transcript of the proceedings: United States Court of Appeals for the Second Circuit, United States ex rel. Lee Kum Hoy v. Edward J. Shaughnessy, Appendix of Respondent-Appellant, Docket number 23972, October term, 1955 [hereafter Appendix], 217a.

4. Board of Special Inquiry Hearing, August 14, 1952, in Appendix, 137a.

5. This wording is from Dorothy E. Roberts, "The Genetic Tie," *University of Chicago Law Review* 62, no. 1 (1995): 209–273.

6. Jacqueline Stevens, *Reproducing the State* (Princeton, NJ: Princeton University Press, 1999); Kristin A. Collins, "Illegitimate Borders: Jus Sanguinis Citizenship

and the Legal Construction of Family, Race, and Nation," *Yale Law Journal* 123 (2014): 2134–2235.

7. This information is culled from Lee Ha's immigration case file (Lee Ha, 111/52, Box 359, Chinese Exclusion Case Files, NARA, regional branch, New York City) as well as from the transcript of the administrative hearings and court records surrounding the children's case (cited in note 3).

8. Ngai, "Legacies of Exclusion," 3.

9. Board of Special Inquiry Hearing, August 14, 1952, in Appendix, 128a.

10. Estelle T. Lau, *Paper Families: Identity, Immigration Administration, and Chinese Exclusion* (Durham, NC: Duke University Press, 2006), 89–90.

11. Lau, *Paper Families,* 100. See also Erika Lee, *At America's Gates: Chinese Immigration during the Exclusion Era, 1882–1943* (Chapel Hill: University of North Carolina Press, 2003).

12. Letter from Edw[in] L. Huff to W. W. Husband, Commissioner-General of Immigration, May 1, 1925, NARA, RG 85, File 55452/385. The Norwegian researcher was Kristine Bonnevie, who conducted the largest ever study of hereditary traits in fingerprints (published in the *Journal of Genetics* the year before). In April 1925, an article about her work appeared in U.S. newspapers, which was probably where the INS agent read about it.

13. Letter from SF District Director [Edwin L. Huff] to INS Commissioner, January 27, 1934, NARA, RG 85, File 55452/385. This voluminous file is frequently cited in histories of paper immigration.

14. Letter from INS Commissioner to SF District Director, February 9, 1934, NARA, RG 85, File 55452/385.

15. Letter from [Edward] Shaughnessy to another INS official [Mr. Henning], March 16, 1925, NARA, RG 85, File 55452/385.

16. A Hong Kong newspaper cited this statistic as originating with Washington authorities; the newspaper is cited in a consular dispatch: Consulate General Despatch [*sic*] 1330, "Hong Kong Press and Governmental Reaction to Consular Fraud Problems at Hong Kong," April 16, 1955, NARA, RG 59, Decimal File, 1955–59, Box 721, Folder 122.4732, 1–1056.

17. Edward Ranzal, "U.S. Drive Aimed at Passport Ring," *New York Times,* February 15, 1956, 1.

18. Everett F. Drumright, "Report on the Problem of Fraud at Hong Kong," NARA, RG 59, Decimal File, 1950–54, Box 0824, Folder 125.4734, 2.

19. 1920 case: *Kwock Gan Fat v. White*; two percent of 6,000: Schatkin, Sussman, and Yarbrough, "Chinese Immigration and Blood Tests," 47; 1,100 actions: *Report of the Commission on Government Security: Pursuant to Public Law 304, 84th Congress, as Amended* (Washington, DC: U.S. Government Printing Office, 1957), 482 (these were so-called Section 503 claims, brought under the relevant section of the 1940 Immigration Code); "practically powerless": "Suits against the Secretary of State under Section 503 of the Nationality Act of 1940" to Argyle Mackey, Commissioner, Central Office, and Bruce Barber, District

Director, San Francisco, CA, February 15, 1952, NARA, RG 85, File 56364/51.6; "obscure Chinese claimant": Letter from Ernest J. Hover, Special Representative and Chief, Examinations Branch, El Paso, TX to Allen C. Devaney. Subject: Declaratory Judgment Cases filed under Section 503, April 21, 1956, NARA, RG 85, File 56364/51.6.

20. 117,000: Ngai, "Legacies of Exclusion," 8; 150 cases per month: Telegraph from HK to Secretary of State, December 28, 1950, NARA, RG 59, Decimal File 1950–54, Box 0824, Folder 125.4735; "incessant demands": Memo to Department of State from Hong Kong, re Status of Visa and Citizenship Work, by Consul K. L. Rankin, January 6, 1950, NARA, RG 59, Decimal File 1950–54, Box 0824, Folder 125.4734; 150 calls: Memo to Department of State from Hong Kong, re Office Procedure, Citizenship and Visa Functions, April 26, 1950, NARA, RG 59, Decimal File 1950–54, Box 0824, Folder 125.4734; "stampede proportions": Memo to Department of State from Hong Kong, re Status of Citizenship and Visa Work, by Consul K. L. Rankin, May 9, 1950, NARA, RG 59, Decimal File 1950–54, Box 0824, Folder 125.4734.

21. The inspectors, Ancel Taylor and Laurence Taylor, were veteran State Department employees. Laurence Taylor had a background in agricultural science, which may suggest he was the source of the idea, although the medicolegal applications of blood group heredity were familiar enough at this time that the idea did not require an expert source. Their visit took place in March–April, 1950; by May the consulate was requesting permission from Washington to implement blood testing. "Report of Action Taken on Inspectors' Recommendation," [June 19, 1950], NARA, RG 59, Decimal File 1950–54, Box 0824, Folder 125.473.

22. Ride was the author of *Genetics and the Clinician* (1938), a textbook that argued for the value of genetics to everyday medical practice, including the application of blood group typing to paternity establishment. He had earlier conducted research on the heredity of blood groups among "races in the Pacific basin."

23. "Persons of Chinese race": Liaison Report, October 9, 1952, from Daniel J. Kelly, Investigator, Investigations Division, NARA, RG 85, File 56267/57B. The report is by an INS official citing a State Department official; 3,000 individuals: "Establish [*sic*] of a laboratory and X-ray machine in the American Consulate General, Hong Kong," September 3, 1953, NARA, RG 59, Decimal File 1950–54, Box 0824, Folder 125.4735; in-house lab: the consulate's doctor convinced the officials it was neither logistically feasible nor cost-effective. "Establish [*sic*] of a laboratory and X-ray machine in the American Consulate General, Hong Kong."

24. Schatkin, Sussman, and Yarbrough, "Chinese Immigration and Blood Tests," 50; Paul M. Scheib, "Fraud in Derivative Citizenship," *Intramural Law Review of New York University* 11 (1955–1956): 271–279; Drumright, "Report on the Problem of Fraud," 43. In fact, at least one scientist disputed this calculation, arguing it overestimated incompatibility and therefore the incidence of fraud:

William C. Boyd, "Chances of Disproof of False Claims of Parent–Child Relationship," *American Journal of Human Genetics* 9, no. 3 (1957): 191–194.

25. In the Matter of L-F-F-, *Administrative Decisions under Immigration & Nationality Laws*, vol. 5, February 1952–June 1954 (Washington, DC: U.S. Government Printing Office, 1952), 153. The observation is made in the context of a discussion of Chinese citizenship cases jamming California dockets.

26. Letter from H. E. Montamat for the Consul General, to HR Landon, District Director, INS, Los Angeles, October 22, 1952, NARA, RG 85, File 56336/205. Also cited in Schatkin, Sussman, and Yarbrough, "Chinese Immigration and Blood Tests," 50.

27. "Report of Activities of the Investigative Unit, American Consulate General, Hong Kong, September 1, 1952 to June 15, 1953," NARA, RG 59, Decimal File, 1950–54, Box 0824, Folder 125.4735, 4a.

28. Their zealousness caused frictions with British colonial officials and local police, however. Various correspondence, NARA, RG 59, Central Decimal File, 1955–59, Box 721, Folder 122.4732, 1–1056.

29. An April 1951 memo from an official in the Passport Office recommends use of tests throughout the State Department: Memo to Mr. Nicholas, Mr. Young and Mrs. Shipley from W. E. Duggan (Passport Office at State Department), April 2, 1951, NARA, RG 85, File 56328/856. Testing in Hong Kong began with the basic ABO blood groups; the State Department advised consular officials to adopt MN and Rh tests if the cost was not excessive.

30. Memo from D. E. Yarbrough, Investigator, to M. F. Fargione, Chief, Investigations Section, New York, Subject: Blood Tests, November 21, 1952, NARA, RG 85, File 98524/528 [hereafter Yarbrough memo].

31. Letter from H. E. Montamat for the Consul General, to HR Landon, District Director, INS, Los Angeles, October 22, 1952.

32. This was the characterization of the INS commissioner to the surgeon general. Letter to Surgeon General Scheele from INS Commissioner [Argyle Mackey], April 7, 1952, NARA, RG 85, File 56328/856.

33. The first mentions of blood testing in Operations Instructions I have found, which relate to payment protocols, date from the fall of 1952.

34. In the Matter of L-F-F-.

35. Letter from W. W. Wiggins, Chief, Examinations Branch to A. C. Devaney, Assistant Commissioner, Inspections and Examinations Division, May 27, 1953, NARA, RG 85, File 56336/205.

36. Coordination occurred across at least a half-dozen USPHS facilities (although not all of the eleven stations were equipped to do blood tests), sixteen INS district offices (although not all regional offices regularly dealt with "Chinese cases"), the consulate in Hong Kong, and the agencies' central offices in Washington, DC.

37. Letter to Commissioner, Central Office, from Joseph Savoretti, District Director, Miami, July 23, 1954, NARA, RG 85, File 56328/856.

38. According to one INS account, there were some ninety actions concerning blood testing pending by November 1952. See Yarbrough memo. These cases are also mentioned in *The Chinese World* beginning in mid-1952.

39. Memo to Mr. Nicholas, Mr. Young and Mrs. Shipley from W. E. Duggan (Passport Office at State Department), April 2, 1951. See also Yarbrough memo.

40. Board of Special Inquiry Hearing, August 14, 1952, in Appendix, 150a. The logic of this decision is reiterated in the BIA's decision of May 11, 1953, Appendix, 157a; and the Decision of the Special Inquiry Officer, December 9, 1953, Appendix, 235a.

41. The BIA upheld this logic in In the Matter of W-K-S- and W-P-S-, *Administrative Decisions under Immigration & Nationality Laws*, vol. 5, February 1952–June 1954 (Washington, DC: U.S. Government Printing Office, 1952), 232–238. The case was from San Francisco.

42. Letter from Willis H. Young, Acting Chief, Passport Division, to Boyd Reynolds, Los Angeles attorney, September 23, 1952, in *Hearings before the President's Commission on Immigration and Naturalization* (Washington, DC: U.S. Government Printing Office, 1952), 1244.

43. In the Matter of L-F-F-. The applicant was given a blood test at the same USPHS office the day before the Lees. Given that he was almost certainly detained at Ellis Island, it is likely that these young people all knew each other. The BIA ruled that the applicant's incompatible blood test results were conclusive, a ruling the applicant did not challenge in court.

44. Benjamin Gim, interview with Mae Ngai, New York City, February 19, 1993. I am very grateful to Mae Ngai for sharing this interview.

45. Edith Cohen, "Benjamin Gim, Founder of Chinatown Firm," *New York Law Journal,* February 5, 1999, 2.

46. "Library": Cohen, "Benjamin Gim."

47. New York contrasted with California—home to the largest Chinese American community—where, as the Chaplin verdict makes clear, case law dating from the early 1940s was unfavorable to blood group evidence.

48. "Not cause for derision": Decision of Board of Immigration Appeals, May 11, 1953, in Appendix, 160a; "third set of tests": 115 F. Supp. 302 (S.D.N.Y. 1953); sweeping views: Brianna Nofil, "Ellis Island's Forgotten Final Act as a Cold War Detention Center," *Atlas Obscura,* February 2, 2016, https://www.atlasobscura.com/articles/ellis-islands-forgotten-final-act-as-a-cold-war-detention-center.

49. Patterns of errors were reported in the San Francisco and New York facilities: In the Matter of D-W-O- and D-W-H-, *Administrative Decisions under Immigration & Nationality Laws*, vol. 5, February 1952–June 1954 (Washington, DC: U.S. Government Printing Office, 1952), 351–369; Yarbrough memo; Letter from R. B. Shipley, Director, Passport Office (State Department), to Devaney, Assistant Commissioner, INS, July 29, 1954, NARA, RG 85, File 56336/341.

50. Reopened Board of Special Inquiry Hearing, November 5, 1953, Examination of Dr. George Cameron by Schatkin, in Appendix, 214a.

51. In the Matter of D-W-O- and D-W-H-.

52. His *Disputed Paternity Proceedings,* first published in 1944, is the American bible of paternity testing. A version of it is still in print today.

53. Reopened Board of Special Inquiry Hearing, November 5, 1953, examination of Sussman by Schatkin, in Appendix, 173a, 178a.

54. Schatkin, Sussman, and Yarbrough, "Chinese Immigration and Blood Tests" (1955) and Sidney Schatkin, Leon Sussman, and Dorris Yarbrough, "Blood Test Evidence in Detecting False Claims of Citizenship," *Criminal Law Review* 3, no. 1 (1956): 45–55. Schatkin was editor of the journal.

55. Leon N. Sussman, "Application of Blood Grouping to Derivative Citizenship," *Journal of Forensic Sciences* 1 (1956): 101–108; Leon N. Sussman, "Blood Grouping Tests in Disputed Paternity Proceedings and Filial Relationship," *Journal of Forensic Sciences* 1 (1956): 25–34; Leon N. Sussman, "Blood Groups in Chinese of New York Area," *American Journal of Clinical Pathology* 26, no. 5 (1956): 471–476.

56. Boyd, "Chances of Disproof." On racial serology, see Jonathan Marks, "The Origins of Anthropological Genetics," *Current Anthropology* 53, no. 5 (2012): 161–172.

57. Yarbrough memo.

58. The test was worthless presumably because of the rarity of Rh-negative individuals of Chinese descent. Sussman, "Blood Groups in Chinese."

59. Yarbrough memo.

60. Quote and statistics are from a memo by George W. Bolin, medical director with the USPHS, stationed in Hong Kong. "Establish [*sic*] of a laboratory and X-ray machine in the American Consulate General, Hong Kong." Bolin cites a maximum 35 percent exclusion rate; others cited 40 percent.

61. Yarbrough memo.

62. The former consul was McConaughy. Letter from H. A. Craves of British Embassy in DC to C. T. Crowe at the Far Eastern Department of the British Foreign Office, March 28, 1956, United States Policy Regarding Issue of Passports to Persons of Chinese Race, 1956, Government Papers, National Archives, Kew, United Kingdom, retrieved from http://www.archivesdirect .amdigital.co.uk/Documents/Details/120983.

63. Yarbrough memo.

64. Drumright made this allegation in his report.

65. In the Matter of L-F-F-, 153.

66. Ethel L. Payne, "How U.S. Makes Enemies Abroad: McCarran-Walker Immigration Act Jim Crows Negroes, Asians," *Chicago Defender,* February 6, 1954, 9; "Payne Asks Ike about Immigration," *Chicago Defender,* March 27, 1954, 1.

67. Reopened Board of Special Inquiry Hearing, November 5, 1953, in Appendix, 179a.

68. Letter to D. D. Boston from J. E. Riley, Acting Assistant Commissioner, Inspections and Examinations Division, Central Office, June 20, 1952, NARA, RG 85, File 56328/856.

69. "Chinese blood typing program": Telegram dated September 14, 1954, NARA, RG 85, File 56328/856; Miami: Letter to Commissioner, Central Office, from Joseph Savoretti, District Director, Miami, July 23, 1954, NARA, RG 85, File 56328/856; Yarbrough: Yarbrough memo; similar references: Letter to District Directors from W. F. Kelly, Assistant Commissioner, Enforcement Division, April 30, 1952; To District Director of Miami [and copied to eight other district directors], from Central Office, Assistant Commissioner, Inspections and Examinations Division, August 19, 1954, NARA, RG 85, File 56328/856; Letter to Commissioner, Central Office, from James W. Butterfield, District Director, Detroit, September 29, 1954, NARA, RG 85, File 56328/856; Letter from A. C. Delaney, Assistant Commissioner, Inspections and Examinations Division, to W. W. Wiggins, Chief, Examinations Branch, July 20, 1953, NARA, RG 85, File 56328/856.

70. The instruction was emitted in September 1953; the draft is from July of that year. Operations Instructions, Sec 205.11, NARA, RG 85, File 56328/856.

71. For its part, the nonethnic U.S. press, other than perfunctorily reporting important judicial decisions regarding the policy, was silent about blood testing and the constitutional issues it raised.

72. "Indignities" / "diabolical scheme": Statement Submitted by Boyd H. Reynolds, Attorney, October 15, 1952, in *Hearings before the President's Commission on Immigration and Naturalization*, 1242. See also "O. P. Stidger Questions Legality of Consulate Demands for Blood Tests," *Chinese World*, September 24, 1953, 2.

73. On the newspaper and its indefatigable editor, see Him Mark Lai, "The Chinese Media in the United States and Canada since World War II," Him Mark Lai Archive, Chinese Historical Society of America, https://himmarklai.org/digitized-articles/2006-201/media-since-wwii/. Dai Ming Lee's editorials were published as *Zhu gang mei zong ling shi bao gao shu de pi pan* [*A Critique of the Report by the U.S. Consul General in Hong Kong*] (San Francisco: Shi Jie Ri Bao, 1956). "Extremely obnoxious": Telegram to Lieutenant General Joseph M. Swing, Commissioner, San Francisco, May 25 [1954], from Dai Ming Lee, NARA, RG 85, File 56336/341. Readers could follow the telegram exchange in "End to Blood Tests for Immigrants Sought," *Chinese World*, May 27, 1954, 1; "Commissioner of Immigration Replies to Chinese World Protest," *Chinese World*, June 8, 1954, 1.

74. "Sweeping victory": "U.S. Supreme Court Decision Outlaws Racial Segregation in Public Schools," *Chinese World*, May 18, 1954, 1.

75. Letter from Commissioner J. M. Swing to Mr. Lee, June 3, 1954, NARA, RG 85, File 56336/341. The fact that Lee's telegram and clippings of the *Chinese World* coverage were preserved in INS files suggests the agency was concerned with public reception of these events.

76. Italics are in the original. The BIA first made this contention in In the Matter of D-W-O- and D-W-H-, which it referred to as a "companion case" to Lee Kum Hoy, emitted three days before the *Brown v. Board* decision. It then repeated itself verbatim a month later in the Lee case: Decision of Board of Immigration Appeals, June 17, 1954, Dismissing Appeal, in Appendix, 239a.

77. Again, this phrasing was invoked both in the BIA's Lee Kum Hoy decision as well as in its companion case, In the Matter of D-W-O- and D-W-H-.

78. "Blood Test Ruling Calls U.S. Biased," *New York Herald Tribune,* August 11, 1955, 15; bond / jail: "NY Chinese Family Declared U.S. Citizens," *Chinese World,* October 19, 1955, 2.

79. Nofil, "Ellis Island's Forgotten Final Act."

80. Reopened Hearing before Special Inquiry Officer, October 28, 1954, in Appendix, 285a–289a.

81. Reopened Hearing before Special Inquiry Officer, October 28, 1954.

82. "Madonna-faced wife": "NY Chinese Family Declared U.S. Citizens." United States *ex rel.* Lee Kum Hoy v. Shaughnessy, 133 F. Supp. 850, 852 (S.D.N.Y. 1955). The decision had occurred a few months earlier, in August 1955: "NY Federal Court Rules Blood Tests Unconstitutional," *Chinese World,* August 12, 1955, 1.

83. "Paternity Tests for Chinese Held Illegal," *Los Angeles Times,* August 11, 1955, 28.

84. United States *ex rel.* Lee Kum Hoy v. Shaughnessy, 237 F.2d 307, 311, no. 3 (2d Cir. 1956).

85. Gabriel Jackson Chin, Cindy Chiang, and Shirley Park, "The Lost *Brown v. Board* of Immigration Law," *North Carolina Law Review* 91, no. 5 (2013): 101–141.

86. General Counsel Opinion no. 16-54, May 13, 1954, re Authority of Service to Compel Persons to Submit to Blood Tests, NARA, RG 85, File 56336/341.

87. "U.S. Citizens of Chinese Ancestry Face Blood Test Demand to Get Passports," *Chinese World,* November 8, 1955, 1; Dai Ming Lee, "U.S. State Department Needs a Housecleaning," *Chinese World,* November 9, 1955, 1; "Immigration Blood Tests Attacked Here," *Chinese World,* November 11, 1955, 2; Dai Ming Lee, "Muddled Thinking Marks State Department Policy," *Chinese World,* November 18, 1955, 1.

88. Dai Ming Lee, "Blood Tests for Passport Applicants," *Chinese World,* November 14, 1955, 1.

89. Ngai, "Legacies of Exclusion."

90. To: J. F. Greene, Chief, General Investigations; From: J. Austin Murphy, Divisional Investigator. Memorandum: Subversive Chinese—Larchmont, New York, May 8, 1956, NARA, RG 85, File 56364/51.6.

91. Affidavits, Ng Mon On, April 3, 1956, and Ng Yee Chor, April 5, 1956, NARA, RG 85, File 56364/51.6.

92. Report of Investigation. Blood test records of Dr. Arthur Liu and Dr. Leon Sussman subpoenaed. New York, May 4, 1956, reported by Investigator D. E. Yarbrough. Several days before, another individual mentioned Dr. Liu in an affidavit in which he testified to having had his blood tested years before. NARA, RG 85, File 56364/51.6. On Dr. Liu's role in the Chinatown community, see Bruce Hall, *Tea That Burns: A Family Memoir of Chinatown* (New York: Simon and Schuster, 2002).

93. Dai Ming Lee, "U.S. State Department Needs a Housecleaning."

94. The telegram exchange was reported in *Chinese World:* Dai Ming Lee, "Muddled Thinking Marks State Department Policy." Shortly thereafter, in a hearing in Washington, DC, a representative of another San Francisco organization, the Chinese Consolidated Benevolent Association, denounced the handling of Chinese citizenship issues, including the use of blood tests. "George Chinn Stresses McCarran Act Inequities at Capital Hearing," *Chinese World,* December 2, 1955.

95. Resolution Register No. 678, Subject: Chinese Persons Seeking Admission to the United States Be Not Subjected to Blood Tests and Letter by Lee R. Pennington to Raymond F. Farrell, Investigations Division, INS, January 25, 1956, NARA, RG 85, File 56364/51.6.

96. Affidavit of Eng Gim Chong before a New York Notary Public, April 30, 1956, NARA, RG 85, File 56364/51.6.

97. On adoption in Chinese kinship practice, see Arthur P. Wolf and Chieh-shan Huang, *Marriage and Adoption in China, 1845–1945* (Stanford, CA: Stanford University Press, 1980).

98. "Report on the Problem of Fraud at Hong Kong," Foreign Service Despatch, December 9, 1955, NARA, RG 85, 56364/51.6; and NARA, RG 59, Central Decimal File 1950–54, Box 0824, Folder 125.4734. Drumright advised that it should be circulated widely within the State Department, the military, the Central Intelligence Agency, the Federal Bureau of Investigation, the Veterans Administration, the Social Security Administration, and other federal agencies. The report is discussed in Ngai, *Impossible Subjects,* chapter 6.

99. "Cultural aspects": Drumright, "Report on the Problem of Fraud," 48; "perfect alibi": Drumright, "Report on the Problem of Fraud," 55. Editor Dai Ming Lee also acknowledged the existence of adoptive practices but notably did not come out in defense of them. Expanded battery: the suggestion was made in an addendum to his original report. Drumright, "Proposals to Better Cope with the Problem of Fraud at Hong Kong," Foreign Service Despatch, December 13, 1955, NARA, RG 85, 56364/51.6. It is also found in the State Department archives: NARA, RG 59, Decimal File 1950–54, Central Decimal File 1955–59, Folder 122.4732. However, available serological tests were too expensive or complicated, or the groups were too rare to make them useful.

100. Letter from James A. Hamilton, Officer in Charge, ORP to Devaney, Assistant Commissioner, Examinations Division, Central Office, Attn Edward Rudnick, Chief Examiner, May 27 1955, NARA, RG 85, File 56336/341.

101. On African American soldiers, see Chapter 7. On the ways that race has shaped the history of U.S. derivative citizenship, see Collins, "Illegitimate Borders."

102. Arissa H. Oh, *To Save the Children of Korea: The Cold War Origins of International Adoption* (Stanford, CA: Stanford University Press, 2015). Ironically, there were even programs to arrange adoptions for children from Hong Kong: Catherine Ceniza Choy, *Global Families: A History of Asian International Adoption in America* (New York: NYU Press, 2013).

103. Chin, Chiang, and Park, "Lost *Brown v. Board*," examines the case and the political context in which it unfolded.

104. As a clerk noted, the government's argument that the peculiar history of fraudulent immigration among Chinese justified blood testing only them could easily be taken up by segregationists to argue, for example, for differential police treatment of African Americans based on assertions of the group's inclination to crime. The clerk is cited in Chin, Chiang, and Park, "Lost *Brown v. Board*," 127.

105. Exhibit "A" Annexed to Supplement of Affidavit of Dorris E. Yarbrough, in Appendix, 96a.

106. Chin, Chiang, and Park, "Lost *Brown v. Board*," finds this argument specious, as the Court could have avoided taking the case in the first place. They discuss possible reasons why the Court saw fit to avoid a ruling on the merits.

107. Conversely, the outcome left open the constitutional question of racial discrimination in immigration policy. The issue was resolved in 1965, when Congress outlawed it.

108. The INS's Operations Instructions guidelines for blood testing continued to be updated at least into the 1960s. INS sources thin out after 1956 due to a reorganization of the archival filing system, but it is probably also because testing had become routine and was unworthy of special comment. Officials continued to look for new serological tests to enhance the power of testing, albeit with limited success: "Legal Division Study" and accompanying letter, from John T. McGill, Chief, Advisory Opinions Division, Visa Office [State Department], to James F. Greene, Deputy Associate Commissioner, Domestic Control, INS, February 23, 1966, United States Citizenship and Immigration Service Library, Washington DC, Vertical Files.

109. There are hints of such testing in the press: "Illegal Entry of Chinese into P.I.," *South China Morning Post,* October 2, 1953, 15; "Request for Admission to Australia Rejected," *South China Morning Post,* November 13, 1963, 12. Several articles in the early 1960s in this same paper discuss cases of Chinese attempting to enter Canada as paper family members; the articles do not clarify whether blood tests were used to uncover these cases.

110. Lucy E. Salyer, *Laws Harsh as Tigers: Chinese Immigrants and the Shaping of Modern Immigration Law* (Chapel Hill: University of North Carolina Press,

2000); Adam M. McKeown, *Melancholy Order: Asian Migration and the Globalization of Borders* (New York: Columbia University Press, 2008).

111. This information comes from an interview that historian Mae Ngai conducted with Benjamin Gim in 1993. Gim died in 2010.

112. Of course, if they were the children of one of Lee Ha's siblings, they would have presumably had a right to U.S. citizenship via descent from the same citizen grandfather.

EPILOGUE

Epigraph: "Smemorato di Collegno, svelato il mistero sull'identità durante 'Chi l'ha visto?,'" *Il Fatto Quotidiano,* July 10, 2014, https://www.ilfattoquotidiano.it /2014/07/10/smemorato-di-collegno-svelato-il-mistero-sullidentita-durante-chi -lha-visto/1056367/.

1. Marc Santora, "Rolling DNA Labs Address the Ultimate Question: 'Who's Your Daddy?'" *New York Times,* November 8, 2016, https://www.nytimes.com/2016 /11/09/nyregion/rolling-dna-labs-address-the-ultimate-question-whos-your -daddy.html.

2. Jack P. Abbott, Kenneth W. Sell, Harry D. Krause, J. B. Miale, E. R. Jennings and W. A. H. Rettberg, "Joint AMA-ABA Guidelines: Present Status of Serologic Testing in Problems of Disputed Parentage," *Family Law Quarterly* 10, no. 3 (1976): 247–285.

3. "Gold standard": Jay Aronson, *Genetic Witness: Science, Law, and Controversy in the Making of DNA Profiling* (New Brunswick, NJ: Rutgers University Press, 2007), 1 (citing defense attorneys Barry Scheck and Peter Neufeld).

4. "Truth machine": Aronson, *Genetic Witness,* 1 (citing Attorney General John Ashcroft).

5. American Association of Blood Banks, Relationship Testing Unit, Annual Report Summary, 2010, 2013; http://www.aabb.org/sa/facilities/Pages /relationshipreports.aspx. This statistic includes only tests performed according to chain-of-command protocols for legal purposes; it does not include the do-it-yourself tests described subsequently in this chapter. One million: Michael Baird, geneticist, DDC DNA Diagnostics Center, interview with the author, January 10, 2019.

6. Jack Curran, "DNA and DNA Forensic Laboratories in the U.S.," *IBISWorld Industry Report,* OD4175, June 2018. The market includes various segments (discussed subsequently), including "discretionary paternity testing," or testing at the behest of private consumers (36.1 percent); legal forensic services, in which courts or police departments outsource DNA testing associated with civil cases of child support or custody and immigration cases as well as forensic testing in criminal cases (42.6 percent); ancestry and "family relationship testing," involving kin other than parents and usually associated with genealogical interest (17.1 percent); and veterinary testing, to determine the pedigree of pets (4.2 percent).

7. Spain: Pilar Benito, "Las pruebas de paternidad se incrementan al caer a la mitad el precio del test," *La Opinión de Murcia,* February 20, 2017, https://www .laopiniondemurcia.es/comunidad/2017/02/20/las-pruebas-de-paternidad-se /807202.html; Italy: Bianca Stancanelli, "Paternità a prova di test," *Panorama,* June 30, 2014, https://www.panorama.it/news/test-di-paternita/; Chile: Patricio Meza S., "10 mil demandas de paternidad se interponen en Chile cada año," *La Segunda,* July 13, 2013, http://www.lasegunda.com/Noticias/Nacional/2013/07 /863382/10-mil-demandas-de-paternidad-se-interponen-en-chile-cada-ano-con -el-adn-es-posible-saber-la-verdad-antes-del-nacimiento; Ecuador: "Demanda de pruebas de ADN se duplicó," *La Hora,* September 30, 2009, https://lahora.com .ec/noticia/938592/demanda-de-pruebas-de-adn-se-duplicc3b3; Honduras: "Se disparan pruebas de paternidad en laboratorios de la UNAH," *La Tribuna,* November 14, 2016, http://www.latribuna.hn/2016/11/14/se-disparan-pruebas -paternidad-laboratorios-la-unah/.

8. Instituto Oscar Freire: Gilka Gattas, geneticist, Departamento de Medicina Legal, Ética Médica e Medicina Social e do Trabalho, Faculdade de Medicina, Universidade de São Paulo, interview with the author, March 2013, São Paulo, Brazil; Brazilian DNA adoption: Sidney Marins, "Brasil já usa código genético para determinar paternidade," *O Globo,* June 12, 1988, 34; Sueann Caulfield and Alexandra Minna Stern, "Shadows of Doubt: The Uneasy Incorporation of Identification Science into Legal Determination of Paternity in Brazil," *Cadernos de Saúde Pública* 33 (2017): 1–14; Rio Grande do Sul: Claudia Fonseca, "Law, Technology, and Gender Relations: Following the Path of DNA Paternity Tests in Brazil," in *Reproduction, Globalization, and the State: New Theoretical and Ethnographic Perspectives,* ed. Carolyn Browner and Carolyn Fishel Sargent (Durham, NC: Duke University Press, 2011), 138–153; popular magazine: "40 Coisas que Mudaram a Sua Vida nos Últimos 40 Anos," *Veja* 2008, special 40th anniversary issue.

9. Nimble Diagnostics: http://www.nimblediagnostics.co.uk/home/pat.html; EasyDNA: https://www.easy-dna.com/contact-us/worldwide-offices/.

10. "Paternity tests growing in China," *Xinhua,* April 12, 2014, http://news .xinhuanet.com/english/video/2014-04/12/c_133256589.htm; Indian Biosciences: http://inbdna.com.

11. This segment also includes forensic testing associated with criminal investigations requested by police departments. Industry reports do not disaggregate what percentage of this segment derives from criminal forensics versus paternity forensics. Curran, "DNA and DNA Forensic Laboratories."

12. Centers for Disease Control and Prevention, "Nonmarital Childbearing in the United States, 1940–99," *National Vital Statistics Reports* 48, no. 16 (October 2000), https://www.cdc.gov/nchs/data/nvsr/nvsr48/nvs48_16.pdf; Angel Castillo, "New Use of Blood Test Is Decisive in Paternity Suits," *New York Times,* June 2, 1981, 1.

13. Ronald Kotulak, "Reliable Paternity Test Unleashes Flood of Suits," *Chicago Tribune,* November 2, 1981, A3.

14. Carmen Solomon-Fears, "Paternity Establishment: Child Support and Beyond," Report for Congress (Congressional Research Service, June 24, 2002), 12.

15. Solomon-Fears, "Paternity Establishment," 9; Mary R. Anderlik and Mark A. Rothstein, "DNA-Based Identity Testing and the Future of the Family: A Research Agenda," *American Journal of Law & Medicine* 28 (2002): 218.

16. Social scientists have challenged any straightforward link between female-headed households and poverty. Sylvia Chant, "Dangerous Equations? How Female-Headed Households Became the Poorest of the Poor: Causes, Consequences and Cautions," *IDS Bulletin* 35, no. 4 (2004): 19–26.

17. Sueann Caulfield, "The Right to a Father's Name: A Historical Perspective on State Efforts to Combat the Stigma of Illegitimate Birth in Brazil," *Law and History Review* 30, no. 1 (2012): 1–36; Nara Milanich, "Daddy Issues: 'Responsible Paternity' as Public Policy in Latin America," *World Policy Journal* 34, no. 3 (2017): 8–14.

18. Cited in Aronson, *Genetic Witness*, 15.

19. Torsten Heinemann, Ilpo Helén, Thomas Lemke, Ursula Naue, and Martin Weiss, *Suspect Families: DNA Analysis, Family Reunification and Immigration Policies* (Farnham: Ashgate, 2016).

20. Dorothy E. Roberts, "The Genetic Tie," *University of Chicago Law Review* 62, no. 1 (1995): 211. On DNA, race, and immigration, see Heide Castañeda, "Paternity for Sale: Anxieties over 'Demographic Theft' and Undocumented Migrant Reproduction in Germany," *Medical Anthropology Quarterly* 22, no. 4 (2008): 340–359; Victoria Degtyareva, "Defining Family in Immigration Law: Accounting for Non-Traditional Families in Citizenship by Descent," *Yale Law Journal* 120 (2011): 862–908; Didier Fassin, "The Mystery Child and the Politics of Reproduction: Between National Imaginaries and Transnational Confrontations," in Browner and Sargent, *Reproduction, Globalization, and the State*, 239–248.

21. "DNA Tests for Would-Be Immigrants on Hold," *RFI,* September 14, 2009, http://www.rfi.fr/actuen/articles/117/article_5119.asp

22. Miriam Jordan, "Refugee Program Halted as DNA Tests Show Fraud," *Wall Street Journal,* August 20, 2008, https://www.wsj.com/articles /SB121919647430755373.

23. U.S. Citizenship and Immigration Service, Senior Policy Council, Options Paper, "Expanding DNA Testing in the Immigration Process," https://www.eff .org/files/filenode/uscis_dna_senior_policy_council_options_paper.pdf. The memo is undated but was issued after the 2008 suspension and before its 2012 reinstatement.

24. The program was delayed in the fall of 2015, ostensibly so that consent forms could be translated. It is unclear if it has since launched. "DHS Delays Refugee Rapid DNA Tests Aimed at Stemming Human Trafficking," http://www.nextgov .com/defense/2015/09/dhs-delays-dna-tests-refugee-camps-aimed-stemming -human-trafficking/120356/.

25. Seth Freed Wessler, "Is Denaturalization the Next Front in the Trump Administration's War on Immigration?," *New York Times Magazine*, December 19, 2018, https://www.nytimes.com/2018/12/19/magazine /naturalized-citizenship-immigration-trump.html.

26. General Counsel Opinion no. 16-54, re Authority of Service to compel persons to submit to blood tests, May 13, 1954, NARA, RG 85, File 56336/341.

27. The report notes that DNA is used to establish individual identity as well as biological relationships. "Adoption of DNA Analysis in Homeland Security to Drive the Global DNA Analysis Market in the Government Sector until 2019, Says Technavio," *Business Wire Report,* February 18, 2016, http://www .businesswire.com/news/home/20160218005047/en/Rising-Adoption-DNA -Analysis-Homeland-Security-Drive.

28. The technique was pioneered by U.S. geneticist Mary-Claire King in conjunction with Argentine colleague Víctor Penchaszadeh and others.

29. "Identigene Turns 20! Identigene DNA Laboratory Celebrates 20 Year Anniversary," http://www.identigene.com/news/identigene-turns-20-identigene-dna -laboratory-celebrates-20-year-anniversary.

30. Curran, "DNA and DNA Forensic Laboratories," 6.

31. Indeed, some Italian doctors voice some of the very reservations their predecessors did sixty years ago: L. Caenazzo, A. Comacchio, P. Tozzo, D. Rodríguez and P. Benciolini, "Paternity Testing Requested by Private Parties in Italy: Some Ethical Considerations," *Journal of Medical Ethics* 34, no. 10 (2008): 735–737, oppose mail-order testing on ethical and legal grounds.

32. Curran, "DNA and DNA Forensic Laboratories," 6.

33. On discourses of identity rights: Anne-Emanuelle Birn, "Uruguay's Child Rights Approach to Health," in *Registration and Recognition: Documenting the Person in World History,* ed. Simon Szreter and Keith Breckenridge (Oxford: Oxford University Press, 2012), 415–447; Article 8 reads: "Children have the right to an identity—an official record of who they are. Governments should respect children's right to a name, a nationality and family ties." United Nations, Convention on the Rights of the Child, https://www.ohchr.org/en /professionalinterest/pages/crc.aspx; Bastard Nation News and Notes, http:// bastardnation.blogspot.com; donor-conceived people: for one example from a public authority in Australia, see State Government of Victoria, Department of Health & Human Services, "A Right to Know Your Identity," 2015, https://www .varta.org.au/sites/varta/files/public/A%20right%20to%20know%20your%20 identity%20-%20DHHS.pdf; examples of the right to identity articulated within Latin American responsible paternity campaigns include, in Mexico, Secretaría de Gobernación, *El derecho a la identidad como derecho humano* (Mexico City: Secretaría de Gobernación, 2011); in Peru, the Ministry of Justice and Human Right's National Campaign for Identity and Paternal Recognition, November 21–22, 2013, https://www.minjus.gob.pe/defensapublica/interna .php?comando=611&codigo=284; and in Paraguay, "Campaña derecho a

identidad promueve la protección de los niños desde su nacimiento," February 5, 2016, http://www.ip.gov.py/ip/campana-derecho-a-identidad-promueve-la -proteccion-de-los-ninos-desde-su-nacimiento/; "political issue": Bastard Nation Mission Statement: http://bastards.org/cgi-sys/suspendedpage.cgi.

34. 23and Me: George Doe, "With Genetic Testing, I Gave My Parents the Gift of Divorce," *Vox,* September 9, 2014, https://www.vox.com/2014/9/9/5975653 /with-genetic-testing-i-gave-my-parents-the-gift-of-divorce-23andme. The story is part of a larger investigative piece: Julia Belluz, "Genetic Testing Brings Families Together, and Sometimes Tears Them Apart," *Vox,* December 18, 2014, https://www.vox.com/2014/9/9/6107039/23andme-ancestry-dna-testing; Irish Catholic woman: Libby Copeland, "She Thought She Was Irish—Until a DNA Test Opened a 100-Year-Old Mystery," *Washington Post,* July 27, 2017, https://www.washingtonpost.com/graphics/2017/lifestyle/she-thought-she-was -irish-until-a-dna-test-opened-a-100-year-old-mystery/?noredirect=on&utm _term=.71fb8e02e720.

35. Sarah Zhang, "When a DNA Test Shatters Your Identity," *Atlantic,* July 17, 2018, https://www.theatlantic.com/science/archive/2018/07/dna-test -misattributed-paternity/562928/.

36. Margaret Talbot, "Family Affair; Separated at Birth," *New York Times Magazine,* April 11, 1999, https://archive.nytimes.com/www.nytimes.com/library /magazine/home/041199wwln-talbot.html.

37. "Southern white girl": Tara Bahrampour, "They Considered Themselves White, But DNA Tests Told A More Complex Story, February 6, 2018, https://www .washingtonpost.com/local/social-issues/they-considered-themselves-white-but -dna-tests-told-a-more-complex-story/2018/02/06/16215d1a-e181-11e7-8679 -a9728984779c_story.html?utm_term=.fab2b7322fe6; Ruth Padawer, "Sigrid Johnson Was Black. A DNA Test Said She Was White," *New York Times,* November 19, 2018, https://www.nytimes.com/2018/11/19/magazine/dna-test -black-family.html; Chana Garcia, "DNA Testing Forced Me To Rethink My Entire Racial Identity," *Huffington Post,* March 3, 2018, https://www .huffingtonpost.com/entry/dna-testing-forced-me-to-rethink-my-entire-racial -identity_us_5a8311b8e4b00ecc923ee61b. On ancestry testing, roots-seeking, and racial identity, Alondra Nelson, *The Social Life of DNA: Race, Reparations, and Reconciliation After the Genome* (New York: Beacon Press, 2016).

38. Doe, "With Genetic Testing."

39. Jim Yardley, "Health Officials Investigating to Determine How Woman Got the Embryo of Another," *New York Times,* March 31, 1999, https://www.nytimes .com/1999/03/31/nyregion/health-officials-investigating-to-determine-how -woman-got-the-embryo-of-another.html; "White Couple Win Black IVF Twins," CNN.com, February 26, 2003, http://edition.cnn.com/2003/WORLD /europe/02/26/britain.twins.reut/index.html; Helen Weathers, "Why Am I Dark, Daddy? The White Couple Who Had Mixed Race Children after IVF Blunder," *Daily Mail,* June 13, 2009, https://www.dailymail.co.uk/news/article

-1192717/Why-I-dark-daddy-The-white-couple-mixed-race-children-IVF
-blunder.html; Kim Bellware, "White Woman Who Sued Sperm Bank over
Black Baby Says It's Not about Race," *Huffington Post,* October 2, 2014,
https://www.huffingtonpost.com/2014/10/02/black-sperm-lawsuit_n_5922180
.html. For a recent baby swap story with a similar theme, see Maïa de la Baume,
"In France, a Baby Switch and a Lesson in Maternal Love," *New York Times,*
February 24, 2015, https://www.nytimes.com/2015/02/25/world/europe/in
-france-a-baby-switch-and-a-test-of-a-mothers-love.html?rref=world
/europe&module=Ribbon&version=context&region=Header&action
=click&contentCollection=Europe&pgtype=article&_r=0.

40. Of the four racial gamete- and embryo-swapping stories cited in the preceding
    note, three involve sperm substitutions.

41. Alessandra Stanley, "So, Who's Your Daddy? In DNA Tests, TV Finds Elixir, to
    Raise Ratings," *New York Times,* March 19, 2002. https://www.nytimes.com
    /2002/03/19/business/media-business-so-who-s-your-daddy-dna-tests-tv-finds
    -elixir-raise-ratings.html.

42. Apophia Agiresaasi, "DNA Testing Trend Brings Joy and Sorrow as Ugandans
    Discover Truth about Families," *Global Press Journal,* July 23, 2016, https://
    globalpressjournal.com/africa/uganda/dna-testing-trend-brings-joy-sorrow
    -ugandans-discover-truth-families/.

43. Cited in Michael Gilding, "DNA Paternity Tests: A Comparative Analysis of the
    U.S. and Australia," *Health Sociology Review* 15 (2006): 92.

44. See https://peaceofmindpaternity.com/contact/.

45. Examples include the following: Identigene: https://dnatesting.com/paternity
    -fraud/ and https://dnatesting.com/paternity-fraud-the-tough-realities-men
    -must-face/ ; Canadian Children's Rights Council: https://canadiancrc.com
    /Newspaper_Articles/Globe_and_Mail_Moms_Little_secret_14DEC02.aspx;
    http://fathersmanifesto.net/paternityfraud.htm. As for scholars, the chief
    offenders, as Michael Gilding notes, are sociobiologists invested in arguments
    about how paternal uncertainty shapes evolutionary strategy. An example of the
    misleading use of statistics is found in Steven M. Platek and Todd K. Shackel-
    ford, *Female Infidelity and Paternal Uncertainty: Evolutionary Perspectives on Male
    Anti-Cuckoldry Tactics* (Cambridge: Cambridge University Press, 2006). See also
    Kermyt G. Anderson, "How Well Does Paternity Confidence Match Actual
    Paternity? Evidence from Worldwide Nonpaternity Rates," *Current Anthropology*
    47, no. 3 (2006): 513–520.

46. A second theory is that it comes from an offhand remark by an English doctor
    in a symposium on artificial insemination in 1972. The doctor never published
    the findings nor was his population sample identified. Michael Gilding traces
    the life of this urban legend in "The Fatherhood Myth," *Inside Story,* July 26,
    2011, http://insidestory.org.au/the-fatherhood-myth/. See also Razib Khan,
    "The Paternity Myth: The Rarity of Cuckoldry," *Discover* blogs, June 20, 2010,
    http://blogs.discovermagazine.com/gnxp/2010/06/the-paternity-myth-the-rarity
    -of-cuckoldry/#.Wm4x3rT83q1.

47. Matt Friedman, "N.J. Legislator Proposes Bill Requiring Genetic Testing for All Newborns, Parents to Verify Paternity," *NJ.com,* March 1, 2012, http://www.nj .com/news/index.ssf/2012/03/nj_legislator_proposes_measure.html.

48. Katharine K. Baker, "Bargaining or Biology? The History and Future of Paternity Law and Parental Status," *Cornell Journal of Law & Public Policy* 14 (2004): 12.

49. Caulfield and Stern, "Shadows of Doubt."

50. Cited in Baker, "Bargaining or Biology," 13.

51. Leh v. Robertson, 463 U.S. 248 (1982).

52. When private parties carry out artificial insemination, the man's intent prior to conception is key to determining whether he is considered a legal parent. Baker, "Bargaining or Biology," 10–11; on surrogacy, 26–28.

53. Kristin A. Collins, "Illegitimate Borders: Jus Sanguinis Citizenship and the Legal Construction of Family, Race, and Nation," *Yale Law Journal* 123 (2014): 2134–2235. The Supreme Court has affirmed this principle multiple times, as in *Nguyen v. INS* (2001) and *Flores-Villar v. United States* (2011). More recently, however, in *Sessions v. Morales Santana* (2017), the Court rejected different residency requirements for unmarried fathers and mothers to transmit their citizenship.

54. "Man Ordered to Pay $65K in Child Support for Kid Who Isn't His," *New York Post,* July 23, 2017, https://nypost.com/2017/07/23/man-ordered-to-pay-65k-in -child-support-for-kid-who-isnt-his/.

55. Janet L. Dolgin, "Family Law and the Facts of Family," in *Naturalizing Power: Essays in Feminist Cultural Analysis,* ed. Sylvia Junko Yanagisako and Carol Lowery Delaney (New York: Routledge, 1995), 47–67 argues that family law inconsistently embraces or rejects biological facts depending on whether they serve to reinforce or undermine an idea of the traditional family; Baker, "Bargaining or Biology," advocates moving from biology to contract as the defining principle of parentage.

56. In addition to the case of Chinese Americans discussed in Chapter 8, and the case of the denaturalized Yemeni American discussed above, see parallel examples in contemporary Germany, Castañeda, "Paternity for Sale," and France, Fassin, "Mystery Child."

57. Nara Milanich, "To Make All Children Equal Is a Change in the Power Structures of Society: The Politics of Family Law in Twentieth Century Chile and Latin America," *Law and History Review* 33, no. 4 (2015): 767–802.

58. Irresponsible paternity: Centro de Estudios para el Adelanto de las Mujeres y la Equidad de Género, "Garantías de cumplimiento de los deberes de paternidad responsable," undated report, http://www.diputados.gob.mx /documentos/CEAMEG/PRESPONSABLE1.pdf; Milanich, "Daddy Issues"; Anjali Thomas, "India's Doubting Fathers and Sons Embrace DNA Paternity Tests," *New York Times,* August 16, 2013, https://india.blogs.nytimes.com /2013/08/16/indias-doubting-fathers-and-sons-embrace-dna-paternity-tests/? _r=0.

59. Bronislaw Malinowski, "Foreword," in Ashley Montagu, *Coming into Being among the Australian Aborigines: The Procreative Beliefs of the Australian Aborigines* (1937; repr., New York: Routledge, 2004), xvi.

60. Baltimore paper: "Has Science Found Answer to Question of Parentage?" *Baltimore Sun,* October 29, 1922, P6P2; Scottish jurist: Alistair R. Brownlie, "Blood and the Blood Groups: A Developing Field for Expert Evidence," *Journal of the Forensic Science Society* 5, no. 3 (1965): 137; American geneticist: Margery Shaw, "Paternity Determination: 1921 to 1983 and Beyond," *JAMA* 250, no. 18 (November 11, 1983): 2537.

61. "Smemorato di Collegno."

ACKNOWLEDGMENTS

This project launched me across three continents, propelled me deep into the history of the twentieth century, and obliged me to trespass across entirely unfamiliar scholarly fields—which is to say it took me well beyond anything resembling a comfort zone. A journey of this sort could not have happened without the intellectual generosity of others. Time and again, insightful interlocutors provided road maps. Supportive colleagues pointed out landmarks as well as pitfalls. Dear friends kept me company along the way.

In the first instance, I am grateful to a number of talented and dedicated librarians and archivists. They include the staff at the Biblioteca Centrale Giuridica, Rome, Italy; the Biblioteca de la Facultad de Medicina at the Universidad de Buenos Aires, Argentina; the Instituto Iberoamericano in Berlin, Germany; and the National Archives and Records Administration (NARA) in Washington, DC, College Park, Maryland, and New York City. I especially want to acknowledge Maria Conforti at the Biblioteca Storia della Medicina of the Università La Sapienza, Rome; Bill Creech, who oversees immigration records at NARA; Lee Hiltzik at the Rockefeller Archive Center; Giovanna Ricci at the Biblioteca Provinciale di Pisa, Italy; David Uhlich of Archives and Special Collections at the University of California, San Francisco; and Zack Wilske at the History Office and Library of U.S. Citizenship and Immigration Services. My stay at the Instituto Oscar Freire at the Universidade de São Paulo, Brazil, was a highlight of my research. I am deeply grateful for the expert assistance and warm welcome of Suely Campos Cardoso, the librarian of the Faculdade de Medicina, and Dr. Gilka Gattas, geneticist at the Departamento de Medicina Legal, Ética Médica e Medicina Social e do Trabalho.

Discussions with geneticists helped clarify the relationship of the past to the present. Gilka Gattas explained not only genetic science but also the ethical quandaries she encounters testing DNA at the Instituto Oscar Freire. Michael Baird, geneticist of DDC DNA Diagnostics Center, shared his impressions of more than three decades working in the U.S. genetic testing industry.

My research and writing were made possible thanks to generous support from the Burkhardt Fellowship of the American Council of Learned Societies and the hospitality of the Kluge Center at the Library of Congress. Barnard College provided financial support and above all that most precious gift—time, in the form of sabbatical leave and course reductions. I also thank the Center for Science and Society, the Heyman Center Society of Fellows, the History Department, and the Institute for Latin American Studies, all of Columbia University, for their support.

I am grateful to colleagues far and wide for engaging with my work and helping me to improve it. They include audiences at the Elizabeth Battle Clark Legal History Workshop at Boston University Law School; the Columbia-Université Paris I Panthéon-Sorbonne Latin American History workshop; the Cuartas Jornadas de Estudios sobre la Infancia and the Seminario Abierto de Discusión at the Universidad San Andrés, in Buenos Aires, Argentina; the Davis Center seminar on "Legalities" at Princeton University; the Family, Kinship, and Household group at the Center for Historical Research, Ohio State University; Georgetown University's History Department; the Kinship and Politics research group at the Zentrum für interdisziplinäre Forschung, Universität Bielefeld, Germany; the Observatorio de Desigualdades of the Universidad Diego Portales, Santiago, Chile; the Preconference Workshop on Latin American Legal History at the American Society of Legal History, Miami, Florida; and the University of New Mexico History Department's International History Group.

Closer to home, I thank participants in the Columbia Law School's Deconstructing and Reconstructing Motherhood conference; the History Department's International History Workshop, the Heyman Society of Fellows; and, of course, the BC-CU Latin American History student-faculty seminar. Audience members at the very first public presentation I gave about my research, at the Barnard Center for Research on Women, raised questions that I have been thinking about ever since. I also thank the (captive!) student audiences in Jonathan Weiner's science writing class at the Journalism School, Pablo Piccato's Latin American graduate historiography course, and my own Family, Race, and Nation seminar at Barnard.

Several research assistants made vital contributions to this project, including Colin Kinniburgh, Christopher Meyer, Sebastian Muñoz, and Jack Neubauer. Without Anne Schult's research and translation, Chapter 6 could never have been written.

I am extremely grateful to friends and colleagues who made bibliographic suggestions, shared archival materials, talked through ideas, read (sometimes reread) proposals and chapters, corrected embarrassing errors, asked pointed questions, offered vigorous critique and equally unstinting encouragement, and brainstormed book titles instead of paying attention at faculty meetings. Those who helped shape this book include Erdmute Alber, Daniel Asen, Gustavo Azenha, Betsy Blackmar, Lila Caimari, Francesco Cassata, Sueann Caulfield, Debbie Coen, Kristin Collins, Isabella Cosse, Angela Creager, Irina Denischenko, Yasmin Ergas, Catherine Fennell, Federico Finchelstein, Claudia Fonseca, Samantha Fox, Alyshia Gálvez, Marianne González Le Saux, Vicky de Grazia, Romeo Guzmán, Jean Howard, Nico Calcina Howson, Elizabeth Quay Hutchison, Robin Judd, Daniel Kevles, Dorothy Ko, Tori Langland, Annick Lempérière, Ricardo López, Jeannett Martin, Alejandra Matus, Sonya Michel, Erika Milam, Steve Mintz, Maxine Molyneux, Celia Naylor, Jack Neubauer, Mae Ngai, Jesse Olszynko-Gryn, Silvana Patriarca, Antonella Pelizzari, Sharon Phillips, Caterina Pizzigoni, Alejandra Ramm,

Anna di Robilant, Stephanie Rupp, Elizabeth Schwall, David Seipp, Shobana Shankar, Birgitte Soland, Rhiannon Stephens, Noah Tamarkin, Molly Tambor, Deborah Valenze, Jonathan Weiner, and John Wertheimer. Paul Katz provided painstaking proofreading. Amy Chazkel and Pablo Piccato read the manuscript and offered terrific feedback, as did my parents, Maxine Margolis and Jerry Milanich. My dad also did expert work on images and permissions. Lisa Tiersten read a late, late draft and in a single conversation reframed the entire argument—and then went on to improve innumerable topic sentences.

I sought out Joyce Seltzer because of her reputation as a uniquely hands-on editor. Sure enough, her incisive feedback made this a better book (and cured me, mostly, of my penchant for rhetorical questions). Janice Audet enthusiastically took up where Joyce left off. Two anonymous readers offered generous and perceptive critiques that forced me to clarify my arguments. Pat Payne salvaged unsalvageable images. Sherry Gerstein graciously tolerated my obsessive-compulsive streak. Angela Baggetta's expertise inspired me to focus on the fun part, the finish line!

This book is my first foray into the history of medicine and science. It is therefore an unfortunate irony that I finished writing it in the waiting rooms of doctors' offices. I thank Drs. Karen Hiotis, Carmen Pérez, Amy Tiersten, and Robin Shafran, as well as Jillian Capodice, for their expert care and for all the times they inquired about this book (incredulous, perhaps, that it was still not done). The best health care providers don't just save your life: they help resuscitate your humanity.

Several individuals generously shared their stories with me. I am deeply grateful to Peter Schiff, son of Dr. Fritz Schiff, for recounting memories of his father and family. Antonio Cipolli, whose birth is the subject of Chapter 7, and his daughter, Dunja Cipolli, helped me to understand perhaps better than anyone else the deep personal stakes of the elusive quest for the father. I thank them for allowing me to participate in their journey.

To Giacomo and Luca: yes, it took seven years, and yes, I promise I'll try to make the next one a novel (and a bestseller). Nicola Cetorelli's most important role in this book was a paternal one: during my innumerable research trips and panicked weekends before deadlines, he cheerfully embraced single fatherhood. He has been an extraordinary partner through thick and thin. This book is dedicated to him.

INDEX

Page numbers followed by n indicate notes. *Italicized* page numbers indicate illustrations.

about among U.S. press and public, 85–87; somatic testing contrasted, 91; Strassman and, 65, 66, 70, 77, 102, 168, 175; as supposed marker of race, 59–60; as test of exclusion, 55–56, 59, 69, 70, 72–73, 78, 85, 219, 223–224, 248; use in Brazilian seduction cases, 76–79; uses in immigration and citizenship proceedings, 208–245; Wiener's lobbying efforts for, 85. *See also* Blood tests; Immigration, citizenship, and blood testing

Blood Groups and Their Areas of Application, The (Schiff), 64–65

Blood tests, 54–88; conflation of distinct kinds of, 80, 87; emergence of in 1920s, 34–35; limitations of blood group tests as inspiring other kinds of, 60, 74; Manoiloff's "race reaction," 65, 66, 70, 77, 102, 168, 175; media and, 75, 87; purporting to uncover both race and paternity, 194–195; Reichert and crystallography, 38–39, 62, 174; search for marker of race in blood, 55, 56–57, 59–61; somatic testing contrasted, 91; Zangemeister and, 74–75, 87. *See also* Abrams, Dr. Albert and oscillophore; Blood groups or types

Board of Immigration Appeals (BIA), in U.S., Chinese immigrants and, 221, 224–226, 234–235

Body: absence of, when father deceased, 135; Chinese immigrants, examination of, 215–216; exhumation of father's, 138; as harboring the secret of kinship and paternity, 5, 23, 55, 62–63, 90–91, 98–100, 190; identity and, 23; Lehmann Nitsche and, 104–105; maternity versus paternity on, 131; race and, 62–63; social and moral meanings of, 143–145, 166; transnational ideas about, 110. *See also* Artists; Eugenics; Resemblance, assessment of; Somatic testing

Brave New World (Huxley), 17

Brazil: Abrams and, 48; blood group testing and, 68, 75, 76–79, 87, 88, 201; DNA testing in twenty-first century, 249, 251, 261; as early adopter of paternity testing,

34, 76–77, 124; forensic testing in, generally, 77–78, 124–126, 140–142, 145–146; lack of scientific consensus about testing methods, 124–126, 135–145; medico-legal community in, 77, 78, 109, 118, 124–126, 140–142, 145–147; seduction and deflowering as applications of paternity testing, 77–79, 146; somatic testing and, 109, 129. *See also* Canella, Giulio (Mario Bruneri); Instituto Oscar Freire (IOF); Silva, Dr. Luiz

Brown v. Board of Education of Topeka, 212, 234

Bruneri, Mario. *See* Canella, Giulio (Mario Bruneri)

Bruneri, Rosa, 117–118

Buffalo Soldiers, 180, 184, 266

Bühler, Engelhard, 161

California: DNA testing in, 249, 261; marital paternity and, 33, 44, 46, 261–262. *See also* Chaplin, Charlie, paternity suit against

Calles, Plutarco, 48

Campbell, Everett, 42

Canella, Giulia, 116–122, *123,* 147–148, 257, 266–267; children of, 120–122, *121, 123*

Canella, Giulio (Mario Bruneri): DNA testing of family of, 256–257, 266–267; Italian cause célèbre surrounding, 116–119; Silva's attempts to determine kinship identity of, 119–123, *121, 123,* 148, 257, 266–267; media and, 297n4

Canella, Julio, 257, 266–267

Caras y Caretas, 34

Cardozo, Benjamin, 197

Caro, Hildegard, 73

Carrozza, Antonio, 190, 191, 196

Catholic Church: Bruneri-Canella case and, 117, 148; canon law, 13, 190, 197; Catholic filmmakers, 193, 201; Catholic jurists, 193–194, 201, 313n53; Catholic professionals' moral reservations about paternity testing, 201–205, 255; Italian paternity laws and, 199; reproductive